Advanced Sorbents for Separation of Metal Ions

Advanced Sorbents for Separation of Metal Ions

Guest Editor
Antonije Onjia

Basel • Beijing • Wuhan • Barcelona • Belgrade • Novi Sad • Cluj • Manchester

Guest Editor
Antonije Onjia
Faculty of Technology and Metallurgy
University of Belgrade
Belgrade
Serbia

Editorial Office
MDPI AG
Grosspeteranlage 5
4052 Basel, Switzerland

This is a reprint of the Special Issue, published open access by the journal *Metals* (ISSN 2075-4701), freely accessible at: https://www.mdpi.com/journal/metals/special_issues/advanced_sorbents_separation_metal_ions.

For citation purposes, cite each article independently as indicated on the article page online and as indicated below:

Lastname, A.A.; Lastname, B.B. Article Title. *Journal Name* **Year**, *Volume Number*, Page Range.

ISBN 978-3-7258-3339-9 (Hbk)
ISBN 978-3-7258-3340-5 (PDF)
https://doi.org/10.3390/books978-3-7258-3340-5

© 2025 by the authors. Articles in this book are Open Access and distributed under the Creative Commons Attribution (CC BY) license. The book as a whole is distributed by MDPI under the terms and conditions of the Creative Commons Attribution-NonCommercial-NoDerivs (CC BY-NC-ND) license (https://creativecommons.org/licenses/by-nc-nd/4.0/).

Contents

About the Editor vii

Antonije Onjia
Advanced Sorbents for Separation of Metal Ions
Reprinted from: *Metals* 2024, 14, 1026, https://doi.org/10.3390/met14091026 1

**Chang Liu, Shichang Zhang, Xinpeng Wang, Lifeng Chen, Xiangbiao Yin,
Mohammed F. Hamza, et al.**
Preparation of Two Novel Stable Silica-Based Adsorbents for Selective Separation of Sr from Concentrated Nitric Acid Solution
Reprinted from: *Metals* 2024, 14, 627, https://doi.org/10.3390/met14060627 6

**Maria del Mar Cerrillo-Gonzalez, Maria Villen-Guzman, Jose Miguel Rodriguez-Maroto and
Juan Manuel Paz-Garcia**
Metal Recovery from Wastewater Using Electrodialysis Separation
Reprinted from: *Metals* 2024, 14, 38, https://doi.org/10.3390/met14010038 20

**Ivana Smičiklas, Bojan Janković, Mihajlo Jović, Jelena Maletaškić, Nebojša Manić and
Snežana Dragović**
Performance Assessment of Wood Ash and Bone Char for Manganese Treatment in Acid Mine Drainage
Reprinted from: *Metals* 2023, 13, 1665, https://doi.org/10.3390/met13101665 38

**Henrik K. Hansen, Claudia Gutiérrez, Natalia Valencia, Claudia Gotschlich, Andrea Lazo,
Pamela Lazo and Rodrigo Ortiz-Soto**
Selection of Operation Conditions for a Batch Brown Seaweed Biosorption System for Removal of Copper from Aqueous Solutions
Reprinted from: *Metals* 2023, 13, 1008, https://doi.org/10.3390/met13061008 60

**Miljan Marković, Milan Gorgievski, Nada Štrbac, Vesna Grekulović, Kristina Božinović,
Milica Zdravković and Milovan Vuković**
Raw Eggshell as an Adsorbent for Copper Ions Biosorption—Equilibrium, Kinetic,
Thermodynamic and Process Optimization Studies
Reprinted from: *Metals* 2023, 13, 206, https://doi.org/10.3390/met13020206 75

**Eleni Mikeli, Danai Marinos, Aikaterini Toli, Anastasia Pilichou, Efthymios Balomenos and
Dimitrios Panias**
Use of Ion-Exchange Resins to Adsorb Scandium from Titanium Industry's Chloride Acidic Solution at Ambient Temperature
Reprinted from: *Metals* 2022, 12, 864, https://doi.org/10.3390/met12050864 93

Aleksandra Nastasović, Bojana Marković, Ljiljana Suručić and Antonije Onjia
Methacrylate-Based Polymeric Sorbents for Recovery of Metals from Aqueous Solutions
Reprinted from: *Metals* 2022, 12, 814, https://doi.org/10.3390/met12050814 106

**Latinka Slavković-Beškoski, Ljubiša Ignjatović, Guido Bolognesi, Danijela Maksin,
Aleksandra Savić, Goran Vladisavljević and Antonije Onjia**
Dispersive Solid–Liquid Microextraction Based on the Poly(HDDA)/Graphene Sorbent
Followed by ICP-MS for the Determination of Rare Earth Elements in Coal Fly Ash Leachate
Reprinted from: *Metals* 2022, 12, 791, https://doi.org/10.3390/met12050791 136

Jonathan Castillo, Norman Toro, Pía Hernández, Patricio Navarro, Cristian Vargas,
Edelmira Gálvez and Rossana Sepúlveda
Extraction of Cu(II), Fe(III), Zn(II), and Mn(II) from Aqueous Solutions with Ionic Liquid R$_4$NCy
Reprinted from: *Metals* **2021**, *11*, 1585, https://doi.org/10.3390/met11101585 . . . . **147**

Jovana Djokić, Dragana Radovanović, Zlatko Nikolovski, Zoran Andjić and
Željko Kamberović
Influence of Electrolyte Impurities from E-Waste Electrorefining on Copper Extraction Recovery
Reprinted from: *Metals* **2021**, *11*, 1383, https://doi.org/10.3390/met11091383 . . . . **157**

About the Editor

Antonije Onjia

Antonije Onjia is a Professor at the Faculty of Technology and Metallurgy, the University of Belgrade, Serbia. With over three decades of academic and research experience, his work focuses on environmental chemistry and materials science. He has made significant contributions to the fields of adsorption processes, environmental remediation, and analytical chemistry, publishing extensively in high-impact journals and contributing to numerous international projects.

As a dedicated educator and researcher, Prof. Onjia has supervised numerous doctoral and master's theses, fostering innovation and collaboration in multidisciplinary research. His expertise also lies in designing novel materials for industrial and environmental applications, particularly for wastewater treatment, hydrometallurgy, and resource recovery.

He has served as a principal investigator on several national and international research projects, focusing on sustainable materials and innovative technologies for pollution control and metal recovery. He has also been actively involved in organizing scientific events and Special Issues, aiming to bridge the gap between academia and industry.

Through his editorial leadership of the "Advanced Sorbents for Separation of Metal Ions" Special Issue, Prof. Onjia continues to drive forward the understanding and application of advanced materials, contributing to sustainable solutions for pressing environmental and industrial challenges.

Editorial

Advanced Sorbents for Separation of Metal Ions

Antonije Onjia

Faculty of Technology and Metallurgy, University of Belgrade, Karnegijeva 4, 11120 Belgrade, Serbia; onjia@tmf.bg.ac.rs

1. Introduction and Scope

Effective, sustainable, and selective methods for recovering or removing metals from various media, such as mining leachates, recycling waste, industrial effluents, and natural water, are necessary due to the increasing demand for metals and stringent environmental constraints [1,2]. Separation of metal ions is an essential stage in recovery and removal procedures [3,4].

Adsorption is considered an effective separation technique that offers excellent workability in process operation and design, and the sorbent can be reused after proper regeneration [5,6]. Adsorption methods for separating metal ions based on traditional sorbents (inorganic clays/zeolites, activated carbon, and polymeric resins) face limitations in terms of selectivity, efficiency, and cost-effectiveness [7].

Recently, advanced sorbents made from new and modified materials, including modified natural minerals [8,9], modified carbons/biochar [10,11], agricultural waste and biosorbent [12,13], metal–organic frameworks (MOFs) [14], synthetic polymers [15], magnetic sorbents [16,17], hydrogels [18], and nanosorbents [19], are promising alternatives for overcoming these challenges. These sorbents have a high surface area, tunable pore structures, and functionalized surfaces, which improve their metal-ion adsorption capacity, selectivity, and kinetics [20]. Various methodologies, including surface modification, hybridization, and template-assisted synthesis, have been employed to develop advanced sorbents [21]. The sorbents with incorporated functional groups, chelating agents, or ion-imprinted polymers may exhibit improved affinity and selectivity toward specific metal ions [22].

Metal-ion adsorption using advanced sorbents is significant in various fields. These sorbents are particularly valuable in water treatment processes, where they are used to remove heavy metals from industrial wastewater [23]. In hydrometallurgy, advanced sorbents are essential in the recovery and purification of valuable metals, such as noble metals [24] and rare earth elements [25], from leachates and process streams in mining and metallurgical operations. Furthermore, the separation of actinides from radioactive waste streams [26] has benefited from radiation-resistant sorbents [27]. Environmental remediation also relies on sorbents to remove metal ions from soil [28] and groundwater [29]. In analytical chemistry, dispersive solid-phase microextraction (DSPME) is a promising technique for the preconcentration and clean-up of trace metal ions in complex matrices [30]. The choice of sorbent in these applications depends on several factors: the metal ions of interest, the composition of the aqueous matrix, and the adsorption characteristics (capacity, selectivity, and reusability) [20].

To optimize metal-ion separation performances, it is essential to understand how the physicochemical properties of both the sorbent and the metal ions influence the adsorption mechanisms. The surface complexation model involves the formation of complexes between the functional groups on the sorbent surface and the metal ions driven by electrostatic interactions, ion exchange, and chelation [31]. Other key processes are precipitation and co-precipitation, where metal hydroxides or other insoluble metal compounds form on the sorbent surface or within its porous structure [32]. Reduction and redox reactions can also

Citation: Onjia, A. Advanced Sorbents for Separation of Metal Ions. *Metals* **2024**, *14*, 1026. https://doi.org/10.3390/met14091026

Received: 23 August 2024
Accepted: 6 September 2024
Published: 10 September 2024

Copyright: © 2024 by the author. Licensee MDPI, Basel, Switzerland. This article is an open access article distributed under the terms and conditions of the Creative Commons Attribution (CC BY) license (https://creativecommons.org/licenses/by/4.0/).

occur, with the sorbent surface catalyzing the reduction of metal ions to their zero-valent or lower oxidation states [33]. Electrostatic interactions, which depend on the solution pH and the point of zero charge of the sorbent, result in either attraction or repulsion between the sorbent surface and metal ions [34]. Additionally, pore and intraparticle diffusion are critical for governing the transport of metal ions within the porous sorbent structure, thereby affecting the adsorption kinetics and isotherms [35,36]. In general, understanding these mechanisms is crucial for the design and optimization of sorbents and for the development of predictive adsorption models.

This Special Issue on "Advanced Sorbents for Separation of Metal Ions" in *Metals* brings together up-to-date research that addresses metal-ion separation challenges through innovative sorbent materials and methodologies. A variety of advanced sorbents, including polymeric materials, silica adsorbents, ion exchange resins, and biosorbents, were investigated.

2. Contributions

The contributed articles in this issue explore novel or modified sorbents based on wood ash, bone char, resins, silica, brown seaweed, eggshells, graphene, ionic liquids, and methacrylate polymers, and discuss the adsorption processes in which these sorbents were used.

Liu et al. presented novel silica-based adsorbents impregnated with crown ethers that demonstrate high selectivity for strontium (Sr) ions in concentrated nitric acid solutions. These adsorbents exhibit improved stability and reduced organic leakage, addressing the significant limitation of crown ethers in acidic environments. This study introduced new materials and provided insights into the adsorption mechanisms, paving the way for their application in nuclear waste management.

Smičiklas et al. evaluated the use of wood ash and bone char for manganese (Mn) removal from acid mine drainage (AMD). Their research revealed that wood ash is highly effective due to its neutralization capacity, whereas bone char shows rapid and efficient Mn separation with minimal interference from other ions. This study emphasizes the potential of using waste materials in sustainable AMD treatment processes.

Hansen et al. investigated copper (Cu) biosorption using brown seaweed. Their work identified optimal conditions for maximum copper uptake and efficient biosorbent regeneration, highlighting the potential of seaweed as a cost-effective and environmentally friendly sorbent for copper removal from wastewater.

Marković et al. explored the use of raw eggshells in copper ion biosorption. This study analyzed the process parameters, equilibrium, kinetics, and thermodynamics, demonstrating that eggshells are effective, low-cost adsorbents for copper removal. The research also optimizes the biosorption process using Response Surface Methodology (RSM).

Mikeli et al. focused on the recovery of scandium (Sc) from titanium industry waste using ion-exchange resins. Their findings indicate that certain resins are highly efficient in extracting scandium without significantly affecting the chloride solution, making this process feasible for industrial applications.

Slavković-Beškoski et al. introduced a novel dispersive solid-phase microextraction method using a poly(HDDA)/graphene sorbent for rare earth element (REE) analysis in coal fly ash leachate. The proposed method provides a fast, sensitive, and efficient method for REE determination, demonstrating significant potential for environmental monitoring and resource recovery.

Castillo et al. investigated the ability of an ionic liquid, R4NCy, to extract multiple metal ions from aqueous solutions. The results show high extraction efficiencies for copper, iron, zinc, and manganese, suggesting the applicability of ionic liquids in pre-treatment processes to remove metal impurities from industrial solutions.

Djokić et al. examined the impact of impurities from e-waste electrorefining on the copper extraction processes. Their optimization of the process parameters revealed strategies to enhance copper recovery while minimizing the co-extraction of other metals, contributing to more efficient e-waste recycling methods.

Cerrillo-Gonzalez et al. summarized an overview of electrodialysis (ED) for metal recovery from wastewater. This review discusses the fundamentals, operational features, and key factors affecting ED performance, highlighting its potential in selective metal recovery and the promotion of a circular economy.

Nastasović et al. reviewed the role of methacrylate-based polymeric sorbents in metal recovery from aqueous solutions. This review emphasizes the versatility, efficiency, and regeneration potential of these sorbents, highlighting their applicability in various environmental and industrial processes.

3. Conclusions and Outlook

The articles in this Special Issue demonstrate significant advancements in the development and application of advanced sorbents for metal-ion separation. From novel silica-based adsorbents and ion-exchange resins to sustainable biosorbents and innovative polymeric materials, the contributions highlight diverse approaches to metal separation. These articles offer prospective options for resource recovery and environmental remediation by shedding light on the synthesis of sorbent materials, elucidating their structure, adsorption mechanism, effectiveness, and process optimization.

Future studies should concentrate on scaling up these sorbent applications, improving their selectivity and efficiency, and incorporating them into encompassing process systems to meet increasing demands for metals and the requirements of environmentally friendly waste management solutions.

Acknowledgments: The Guest Editor thanks all of the authors who submitted manuscripts to this Special Issue and the reviewers who put in an immense amount of effort behind the scenes to ensure the Special Issue was of high quality.

Conflicts of Interest: The author declares no conflicts of interest.

List of Contributions

1. Liu, C.; Zhang, S.; Wang, X.; Chen, L.; Yin, X.; Hamza, M.F.; Wei, Y.; Ning, S. Preparation of Two Novel Stable Silica-Based Adsorbents for Selective Separation of Sr from Concentrated Nitric Acid Solution. *Metals* **2024**, *14*, 627. https://doi.org/10.3390/met14060627.
2. Smičiklas, I.; Janković, B.; Jović, M.; Maletaškić, J.; Manić, N.; Dragović, S. Performance Assessment of Wood Ash and Bone Char for Manganese Treatment in Acid Mine Drainage. *Metals* **2023**, *13*, 1665. https://doi.org/10.3390/met13101665.
3. Hansen, H.K.; Gutiérrez, C.; Valencia, N.; Gotschlich, C.; Lazo, A.; Lazo, P.; Ortiz-Soto, R. Selection of Operation Conditions for a Batch Brown Seaweed Biosorption System for Removal of Copper from Aqueous Solutions. *Metals* **2023**, *13*, 1008. https://doi.org/10.3390/met13061008.
4. Marković, M.; Gorgievski, M.; Štrbac, N.; Grekulović, V.; Božinović, K.; Zdravković, M.; Vuković, M. Raw Eggshell as an Adsorbent for Copper Ions Biosorption—Equilibrium, Kinetic, Thermodynamic and Process Optimization Studies. *Metals* **2023**, *13*, 206. https://doi.org/10.3390/met13020206.
5. Mikeli, E.; Marinos, D.; Toli, A.; Pilichou, A.; Balomenos, E.; Panias, D. Use of Ion-Exchange Resins to Adsorb Scandium from Titanium Industry's Chloride Acidic Solution at Ambient Temperature. *Metals* **2022**, *12*, 864. https://doi.org/10.3390/met12050864.
6. Slavković-Beškoski, L.; Ignjatović, L.; Bolognesi, G.; Maksin, D.; Savić, A.; Vladisavljević, G.; Onjia, A. Dispersive Solid–Liquid Microextraction Based on the Poly(HDDA)/Graphene Sorbent Followed by ICP-MS for the Determination of Rare Earth Elements in Coal Fly Ash Leachate. *Metals* **2022**, *12*, 791. https://doi.org/10.3390/met12050791.
7. Castillo, J.; Toro, N.; Hernández, P.; Navarro, P.; Vargas, C.; Gálvez, E.; Sepúlveda, R. Extraction of Cu(II), Fe(III), Zn(II), and Mn(II) from Aqueous Solutions with Ionic Liquid R4NCy. *Metals* **2021**, *11*, 1585. https://doi.org/10.3390/met11101585.
8. Djokić, J.; Radovanović, D.; Nikolovski, Z.; Andjić, Z.; Kamberović, Ž. Influence of Electrolyte Impurities from E-Waste Electrorefining on Copper Extraction Recovery. *Metals* **2021**, *11*, 1383. https://doi.org/10.3390/met11091383.

9. Cerrillo-Gonzalez, M.D.M.; Villen-Guzman, M.; Rodriguez-Maroto, J.M.; Paz-Garcia, J.M. Metal Recovery from Wastewater Using Electrodialysis Separation. *Metals* **2023**, *14*, 38. https://doi.org/10.3390/met14010038.
10. Nastasović, A.; Marković, B.; Suručić, L.; Onjia, A. Methacrylate-Based Polymeric Sorbents for Recovery of Metals from Aqueous Solutions. *Metals* **2022**, *12*, 814. https://doi.org/10.3390/met12050814.

References

1. Hagelüken, C.; Goldmann, D. Recycling and Circular Economy—Towards a Closed Loop for Metals in Emerging Clean Technologies. *Miner. Econ.* **2022**, *35*, 539–562. [CrossRef]
2. Stanković, S.; Kamberović, Ž.; Friedrich, B.; Stopić, S.R.; Sokić, M.; Marković, B.; Schippers, A. Options for Hydrometallurgical Treatment of Ni-Co Lateritic Ores for Sustainable Supply of Nickel and Cobalt for European Battery Industry from South-Eastern Europe and Turkey. *Metals* **2022**, *12*, 807. [CrossRef]
3. Gunarathne, V.; Rajapaksha, A.U.; Vithanage, M.; Alessi, D.S.; Selvasembian, R.; Naushad, M.; You, S.; Oleszczuk, P.; Ok, Y.S. Hydrometallurgical Processes for Heavy Metals Recovery from Industrial Sludges. *Crit. Rev. Environ. Sci. Technol.* **2022**, *52*, 1022–1062. [CrossRef]
4. Castro, L.; Blázquez, M.L.; Muñoz, J.Á. Leaching/Bioleaching and Recovery of Metals. *Metals* **2021**, *11*, 1732. [CrossRef]
5. Shrestha, R.; Ban, S.; Devkota, S.; Sharma, S.; Joshi, R.; Tiwari, A.P.; Kim, H.Y.; Joshi, M.K. Technological Trends in Heavy Metals Removal from Industrial Wastewater: A Review. *J. Environ. Chem. Eng.* **2021**, *9*, 105688. [CrossRef]
6. Tadić, T.; Marković, B.; Vuković, Z.; Stefanov, P.; Maksin, D.; Nastasović, A.; Onjia, A. Fast Gold Recovery from Aqueous Solutions and Assessment of Antimicrobial Activities of Novel Gold Composite. *Metals* **2023**, *13*, 1864. [CrossRef]
7. Chai, W.S.; Cheun, J.Y.; Kumar, P.S.; Mubashir, M.; Majeed, Z.; Banat, F.; Ho, S.-H.; Show, P.L. A Review on Conventional and Novel Materials towards Heavy Metal Adsorption in Wastewater Treatment Application. *J. Clean. Prod.* **2021**, *296*, 126589. [CrossRef]
8. Marjanovic, V.; Peric-Grujic, A.; Ristic, M.; Marinkovic, A.; Markovic, R.; Onjia, A.; Sljivic-Ivanovic, M. Selenate Adsorption from Water Using the Hydrous Iron Oxide-Impregnated Hybrid Polymer. *Metals* **2020**, *10*, 1630. [CrossRef]
9. Godage, N.H.; Gionfriddo, E. Use of Natural Sorbents as Alternative and Green Extractive Materials: A Critical Review. *Anal. Chim. Acta* **2020**, *1125*, 187–200. [CrossRef]
10. Deshwal, N.; Singh, M.B.; Bahadur, I.; Kaushik, N.; Kaushik, N.K.; Singh, P.; Kumari, K. A Review on Recent Advancements on Removal of Harmful Metal/Metal Ions Using Graphene Oxide: Experimental and Theoretical Approaches. *Sci. Total Environ.* **2023**, *858*, 159672. [CrossRef]
11. Liu, Z.; Xu, Z.; Xu, L.; Buyong, F.; Chay, T.C.; Li, Z.; Cai, Y.; Hu, B.; Zhu, Y.; Wang, X. Modified Biochar: Synthesis and Mechanism for Removal of Environmental Heavy Metals. *Carbon Res.* **2022**, *1*, 8. [CrossRef]
12. Imran-Shaukat, M.; Wahi, R.; Ngaini, Z. The Application of Agricultural Wastes for Heavy Metals Adsorption: A Meta-Analysis of Recent Studies. *Bioresour. Technol. Rep.* **2022**, *17*, 100902. [CrossRef]
13. Syeda, H.I.; Sultan, I.; Razavi, K.S.; Yap, P.-S. Biosorption of Heavy Metals from Aqueous Solution by Various Chemically Modified Agricultural Wastes: A Review. *J. Water Process Eng.* **2022**, *46*, 102446. [CrossRef]
14. Lin, G.; Zeng, B.; Li, J.; Wang, Z.; Wang, S.; Hu, T.; Zhang, L. A Systematic Review of Metal Organic Frameworks Materials for Heavy Metal Removal: Synthesis, Applications and Mechanism. *Chem. Eng. J.* **2023**, *460*, 141710. [CrossRef]
15. Berber, M.R. Current Advances of Polymer Composites for Water Treatment and Desalination. *J. Chem.* **2020**, *2020*, 7608423. [CrossRef]
16. Marković, B.M.; Vuković, Z.M.; Spasojević, V.V.; Kusigerski, V.B.; Pavlović, V.B.; Onjia, A.E.; Nastasović, A.B. Selective Magnetic GMA Based Potential Sorbents for Molybdenum and Rhenium Sorption. *J. Alloys Compd.* **2017**, *705*, 38–50. [CrossRef]
17. Suručić, L.; Tadić, T.; Janjić, G.; Marković, B.; Nastasović, A.; Onjia, A. Recovery of Vanadium (V) Oxyanions by a Magnetic Macroporous Copolymer Nanocomposite Sorbent. *Metals* **2021**, *11*, 1777. [CrossRef]
18. Sun, R.; Gao, S.; Zhang, K.; Cheng, W.-T.; Hu, G. Recent Advances in Alginate-Based Composite Gel Spheres for Removal of Heavy Metals. *Int. J. Biol. Macromol.* **2024**, *268*, 131853. [CrossRef]
19. Naseer, A. Role of Nanocomposites and Nano Adsorbents for Heavy Metals Removal and Dyes. An Overview. *Desalination Water Treat.* **2024**, *320*, 100662. [CrossRef]
20. Rajendran, S.; Priya, A.K.; Senthil Kumar, P.; Hoang, T.K.A.; Sekar, K.; Chong, K.Y.; Khoo, K.S.; Ng, H.S.; Show, P.L. A Critical and Recent Developments on Adsorption Technique for Removal of Heavy Metals from Wastewater-A Review. *Chemosphere* **2022**, *303*, 135146. [CrossRef]
21. Kaur, A.; Bajaj, B.; Kaushik, A.; Saini, A.; Sud, D. A Review on Template Assisted Synthesis of Multi-Functional Metal Oxide Nanostructures: Status and Prospects. *Mater. Sci. Eng. B* **2022**, *286*, 116005. [CrossRef]
22. Zhang, X.; Zhang, K.; Shi, Y.; Xiang, H.; Yang, W.; Zhao, F. Surface Engineering of Multifunctional Nanostructured Adsorbents for Enhanced Wastewater Treatment: A Review. *Sci. Total Environ.* **2024**, *920*, 170951. [CrossRef] [PubMed]
23. Krishnan, S.; Zulkapli, N.S.; Kamyab, H.; Taib, S.M.; Din, M.F.B.M.; Majid, Z.A.; Chaiprapat, S.; Kenzo, I.; Ichikawa, Y.; Nasrullah, M.; et al. Current Technologies for Recovery of Metals from Industrial Wastes: An Overview. *Environ. Technol. Innov.* **2021**, *22*, 101525. [CrossRef]

24. Zupanc, A.; Install, J.; Jereb, M.; Repo, T. Sustainable and Selective Modern Methods of Noble Metal Recycling. *Angew. Chem. Int. Ed.* **2023**, *62*, e202214453. [CrossRef] [PubMed]
25. Bishop, B.A.; Alam, M.S.; Flynn, S.L.; Chen, N.; Hao, W.; Ramachandran Shivakumar, K.; Swaren, L.; Gutierrez Rueda, D.; Konhauser, K.O.; Alessi, D.S.; et al. Rare Earth Element Adsorption to Clay Minerals: Mechanistic Insights and Implications for Recovery from Secondary Sources. *Environ. Sci. Technol.* **2024**, *58*, 7217–7227. [CrossRef]
26. Bao, L.; Cai, Y.; Liu, Z.; Li, B.; Bian, Q.; Hu, B.; Wang, X. High Sorption and Selective Extraction of Actinides from Aqueous Solutions. *Molecules* **2021**, *26*, 7101. [CrossRef]
27. Hashim, K.S.; Shaw, A.; AlKhaddar, R.; Kot, P.; Al-Shamma'a, A. Water Purification from Metal Ions in the Presence of Organic Matter Using Electromagnetic Radiation-Assisted Treatment. *J. Clean. Prod.* **2021**, *280*, 124427. [CrossRef]
28. Campillo-Cora, C.; Conde-Cid, M.; Arias-Estévez, M.; Fernández-Calviño, D.; Alonso-Vega, F. Specific Adsorption of Heavy Metals in Soils: Individual and Competitive Experiments. *Agronomy* **2020**, *10*, 1113. [CrossRef]
29. Al-Hashimi, O.; Hashim, K.; Loffill, E.; Marolt Čebašek, T.; Nakouti, I.; Faisal, A.A.H.; Al-Ansari, N. A Comprehensive Review for Groundwater Contamination and Remediation: Occurrence, Migration and Adsorption Modelling. *Molecules* **2021**, *26*, 5913. [CrossRef]
30. Uzcan, F.; Soylak, M. Magnetic Dispersive Solid Phase Microextraction of Cadmium on Fe_3O_4@MIL-53-(Fe)@Ti_3AlC_2 from Edible Offal and Water Samples. *J. Food Compos. Anal.* **2024**, *135*, 106636. [CrossRef]
31. Haoli, Q.; Yan, L.; Mei, L.; Yang, Y.; Ya, A. Adsorption Behavior of Cadmium in Argillaceous Limestone Yellow Soils Simulated by The Surface Complexation Model. *Water Air Soil Pollut.* **2024**, *235*, 497. [CrossRef]
32. Da Conceição, F.T.; Da Silva, M.S.G.; Menegário, A.A.; Antunes, M.L.P.; Navarro, G.R.B.; Fernandes, A.M.; Dorea, C.; Moruzzi, R.B. Precipitation as the Main Mechanism for Cd(II), Pb(II) and Zn(II) Removal from Aqueous Solutions Using Natural and Activated Forms of Red Mud. *Environ. Adv.* **2021**, *4*, 100056. [CrossRef]
33. Sepehri, S.; Kanani, E.; Abdoli, S.; Rajput, V.D.; Minkina, T.; Asgari Lajayer, B. Pb(II) Removal from Aqueous Solutions by Adsorption on Stabilized Zero-Valent Iron Nanoparticles—A Green Approach. *Water* **2023**, *15*, 222. [CrossRef]
34. Akpomie, K.G.; Conradie, J.; Adegoke, K.A.; Oyedotun, K.O.; Ighalo, J.O.; Amaku, J.F.; Olisah, C.; Adeola, A.O.; Iwuozor, K.O. Adsorption Mechanism and Modeling of Radionuclides and Heavy Metals onto ZnO Nanoparticles: A Review. *Appl. Water Sci.* **2023**, *13*, 20. [CrossRef]
35. Wang, J.; Guo, X. Adsorption Kinetics and Isotherm Models of Heavy Metals by Various Adsorbents: An Overview. *Crit. Rev. Environ. Sci. Technol.* **2023**, *53*, 1837–1865. [CrossRef]
36. Marković, D.D.; Lekić, B.M.; Rajaković-Ognjanović, V.N.; Onjia, A.E.; Rajaković, L.V. A New Approach in Regression Analysis for Modeling Adsorption Isotherms. *Sci. World J.* **2014**, *2014*, 930879. [CrossRef]

Disclaimer/Publisher's Note: The statements, opinions and data contained in all publications are solely those of the individual author(s) and contributor(s) and not of MDPI and/or the editor(s). MDPI and/or the editor(s) disclaim responsibility for any injury to people or property resulting from any ideas, methods, instructions or products referred to in the content.

Article

Preparation of Two Novel Stable Silica-Based Adsorbents for Selective Separation of Sr from Concentrated Nitric Acid Solution

Chang Liu [1], Shichang Zhang [2], Xinpeng Wang [1,*], Lifeng Chen [3], Xiangbiao Yin [3], Mohammed F. Hamza [3], Yuezhou Wei [3,4] and Shunyan Ning [3,*]

1. State Key Laboratory of Featured Metal Materials and Life-Cycle Safety for Composite Structures, MOE Key Laboratory of New Processing Technology for Nonferrous Metals and Materials, School of Resources, Environment and Materials, Guangxi University, Nanning 530004, China; liuchangguangxiedu@st.gxu.edu.cn
2. School of Nuclear Science and Technology, University of Science and Technology of China, Hefei 230026, China; zhangshichang@mail.ustc.edu.cn
3. School of Nuclear Science and Technology, University of South China, 28 Changsheng West Road, Hengyang 421001, China; chenlf@usc.edu.cn (L.C.); yinxb@usc.edu.cn (X.Y.); m_fouda21@usc.edu.cn (M.F.H.); yzwei@usc.edu.cn (Y.W.)
4. School of Nuclear Science and Engineering, Shanghai Jiao Tong University, Shanghai 200240, China
* Correspondence: wangxinpeng@gxu.edu.cn (X.W.); ningshunyan@usc.edu.cn (S.N.); Tel.: +86-0771-3392507 (X.W.)

Citation: Liu, C.; Zhang, S.; Wang, X.; Chen, L.; Yin, X.; Hamza, M.F.; Wei, Y.; Ning, S. Preparation of Two Novel Stable Silica-Based Adsorbents for Selective Separation of Sr from Concentrated Nitric Acid Solution. *Metals* 2024, 14, 627. https://doi.org/10.3390/met14060627

Academic Editor: Antonije Onjia

Received: 26 April 2024
Revised: 19 May 2024
Accepted: 21 May 2024
Published: 25 May 2024

Copyright: © 2024 by the authors. Licensee MDPI, Basel, Switzerland. This article is an open access article distributed under the terms and conditions of the Creative Commons Attribution (CC BY) license (https://creativecommons.org/licenses/by/4.0/).

Abstract: Crown ethers are famous for the highly selectively grab Sr(II) from concentrated nitric acid solution due to the size match, but they suffer from the high leakage into the liquid phase caused by the presence of a large number of hydrophilic groups. To reduce their leakage, two novel porous silica-based adsorbents, (DtBuCH18C6 + Dodec)/SiAaC-g-ABSA and (DtBuCH18C6 + Dodec)/SiAaC-g-3-ABSA, were prepared by vacuum impregnation with organic contents of about 55.9 wt.% and 56.1 wt.%, respectively. The two adsorbents have good reusability and structural stability, and the total organic carbon leakage rates in 2 M HNO_3 solution are lower than 0.56 wt.% and 0.29 wt.%, respectively. Batch adsorption experiments revealed that the two adsorbents possessed good adsorption selectivity towards Sr, with $SF_{Sr/M}$ over 40, except that of $SF_{Sr/Ba}$ in 2 M HNO_3 solution. The adsorption equilibrium of Sr in 2 M HNO_3 solution was reached within 1 h, with saturated adsorption capacities of 36.9 mg/g and 37.5 mg/g, respectively. Furthermore, the XPS results indicate that the adsorption mechanism is the coordination of the crown ether ring with Sr. This work not only develops two novel adsorbents for the separation of Sr in nitric acid environments; it also provides a method for effectively reducing the water solubility of crown ethers.

Keywords: strontium; adsorption; crown ethers; high-level liquid waste

1. Introduction

Nuclear energy plays an increasingly important role in the restructuring of energy sources because of its cleanliness and high efficiency [1]. The development of nuclear energy will inevitably produce large quantities of nuclear spent fuel. The PUREX process (plutonium uranium recovery by extraction) effectively separates U and Pu from nuclear spent fuel, while other fission products, including minor actinides (Np, Am, and Cm), long-lived fission products (^{99}Tc, ^{129}I, etc.), and high heat-generating elements (^{90}Sr and ^{137}Cs), are retained in the high-level liquid waste (HLLW) [2,3]. The minor actinides and long-lived fission products can be converted into short-lived or stabilized nuclides through partitioning and transmutation strategies, which can effectively shorten the potential threat time of HLLW [4]. However, the radiological and biochemical toxicity of other nuclides, such as ^{90}Sr, still exists in the HLLW, which will have a significant negative impact on the vitrification of HLLW. Therefore, it is necessary to remove ^{90}Sr from HLLW prior to the vitrification process.

HLLWs are characterized by high radioactivity, multiple components, and high acidity, which makes the efficient separation of ^{90}Sr ($T_{1/2}$ = 28.8 a) from it extremely challenging [5–7]. Crown ethers and their derivatives carry cavities that can coordinate with Sr in high-concentration nitric acid environments and have the potential to separate Sr from HLLW [8–10]. However, the extraction of Sr by crown ethers and their derivatives usually requires the consumption of large quantities of organic diluents, which will inevitably generate a large amount of organic waste. In contrast, organic–inorganic hybrid adsorbent materials prepared by impregnating organic ligands into stabilized carriers combine the excellent properties of organic ligands with the stability of carriers and produce almost no organic waste [11–13]. Chen et al. [14] prepared a novel silica-based adsorbent by impregnating 4′,4″(5″)-di-tert-butyldicyclohexano-18-crown-6 (DtBuCH18C6) into the interior of the porous carrier (SiO$_2$-P), which had an adsorption capacity of 0.43 mmol/g of Sr in 2 M HNO$_3$ solution. However, the presence of a large number of oxygen atoms in the structure of the crown ether leads to its high hydrophilicity, which makes it prone to leakage when adsorbing Sr (II) [14].

To address the above issues, researchers have modified crown ethers to reduce their leakage in the aqueous phase. Several organic ligands, such as 1-dodecanol, tri-n-butyl phosphate, and dodecyl benzenesulfonic acid, are prone to form hydrogen bonds with crown ethers, and impregnating them into porous carriers can effectively reduce the leakage of crown ethers and thus improve the adsorption efficiency of Sr [15–17]. The modification of crown ethers' adsorbents using silica-based hybrid carriers could be applied to separate Sr in nitric acid environments.

But as with the existence of hydrophilic sulfonic acid group, the total organic carbon (TOC) in the liquid phase was still high due to the leakage of organic content from the adsorbent during the adsorption process, which needs to be further improved [17]. In this work, we investigated the static and dynamic adsorption properties of two novel stable silica-based adsorbents, (DtBuCH18C6 + Dodec)/SiAaC-g-ABSA and (DtBuCH18C6 + Dodec)/SiAaC-g-3-ABSA, by batch and column experiments on the two adsorbents on Sr. This material is different from the material in previous studies that was only impregnated with DtBuCH18C6 and dodecanol to the carrier SiO$_2$-P [14]. In this work, aminobenzene sulfonic acid (ABSA) and 3-aminobenzene sulfonic acid (3-ABSA) were grafted onto the previously synthesized SiAaC and then impregnated with DtBuCH18C6 and dodecanol to further improve the stability and adsorption properties of the materials [18]. The experimental data were fitted and analyzed to obtain the main adsorption parameters. The reusability and structural stability of the two adsorbents were studied. Finally, the XPS technique was used to study the change in binding energy of functional groups before and after adsorption to reveal the adsorption mechanism.

2. Experimental

2.1. Chemicals

4′,4″(5″)-di-tert-butyldicyclohexano-18-crown-6 (DtBuCH18C6, 90%) was purchased from Sigma-Aldrich (Shanghai, China). Dimethyl sulfoxide (DMSO, AR), 3-(diethoxyphosphoryloxy)-1,2,3-benzotriazin-4(3H)-one (DEPBT, 98%), 1-dodecanol (Dodec, 98%), aminobenzene sulfonic acid (ABSA, AR), and 3-aminobenzene sulfonic acid (3-ABSA, 98%) were purchased from Shanghai Aladdin Biochemical Technology Co., Ltd., (Aladdin Industrial, Inc., Shanghai, China). Other chemicals, including Sr(NO$_3$)$_2$·6H$_2$O, used in this work were analytical grade and purchased from Shanghai Macklin Biochemical Technology Co., Ltd., (Aladdin Industrial, Inc., Shanghai, China).

2.2. Synthesis

The porous silica-based carrier SiAaC containing -COOH group was functionalized and modified to prepare two novel adsorbents, named (DtBuCH18C6 + Dodec)/SiAaC-g-ABSA and (DtBuCH18C6 + Dodec)/SiAaC-g-3-ABSA, respectively. The preparation process of SiAaC was provided in our previous work [18]. Figure 1 shows the synthesis processes of the two adsorbents. More details follow below.

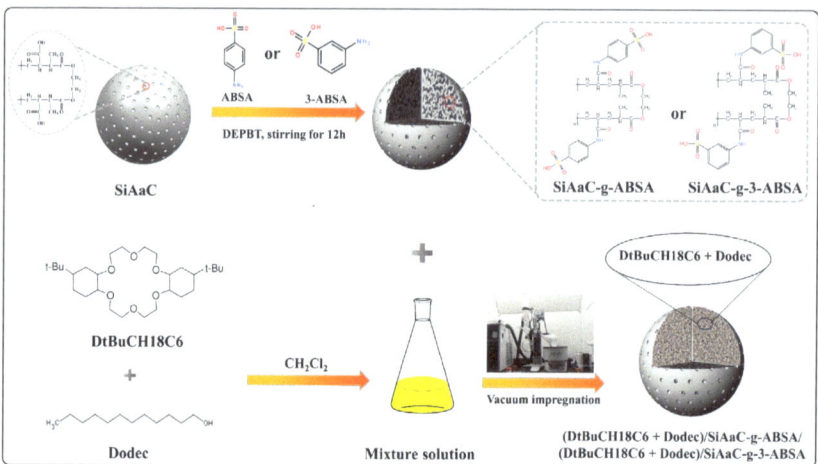

Figure 1. Schematic diagram of the preparation processes of the adsorbents.

SiAaC was amino-modified according to the following procedure. Firstly, 5 g of SiAaC and 50 mL of DMSO were placed in a beaker and mixed at room temperature, after which 5 g of ASAB or 3-ABSA was added. Then, 3 g catalyst of DEPBT was added to improve the grafting efficiency. The above mixed solution was stirred at 200 rpm for 12 h at room temperature. The collected products were named SiAaC-g-ABSA and SiAaC-g-3-ABSA, respectively.

(DtBuCH18C6 + Dodec)/SiAaC-g-ABSA and (DtBuCH18C6 + Dodec)/SiAaC-g-3-ABSA were prepared by the vacuum impregnation method, as detailed below. In total, 5 g of SiAaC-g-ABSA or SiAaC-g-3-ABSA was weighted and placed into a glass flask. Then, 0.9 g DtBuCH$_{18}$C$_6$ and 0.3 g Dodec were dissolved into 200 mL of CH$_2$Cl$_2$, and the mixture solution was added to the flask. The glass flask was fixed to a rotary evaporator (EYELA, N-300 V-WB, Tokyo, Japan) and rotated at 200 rpm. Decompression was carried out at a rate of 20 pha/10 min until the CH$_2$Cl$_2$ was completely evaporated. The collected products are named (DtBuCH18C6 + Dodec)/SiAaC-g-ABSA and (DtBuCH18C6 + Dodec)/SiAaC-g-3-ABSA, respectively.

2.3. Characterization

The SEM technique (HITACHI SU8200, Tokyo, Japan) was used to obtain the surface morphology of the adsorbents. TG-DSC analyses were conducted under an oxygen environment, using a NETZSCH STA 449F3 analyzer at a heating rate of 10 °C/min to analyze the thermal stability and component of the adsorbents (Selb, Germany). The changes in functional groups and chemical state were studied by FTIR (SHIMADZU, IRTracer-100, Tokyo, Japan) and XPS (Scalab250XI, C1s: 284.6 eV, Paris, France) analysis. The content of total organic carbon in solution was measured by a TOC analyzer (SHIMADZU, VCPH).

2.4. Batch Experiments

The adsorption performances of (DtBuCH18C6 + Dodec)/SiAaC-g-ABSA and (DtBuCH18C6 + Dodec)/SiAaC-g-3-ABSA on Sr were evaluated by batch adsorption experiments. In total, 0.1 g of the adsorbent and 5 mL HNO$_3$ solution containing Sr (NO$_3$)$_2$ were mixed in a glass vial, respectively. The vial was secured to a water-bath shaker oscillating at a rate of 160 rpm. The concentrations of Sr and other metal ions in the solution before and after adsorption were determined by atomic absorption spectrometry (AAS, SHIMADZU

AA-7000, Japan) and inductively coupled plasma optical emission spectroscopy (ICP-OES, Thermo ICAP 7000, Waltham, MA, USA), respectively. The adsorption capacity Q (mg/g), adsorption efficiency E (%), distribution coefficient K_d (mL/g) and separation factor (SF) are calculated by Equations (1)–(4) [19–21]:

$$Q = \frac{(C_o - C)}{m} \times V \tag{1}$$

$$E = \frac{(C_o - C)}{C_o} \times 100\% \tag{2}$$

$$K_d = \frac{(C_o - C)}{C} \times \frac{V}{m} \tag{3}$$

$$SF_{A/B} = \frac{K_{d\,A}}{K_{d\,B}} \tag{4}$$

where C_o and C (mg/L) are the initial and equilibrium concentrations of the metal ions in the solutions, respectively; and V (mL) and m (g) are the aqueous phase volume and adsorbent mass, respectively.

2.5. Column Experiments

The dynamic adsorption behavior of (DtBuCH18C6 + Dodec)/SiAaC-g-ABSA and (DtBuCH18C6 + Dodec)/SiAaC-g-3-ABSA on Sr was investigated using column experiments, respectively. A total of 1.9981 g of the adsorbent was weighed and packed into a glass column ($\phi \times h$ = 10 mm × 100 mm). A peristaltic pump (EYELA MP 2000, Japan) was used to pump the feed solution (C_o = 461 mg Sr/L) into the packed column at a rate of 0.5 mL/min, and the effluent was collected using a fraction collector (EYELA DC1500C, Japan). The concentration of Sr in the solution was determined by atomic absorption spectrometry (AAS, SHIMADZU AA-7000, Japan). The experimental data were fitted by the Thomas model (Equation (5)) [22,23].

$$\frac{C}{C_o} = \frac{1}{1 + exp(\frac{K_{Th}}{v}(q_o m - C_o V))} \tag{5}$$

where C_o and C are the Sr concentrations (mg/L) in the feed solution and effluent, respectively; K_{Th} (mL·g^{-1}·min^{-1}) and v (mL/min) are the Thomas constant and flow rate, respectively; q_o (mg/g) is the column adsorption capacity; and m (g) and V (mL) are the amount of adsorbent and the effluent volume.

3. Results and Discussion

3.1. Characterization of the Materials

The SEM technique was used to study the surface morphology of (DtBuCH18C6 + Dodec)/SiAaC-g-ABSA and (DtBuCH18C6 + Dodec)/SiAaC-g-3-ABSA, and the results are shown in Figure 2a,b. The two adsorbents have regular spherical morphology with a particle size of 40–100 µm. Figure 2c,d show the TG-DSC results of the two adsorbents. As the temperature increases from 25 °C to 800 °C, from the TG curves, the two adsorbents undergo mass loss until about 450 °C and then remain stable. The organic contents of (DtBuCH18C6 + Dodec)/SiAaC-g-ABSA and (DtBuCH18C6 + Dodec)/SiAaC-g-3-ABSA are 55.9 wt.% and 56.1 wt.%, respectively. According to the DSC curves, several characteristic peaks resulting from mass decomposition are clearly observed. The characteristic peak near 100 °C originates from the evaporation of water in the adsorbents. The broad characteristic peaks between 100 °C and 496 °C derive from the thermal decomposition of organic matter in the adsorbents.

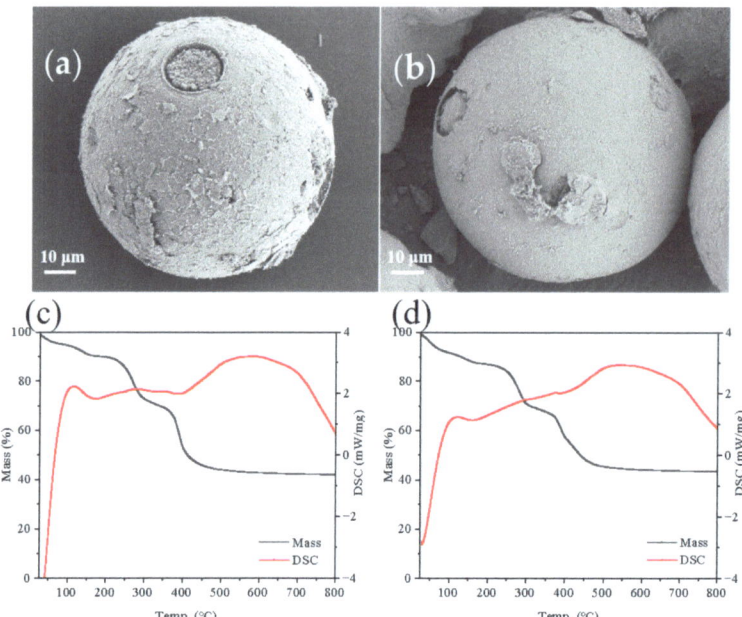

Figure 2. (**a**,**b**) SEM and (**c**,**d**) TG-DSC results of (DtBuCH18C6 + Dodec)/SiAaC-g-ABSA and (DtBuCH18C6 + Dodec)/SiAaC-g-3-ABSA.

The FTIR technique was employed to investigate the changes in functional groups during the preparation of the adsorbents, and the results are shown in Figure 3. The peaks at 469 cm^{-1}, 803 cm^{-1}, and 1096 cm^{-1} originate from the stretching vibration of the Si-O-Si [24]. After sulfonation modification of SiAaC using 3-ABSA and ABSA, the characteristic peak of S-O is observed at around 568 cm^{-1} [17]. Due to the partial overlap of the peak positions of S=O and Si-O, the S=O peak in the FTIR spectra of SiAaC-g-3-ABSA and SiAaC-g-ABSA cannot be observed as an independent peak. After the modification of SiAaC-g-3-ABSA and SiAaC-g-ABSA using DtBuCH18C6 and Dodec, the C-O-C characteristic peaks (around 1260 cm^{-1}) are significantly enhanced. The peak at 2929–2961 cm^{-1} derives from the C-H vibration [25]. The peak around 3461 cm^{-1} is from the adsorbed water [26].

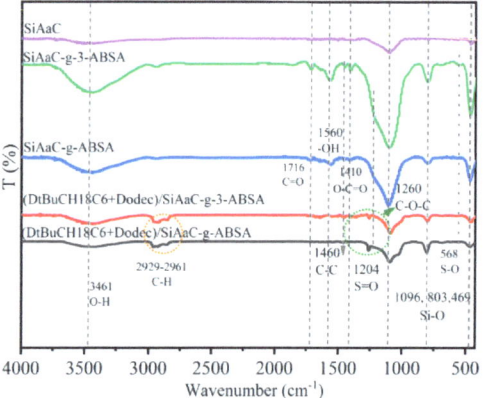

Figure 3. FTIR results of SiAaC, SiAaC-g-3-ABSA, SiAaC-g-ABSA, (DtBuCH18C6 + Dodec)/SiAaC-g-3-ABSA and (DtBuCH18C6 + Dodec)/SiAaC-g-ABSA.

3.2. Batch Adsorption Experiments

3.2.1. Adsorption Selectivity

Figure 4 shows the adsorption selectivity of Sr by (DtBuCH18C6 + Dodec)/SiAaC-g-ABSA and (DtBuCH18C6 + Dodec)/SiAaC-g-3-ABSA in simulated high-level liquid waste. At a nitric acid concentration of 0.5 M, the sulfonic acid group promotes the coordination of the crown ether ring with Sr, resulting in the two adsorbents exhibiting good adsorption of Sr [17]. As the nitric acid concentration increases to 1 M, the coordination promotion of the sulfonic acid groups to the crown ether ring is inhibited, leading to a weakening of the adsorption of Sr by the two adsorbents. As the nitric acid concentration increases to 2 M, the coordination of the crown ether ring to Sr is enhanced, leading to a rise in Sr adsorption. With the further increase in nitric acid concentration, there is competition adsorption of nitric acid with Sr, resulting in the weakening of the adsorption of Sr by the two adsorbents (Equations (6)–(8)). In addition, the two adsorbents show weak adsorption on Ba and Pd in the simulated high-level liquid waste in Figure 4, and almost no adsorption for other metal ions, with $SF_{Sr/M}$ over 40, except that of $SF_{Sr/Ba}$ in the simulated high-level liquid waste of 2 M HNO_3. It indicates good adsorption selectivity of the two adsorbents on Sr.

$$Sr^{2+} + 2RSO_3^- + DtBuCH18C6 \leftrightarrow Sr(RSO_3)_2 \cdot DtBuCH18C6 \tag{6}$$

$$Sr^{2+} + 2NO_3^- + DtBuCH18C6 \leftrightarrow Sr(NO_3)_2 \cdot DtBuCH18C6 \tag{7}$$

$$HNO_3 + DtBuCH18C6 \leftrightarrow HNO_3 \cdot DtBuCH18C6 \tag{8}$$

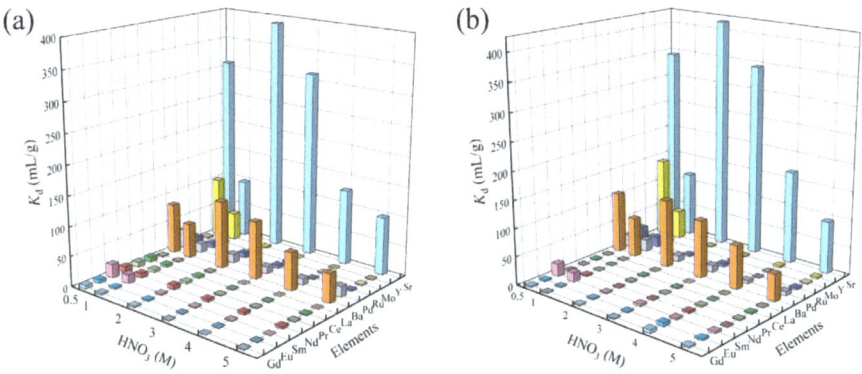

Figure 4. Adsorption selectivity of (**a**) (DtBuCH18C6 + Dodec)/SiAaC-g-ABSA and (**b**) (DtBuCH18C6 + Dodec)/SiAaC-g-3-ABSA on Sr in different HNO_3 concentrations ([M] = 1 mM, m/V = 0.1 g/5 mL, t = 6 h, T = 298 K).

3.2.2. Kinetics

The adsorption kinetics of (DtBuCH18C6 + Dodec)/SiAaC-g-ABSA and (DtBuCH18C6 + Dodec)/SiAaC-g-3-ABSA on Sr in 2 M HNO_3 solution were studied. According to Figure 5, with the increase in adsorption time, the adsorption of Sr by the adsorbent first increased rapidly and then remained stable. The adsorption equilibriums of Sr by the two adsorbents obtained within 1 h with the adsorption capacities were about 36.9 mg/g and 37.5 mg/g.

Moreover, a pseudo-second-order kinetics model (Equation (9)) was adopted to analyze the experimental data [27]. The results are shown in Figure 5 and Table 1. The correlation coefficients (R^2) were higher than 0.99, indicating the good applicability of the pseudo-second-order model to the adsorption processes. Therefore, it can be hypothesized that the adsorption of Sr by the two adsorbents is chemisorption [28,29].

$$Q_t = \frac{k_2 Q_e^2 t}{1 + k_2 Q_e t} \qquad (9)$$

where Q_e and Q_t (mg/g) are equilibrium and adsorption capacities at time, t (min), respectively; k_2 (min^{-1}) is the adsorption rate constants of pseudo-second order.

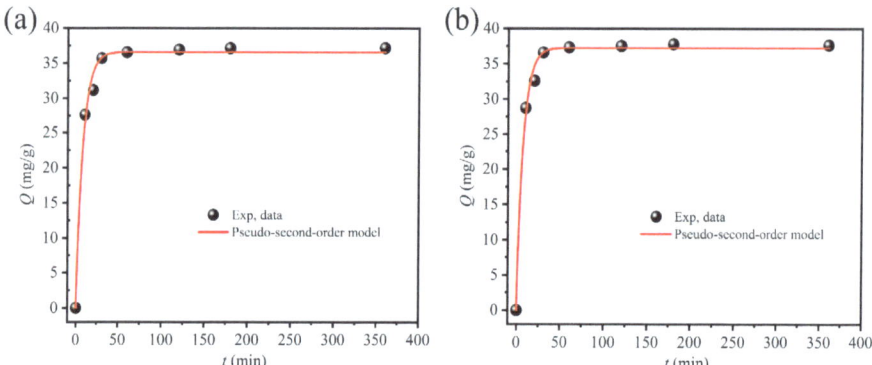

Figure 5. Adsorption kinetics of (**a**) (DtBuCH18C6 + Dodec)/SiAaC-g-ABSA and (**b**) (DtBuCH18C6 + Dodec)/SiAaC-g-3-ABSA on Sr in 2 M HNO$_3$ concentration (C_o = 5 mM, m/V = 0.1 g/5 mL, T = 298 K).

Table 1. Kinetics parameters of Sr by the two adsorbents.

T (K)	Adsorbents	Pseudo-Second-Order Model		
		K_2 (g·h/mg)	Q_e (mg/g)	R^2
298	(DtBuCH18C6 + Dodec)/SiAaC-g-ABSA	0.13	36.6	0.99
	(DtBuCH18C6 + Dodec)/SiAaC-g-3-ABSA	0.14	37.4	0.99

3.2.3. Isotherms

The adsorption isotherms of Sr by (DtBuCH18C6 + Dodec)/SiAaC-g-ABSA and (DtBuCH18C6 + Dodec)/SiAaC-g-3-ABSA in 2 M HNO$_3$ solution were studied by varying the initial Sr concentration of the solution, with the results shown in Figure 6. The adsorption of Sr by the two adsorbents increased significantly with the increase in the equilibrium ion concentration, and the maximum adsorption capacity was about 39.1 mg/g and 39.2 mg/g, respectively.

Moreover, Langmuir and Freundlich isotherm models were applied to analyze the experimental data [30–32]. The results are shown in Figure 6 and Table 2. According to the fitting results, the Langmuir model has a higher correlation coefficient (R^2 = 0.99) compared to the Freundlich model, indicating that the Langmuir model has better applicability to the adsorption process. Therefore, it can be concluded that the adsorption of Sr by the two adsorbents is monomolecular layer chemisorption [24,33,34].

$$Q_e = \frac{q_m \times K_L \times C_e}{1 + K_L \times C_e} \qquad (10)$$

$$Q_e = K_F \times C_e^{\frac{1}{n}} \qquad (11)$$

where q_m (mg/g) is the calculated saturation adsorption capacity; C_e (mM) is the equilibrium ions concentration; K_L (L/mg) and K_F (mg$^{1-n}\cdot$L^n/g) are the Langmuir and Freundlich model constants, respectively; and n is the adsorption intensity.

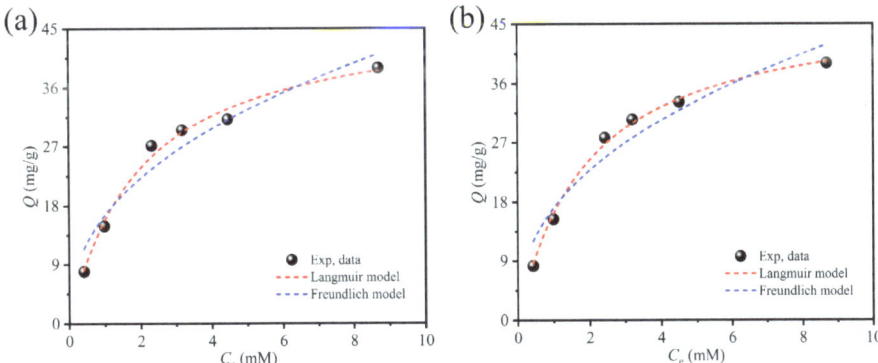

Figure 6. Adsorption isotherms of (**a**) (DtBuCH18C6 + Dodec)/SiAaC-g-ABSA and (**b**) (DtBuCH18C6 + Dodec)/SiAaC-g-3-ABSA on Sr in 2 M HNO$_3$ concentration (m/V = 0.1 g/5 mL, T = 298 K).

Table 2. Isotherms parameters of Sr by the two adsorbents.

Adsorbents	Langmuir Model			Freundlich Model		
	K_L (L/mg)	q_m (mg/g)	R^2	n	K_F (mg$^{1-n}\cdot$L^n/g)	R^2
(DtBuCH18C6 + Dodec)/SiAaC-3-ABSA	0.50	47.4	0.99	0.50	16.6	0.94
(DtBuCH18C6 + Dodec)/SiAaC-g-3-ABSA	0.51	48.2	0.99	0.4	17.1	0.93

3.3. Reusability and Stability

To effectively assess the utility of (DtBuCH18C6 + Dodec)/SiAaC-g-ABSA and (DtBuCH18C6 + Dodec)/SiAaC-g-3-ABSA, the reusability of the two adsorbents was studied. The desorption capacity, Q_d (mg/g), and desorption efficiency, E_d (%), are calculated by Equation (12) and Equation (13), respectively [28]:

$$Q_d = C_d \times \frac{V}{m} \tag{12}$$

$$E_d = Q_d/Q \times 100\% \tag{13}$$

where C_d (mg/L) means the concentration of Sr in the desorption solution.

According to Figure 7, pure water is able to effectively desorb the adsorbed Sr by the two adsorbents with the desorption efficiencies of about 77% and 82%, respectively. The adsorption efficiency of the adsorbent for Sr decreases a little as the number of adsorption–desorption cycles increases. After five adsorption–desorption cycles, the adsorption efficiencies of the adsorbent for Sr are still kept at about 57% and 75%, and the desorption efficiencies are 78% and 85%, respectively. The above results indicate that the two adsorbents have good reusability.

The structural stability of the two adsorbents was investigated by determining the total organic carbon (TOC) content in the aqueous phase. The TOC leakage of the two adsorbents in different concentrations of nitric acid solutions is shown in Figure 8. In 0.1–5 M nitric acid solutions, the TOC leakages of the two adsorbents are less than 188 ppm and 95 ppm, respectively. The calculated TOC leakage rates for (DtBuCH18C6 + Dodec)/SiAaC-g-ABSA and (DtBuCH18C6 + Dodec)/SiAaC-g-3-ABSA are less than 0.56 wt.% and 0.29 wt.%, respectively, indicating the good structural stability of the two adsorbents. In this study, the leakage rate of organic compounds is significantly reduced compared to that in previous work [17].

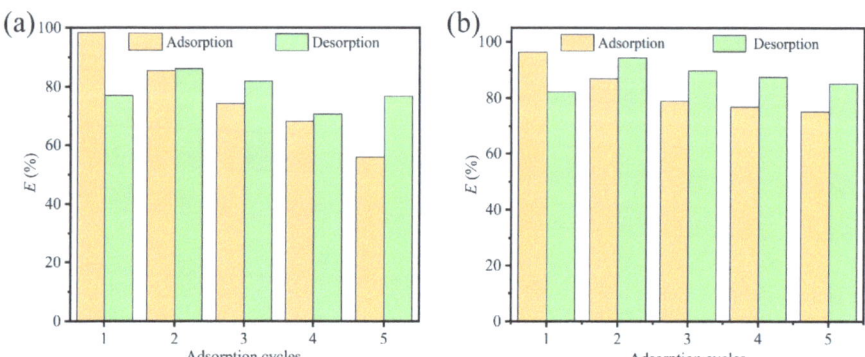

Figure 7. Reusability of (**a**) (DtBuCH18C6 + Dodec)/SiAaC-g-ABSA and (**b**) (DtBuCH18C6 + Dodec)/SiAaC-g-3-ABSA on Sr in 2 M HNO$_3$ concentration (C_o = 5 mM, m/V = 0.1 g/5 mL, t = 180 min, T = 298 K).

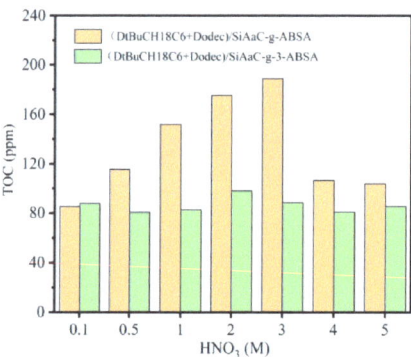

Figure 8. TOC leakage of the two adsorbents in different concentrations of HNO$_3$ solutions. (Contact time: 24 h, m/V = 0.1 g/5 mL, and T = 298 K).

Figure 9 explains to us that this reduction in leakage rate is attributed to the intermolecular bonding between the -SO3H group introduced through grafting and DtBuCH18C6 through hydrogen bonding in our study. Modifiers that can form hydrogen bonds with DtBuCH18C6 can effectively reduce the leakage of the latter in the solution, which has been confirmed in previous studies [16]. Furthermore, (DtBuCH18C6 + Dodec)/SiAaC-g-3-ABSA exhibits a lower leakage rate compared to (DtBuCH18C6 + Dodec)/SiAaC-g-ABSA. This is attributed to the higher grafting efficiency of 3-ABSA in the former, which enhances the protection of the modifier on DtBuCH18C6 and further improves the reusability of the material.

3.4. Column Experiments

The dynamic adsorption behavior of Sr by two adsorbents was investigated, and the breakthrough curves were fitted using the Thomas model [35]. According to Figure 9, the Sr in the effluent is observed when the effluent volume exceeds 70 mL and 85 mL, respectively. After that, the breakthrough curves climb rapidly, and the two adsorbents reach saturation adsorption when the effluent volumes are higher than 139 mL and 154 mL, respectively. Based on the fitting results of the Thomas model in Figure 10, the correlation coefficients, R^2, are close to 1, indicating that the model has good applicability to the dynamic adsorption process. The dynamic adsorption capacities of the two adsorbents for Sr are about 30.2 mg/g and 35.5 mg/g, respectively, while the theoretical adsorption capacities fitted by the Thomas model are 41.6 mg/g and 42.4 mg/g, respectively. The discrepancy between the theoretical Q and actual Q is likely caused by the fluctuation of the flow rate during the adsorption process [36].

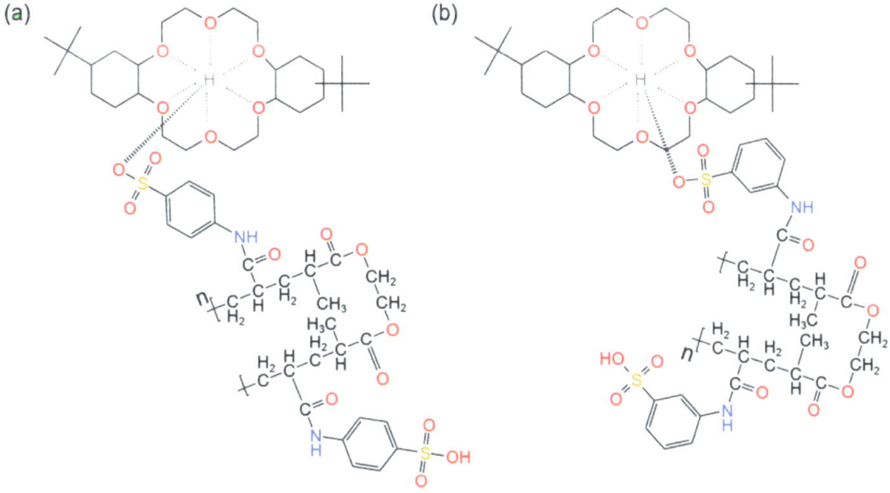

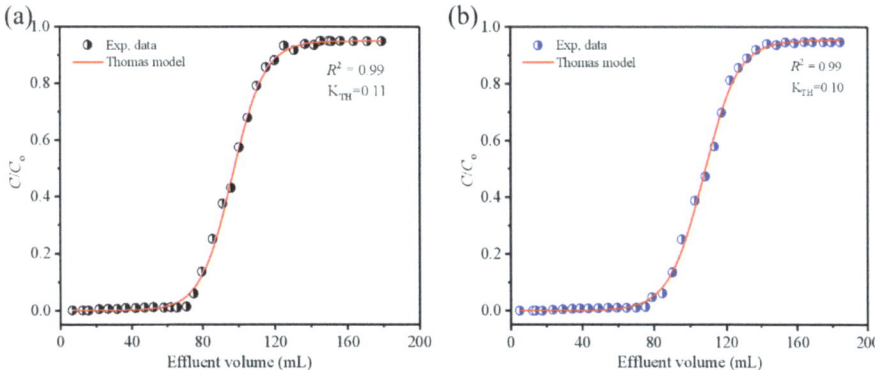

Figure 9. The intermolecular binding of DtBuCH18C6 to (**a**) (DtBuCH18C6 + Dodec)/SiAaC-g-ABSA and (**b**) (DtBuCH18C6 + Dodec)/SiAaC-g-3-ABSA via hydrogen bonds, respectively.

Figure 10. Breakthrough curves of (**a**) (DtBuCH18C6 + Dodec)/SiAaC-3-ABSA and (**b**) (DtBuCH18C6 + Dodec)/SiAaC-g-3-ABSA on Sr in 2 M HNO_3 concentration (C_o = 461 mg/L, m = 1.9981 g, $\phi \times h$ = 10 mm × 100 mm, flow speed = 0.5 mL/min, T = 298 K).

3.5. Mechanism Study

XPS analysis was used to study the adsorption mechanism. Figure 11 shows the XPS results of the (DtBuCH18C6 + Dodec)/SiAaC-g-ABSA and (DtBuCH18C6 + Dodec)/SiAaC-g-3-ABSA before and after adsorption of Sr from 2 M HNO_3 solution. After the adsorption of Sr by the two adsorbents, the characteristic peak of Sr 3d was detected in the full XPS spectra (Figure 11a,d). According to the fitting results in Figure 11b,e, Sr 3d can be divided into Sr 3d 3/2 and Sr 3d 5/2, with the binding energies of about 135.6 eV and 133. 8 eV, respectively [37–39]. According to previous reports [40,41], crown ethers are susceptible to coordination interactions, with Sr leading to more pronounced changes in C 1s binding energy. Therefore, in the present work, the changes in C 1s binding energy before and after the adsorption of Sr by the two adsorbents were investigated. According to the XPS results in Figure 11c,f, C 1s can be divided into three forms, i.e., C-C, C-O-C, and C=O, with the binding energies of 284.79 eV, 285.17 eV, and 286.22 eV for (DtBuCH18C6 + Dodec)/SiAaC-g-ABSA and 284.81 eV, 285.41 eV, and 286.41 eV for (DtBuCH18C6 + Dodec)/SiAaC-g-3-ABSA, respectively [18,27]. After the adsorption of Sr by the two adsorbents, the binding energy of C-O-C shifted to 284.88 eV and 285.41 eV, respectively. The change in C-O-C binding energy after adsorption is presumed to be caused by the coordination between the crown ether ring and Sr [27].

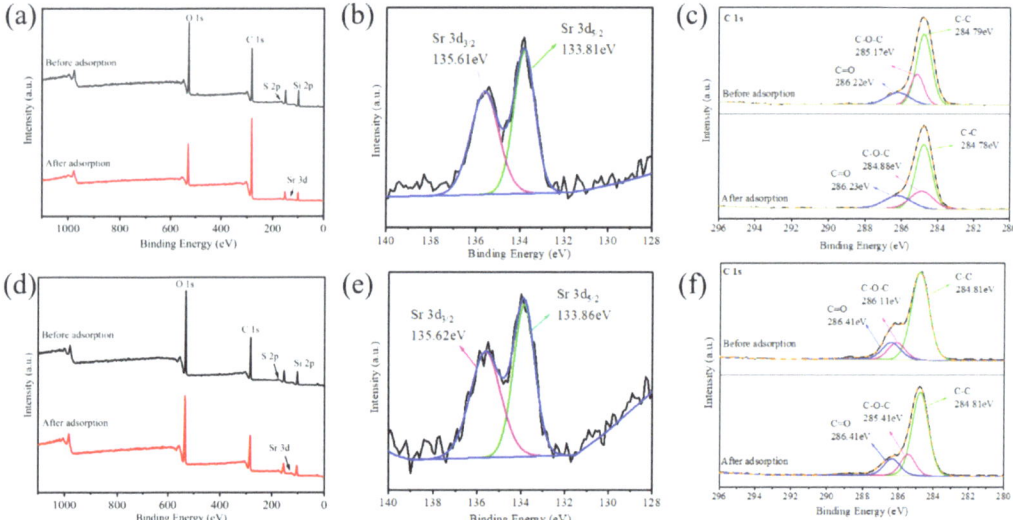

Figure 11. XPS results of (**a–c**) (DtBuCH$_{18}$C$_6$ + Dodec)/SiAaC-3-ABSA and (**d–f**) (DtBuCH$_{18}$C$_6$ + Dodec)/SiAaC-g-3-ABSA before and after adsorption of Sr.

3.6. Comparison of (DtBuCH18C6 + Dodec)/SiAaC-g-ABSA and (DtBuCH18C6 + Dodec) /SiAaC-g-3-ABSA with Other Materials

In practical applications, utility and cost-effectiveness are paramount. A comparison of various solid adsorbent materials is presented in Table 3. The prepared (DtBuCH18C6 + Dodec)/SiAaC-g-ABSA and (DtBuCH18C6 + Dodec)/SiAaC-g-3-ABSA exhibit high K_d values, complete desorption, excellent reusability, and high selectivity. In summary, both (DtBuCH18C6 + Dodec)/SiAaC-g-ABSA and (DtBuCH18C6 + Dodec)/SiAaC-g-3-ABSA are well-suited for the treatment of Sr.

Table 3. Comparison of (DtBuCH18C6 + Dodec)/SiAaC-3-ABSA and (DtBuCH18C6 + Dodec)/SiAaC-g-3-ABSA with other materials.

Adsorbents	Other Species	C(HNO$_3$)	K_d (cm^3/g)	Reusability	Ref.
(DtBuCH18C6 + [C2mim][NTf2])/SiO$_2$-P	Sr, Ba, Na, Ca, La, Nd, Sm, Gd, Ru, Pd, Zr, Mo	3 M	30	-	[42]
(DtBuCH18C6 + Oct)/SiO$_2$-P	Ru, Pd, Ba, Mo, La, Y, Cs, Na, K	2 M	<200	-	[43]
(DtBuCH18C6 + Dodec)/SiO$_2$-P	Cs, Ru, Pd, La, Nd, Sm, Gd, Zr, Mo	3 M	182.0	-	[11]
(DtBuCH18C6 + Dodec + DBS)/SiO$_2$-P	-	3 M	260.3	-	[17]
(DtBuCH18C6 + Dodec)/SiAaC-g-ABSA	Gd, Eu, Sm, Nd, Pr, Ce, La, Ba, Pd, Ru, Mo, Y	2 M	389.68	4	This study
(DtBuCH18C6 + Dodec)/SiAaC-g-3-ABSA		2 M	416.68	≥5	

4. Conclusions

In this work, two novel silica-based adsorbents (DtBuCH18C6 + Dodec)/SiAaC-g-ABSA and (DtBuCH18C6 + Dodec)/SiAaC-g-3-ABSA were prepared successively by chemical grafting and vacuum impregnation for the selective separation of Sr in nitric acid environment. Different from the previous research work (DtBuCH18C6 + dodecanol)/SiO$_2$-P materials, SiAaC-g-ABSA and SiAaC-g-3-ABSA were used as carriers in this work, respectively. And the new supports can further improve the stability and adsorption properties of the materials. The formation of hydrogen bonds between the -SO3H groups introduced through grafting in our study and DtBuCH18C6 further reduces the leakage of the latter in the solution. The two adsorbents exhibited good adsorption selectivity for Sr in 0.5–5 M HNO$_3$ solutions with the separation factor, $SF_{Sr/M}$, over 40, except that of $SF_{Sr/Ba}$. The adsorption of Sr matched well with pseudo-second-order and Langmuir model with the saturated adsorption capacities of 36.9 mg/g and 37.5 mg/g obtained within 1 h, respectively. The adsorbed Sr could be desorbed efficiently using pure water with desorption efficiencies of 77% and 82%, respectively. They possess good reusability with Sr adsorption efficiencies of about 57% and 75%, respectively, after five adsorption–desorption cycles. The leakage rates for the two adsorbents in the liquid phase were less than 0.56 wt.% and 0.29 wt.%, respectively, according to the TOC analysis, indicating the good structural stability of the two adsorbents. In column experiments, the dynamic adsorption capacities of the two adsorbents for Sr were about 30.2 mg/g and 35.5 mg/g, respectively. Finally, the XPS results indicate that the adsorption mechanism is the coordination between the adsorbents and Sr.

Author Contributions: Conceptualization, C.L., X.Y., L.C. and M.F.H.; Methodology, C.L. and X.W.; Validation, Y.W.; Formal analysis, C.L. and S.Z.; Investigation, C.L., L.C., X.W.; Resources, X.W.; Data curation, C.L.; Writing—original draft preparation, C.L.; Writing—review and editing, S.Z., X.W., L.C., M.F.H., Y.W. and S.N.; Visualization, C.L.; Supervision, L.C., X.Y., M.F.H., Y.W. and S.N.; Project administration, S.N.; Funding acquisition, S.N. All authors have read and agreed to the published version of the manuscript.

Funding: This work was supported by the National Natural Science Foundation of China (12275124) and the Science and Technology Innovation Program of Hunan Province (2023RC1067).

Data Availability Statement: The data presented in this study are available on request from the corresponding author. As the project is still ongoing, the data will not be disclosed due to the needs of the project.

Conflicts of Interest: The authors declare that they have no known competing financial interests or personal relationships that could have appeared to influence the work reported in this paper.

References

1. Xu, Y.; Kim, S.Y.; Ito, T.; Nakazawa, K.; Funaki, Y.; Tada, T.; Hitomi, K.; Ishii, K. Adsorption and separation behavior of yttrium and strontium in nitric acid solution by extraction chromatography using a macroporous silica-based adsorbent. *J. Chromatogr. A* **2012**, *1263*, 28–33. [CrossRef] [PubMed]
2. Venkatesan, K.A.; Selvan, B.R.; Antony, M.P.; Srinivasan, T.G.; Rao, P.R.V. Extraction of palladium from nitric acid medium by commercial resins with phosphinic acid, methylene thiol and isothiouronium moieties attached to polystyrene-divinylbenzene. *J. Radioanal. Nucl. Chem.* **2005**, *266*, 431–440. [CrossRef]
3. Khan, P.N.; Bhattacharyya, A.; Banerjee, D.; Sugilal, G.; Kaushik, C.P. Partitioning of heat generating fission product (^{137}Cs & ^{90}Sr) from acidic medium by 1,3-dioctyloxy-calix[4]arenecrown-6 (CC6) & Octabenzyloxyoctakis[[[(N,N-diethylamino)carbonyl)]methyl]oxy]calix[8]arene (BOC8A) in nitro octane diluent: Batch scale study & process parameter optimization. *Sep. Purif. Technol.* **2021**, *274*, 119102. [CrossRef]
4. Xiao, C.L.; Wang, C.Z.; Yuan, L.Y.; Li, B.; He, H.; Wang, S.; Zhao, Y.L.; Chai, Z.F.; Shi, W.Q. Excellent selectivity for actinides with a tetradentate 2,9-diamide-1,10-phenanthroline ligand in highly acidic solution: A hard–soft donor combined strategy. *Inorg. Chem.* **2014**, *53*, 1712–1720. [CrossRef] [PubMed]
5. Xu, L.; Zhang, A.; Pu, N.; Xu, C.; Chen, J. Development of Two novel silica based symmetric triazine-ring opening N-donor ligands functional adsorbents for highly efficient separation of palladium from HNO_3 solution. *J. Hazard. Mater.* **2019**, *376*, 188–199. [CrossRef] [PubMed]
6. Datta, S.J.; Oleynikov, P.; Moon, W.K.; Ma, Y.; Mayoral, A.; Kim, H.; Dejoie, C.; Song, M.K.; Terasaki, O.; Yoon, K.B. Removal of ^{90}Sr from highly Na^+-rich liquid nuclear waste with a layered vanadosilicate. *Energy Environ. Sci.* **2019**, *12*, 1857–1865. [CrossRef]
7. Zhang, M.; Gu, P.; Yan, S.; Pan, S.; Dong, L.; Zhang, G. A novel nanomaterial and its new application for efficient radioactive strontium removal from tap water: KZTS-NS metal sulfide adsorbent versus CTA-F-MF process. *Chem. Eng. J.* **2020**, *391*, 123486. [CrossRef]
8. Song, Y.; Du, Y.; Lv, D.; Ye, G.; Wang, J. Macrocyclic receptors immobilized to monodisperse porous polymer particles by chemical grafting and physical impregnation for strontium capture: A comparative study. *J. Hazard. Mater.* **2014**, *274*, 221–228. [CrossRef]
9. Saha, D.; Vithya, J.; Kumar, R.; Joseph, M. Studies on purification of ^{89}Sr from irradiated yttria target by multi-column extraction chromatography using DtBuCH18-C-6/XAD-7 resin. *Radiochim. Acta* **2019**, *107*, 479–487. [CrossRef]
10. Yi, R.; Xu, C.; Sun, T.; Wang, Y.; Ye, G.; Wang, S.; Chen, J. Improvement of the extraction ability of bis(2-propyloxy)calix[4]arene-crown-6 toward cesium cation by introducing an intramolecular triple cooperative effect. *Sep. Purif. Technol.* **2018**, *199*, 97–104. [CrossRef]
11. Wu, Y.; Kim, S.-Y.; Tozawa, D.; Ito, T.; Tada, T.; Hitomi, K.; Kuraoka, E.; Yamazaki, H.; Ishii, K. Equilibrium and kinetic studies of selective adsorption and separation for strontium using DtBuCH18C6 loaded resin. *J. Nucl. Sci. Technol.* **2012**, *49*, 320–327. [CrossRef]
12. Zhang, A.; Xiao, C.; Chai, Z. SPEC Process II. Adsorption of strontium and some typical co-existent elements contained in high level liquid waste onto a macroporous silica-based crown ether impregnated functional composite. *J. Radioanal. Nucl. Chem.* **2009**, *280*, 181–191. [CrossRef]
13. Guo, C.; Yuan, M.; He, L.; Cheng, L.; Wang, X.; Shen, N.; Ma, F.; Huang, G.; Wang, S. Efficient capture of Sr^{2+} from acidic aqueous solution by an 18-crown-6-etherbased metal organic framework. *CrystEngComm* **2021**, *23*, 3349–3355. [CrossRef]
14. Chen, Z.; Wu, Y.; Wei, Y. Adsorption characteristics and radiation stability of a silica-based DtBuCH18C6 adsorbent for Sr (II) separation in HNO_3 medium. *J. Radioanal. Nucl. Chem.* **2014**, *299*, 485–491. [CrossRef]
15. Sharma, J.N.; Khan, P.N.; Dhami, P.S.; Jagasia, P.; Tessy, V.; Kaushik, C.P. Separation of strontium-90 from a highly saline high level liquid waste solution using 4,4′(5′)-[di-tert-butyldicyclohexano]-18-crown-6 + isodecyl alcohol/n-dodecane solvent. *Sep. Purif. Technol.* **2019**, *229*, 115502. [CrossRef]
16. Zhang, A.; Xiao, C.; Kuraoka, E.; Kumagai, M. Preparation of a novel silica-based DtBuCH18C6 impregnated polymeric composite modified by tri-n-butyl phosphate and its application in chromatographic partitioning of strontium from high level liquid waste. *Ind. Eng. Chem. Res.* **2007**, *46*, 2164–2171. [CrossRef]
17. Wang, Y.; Wen, Y.; Mao, C.; Sang, H.; Wu, Y.; Li, H.; Wei, Y. Development of chromatographic process for the dynamic separation of ^{90}Sr from high level liquid waste through breakthrough curve simulation and thermal analysis. *Sep. Purif. Technol.* **2022**, *282*, 120103. [CrossRef]
18. Liu, H.; Ning, S.; Zhang, S.; Wang, X.; Chen, L.; Fujita, T.; Wei, Y. Preparation of a mesoporous ion-exchange resin for efficient separation of palladium from simulated electroplating wastewater. *J. Environ. Chem. Eng.* **2022**, *10*, 106966. [CrossRef]
19. Ma, F.Y.; Li, Z.; Zhou, W.; Li, Q.N.; Zhang, L. Application of polyantimonic acid-polyacrylonitrile for removal of strontium (II) from simulated high-level liquid waste. *J. Radioanal. Nucl. Chem.* **2017**, *311*, 2007–2013. [CrossRef]
20. Ning, S.Y.; Wang, X.P.; Zou, Q.; Shi, W.Q.; Tang, F.D.; He, L.F.; Wei, Y.Z. Direct separation of minor actinides from high level liquid waste by Me_2-CA-BTP/SiO_2-P adsorbent. *Sci. Rep.* **2017**, *7*, 14679. [CrossRef]
21. Liu, J.Q.; Liu, Y.J.; Talay, D.K.; Calverley, E.; Brayden, M.; Martinez, M. A new carbon molecular sieve for propylene/propane separations. *Carbon* **2015**, *85*, 201–211. [CrossRef]
22. Chang, K.C.; Lo, H.-M.; Lin, K.-L.; Liu, M.-H.; Chiu, H.-Y.; Lo, F.-C.; Chang, J.H. Cu adsorption in fixed bed column with three different influent concentration. *E3S Web Conf.* **2019**, *120*, 03003. [CrossRef]

23. Unuabonah, E.I.; Omorogie, M.O.; Oladoja, N.A. Modeling in adsorption: Fundamentals and applications, Compos. *Nanoadsorbents* **2018**, 85–118. [CrossRef]
24. Tian, X.; Wang, S.; Li, J.S.; Liu, F.X.; Wang, X.; Chen, H.; Ni, H.Z.; Wang, Z. Composite membranes based on polybenzimidazole and ionic liquid functional Si-O-Si network for HT-PEMFC applications. *Int. J. Hydrogen Energy* **2017**, *42*, 21913–21921. [CrossRef]
25. Kozubal, J.; Heck, T.; Metz, R.B. Vibrational Spectroscopy of Intermediates and C-H Activation Products of Sequential Zr^+ Reactions with CH_4. *J. Phys. Chem. A* **2020**, *124*, 8235–8245. [CrossRef]
26. Dai, Y.; Liu, Y.; Zhang, A.Y. Preparation and characterization of a mesoporous polycrown impregnated silica and its adsorption for palladium from highly acid medium. *J. Porous Mater.* **2017**, *24*, 1037–1045. [CrossRef]
27. Tang, J.; Liao, L.; He, X.; Lv, L.; Yin, X.; Li, W.; Wei, Y.; Ning, S.; Chen, L. Efficient separation of radium from natural thorium using a mesoporous silica-supported composite resin with sulfonic acid groups for the acquisition of targeted α-nuclides ^{212}Pb. *Chem. Eng. J.* **2024**, *485*, 150022. [CrossRef]
28. Wang, J.; Guo, X. Adsorption kinetic models: Physical meanings, applications, and solving methods. *J. Hazard. Mater.* **2020**, *390*, 122156. [CrossRef]
29. Zhang, A.Y.; Wang, W.H.; Chai, Z.F.; Kumagai, M. Separation of strontium ions from a simulated highly active liquid waste using a composite of silica-crown ether in a polymer. *J. Sep. Sci.* **2008**, *31*, 3148–3155. [CrossRef]
30. Lu, L.; Na, C.Z. Gibbsian interpretation of Langmuir, Freundlich and Temkin isotherms for adsorption in solution. *Philos. Mag. Lett.* **2022**, *102*, 239–253. [CrossRef]
31. Ezzati, R.; Pseudo-First-Order, D.O. Derivation of Pseudo-First-Order, Pseudo-Second-Order and Modified Pseudo-First-Order rate equations from Langmuir and Freundlich isotherms for adsorption. *Chem. Eng. J.* **2020**, *392*, 123705. [CrossRef]
32. Martín, F.S.; Kracht, W.; Vargas, T. Attachment of *Acidithiobacillus ferrooxidans* to pyrite in fresh and saline water and fitting to Langmuir and Freundlich isotherms. *Biotechnol. Lett.* **2020**, *42*, 957–964. [CrossRef] [PubMed]
33. Ning, S.; Zhang, S.; Zhang, W.; Zhou, J.; Wang, S.; Wang, X.; Wei, Y. Separation and recovery of Rh, Ru and Pd from nitrate solution with a silica-based isoBu-BTP/SiO_2-P adsorbent. *Hydrometallurgy* **2019**, *191*, 105207. [CrossRef]
34. Baseri, H.; Tizro, S. Treatment of nickel ions from contaminated water by magnetit based nanocomposite adsorbents: Effects of thermodynamic and kinetic parameters and modeling with Langmuir and Freundlich isotherms. *Process Saf. Environ. Prot.* **2017**, *109*, 465–477. [CrossRef]
35. Su, Z.; Ning, S.; Li, Z.; Zhang, S. High-efficiency separation of palladium from nitric acid solution using a silica-polymer-based adsorbent isoPentyl-BTBP/SiO_2-P. *J. Environ. Chem. Eng.* **2022**, *10*, 107928. [CrossRef]
36. Wang, W.; Zhang, S.; Chen, L.; Li, J.; Wu, K.; Zhang, Y.; Su, Z.; Yin, X.; Hamza, M.F.; Wei, Y.; et al. Efficient separation of palladium from nitric acid solution by a novel silica-based ion exchanger with ultrahigh adsorption selectivity. *Sep. Purif. Technol.* **2023**, *322*, 124326. [CrossRef]
37. Adimule, V.; Nandi, S.S.; Yallur, B.C.; Bhowmik, D.; Jagadeesha, A.H. Optical, Structural and Photoluminescence Properties of Gd_x SrO: CdO Nanostructures Synthesized by Co Precipitation Method. *J. Fluoresc.* **2021**, *31*, 487–499. [CrossRef] [PubMed]
38. Zhang, Z.; Gu, P.; Zhang, M.; Yan, S.; Dong, L.; Zhang, G. Synthesis of a robust layered metal sulfide for rapid and effective removal of Sr^{2+} from aqueous solutions. *Chem. Eng. J.* **2019**, *372*, 1205–1215. [CrossRef]
39. Gupta, K.; Yuan, B.; Chen, C.; Varnakavi, N.; Fu, M.-L. $K_{2x}Mn_xSn_{3-x}S_6$ (x = 0.5–0.95) (KMS-1) immobilized on the reduced graphene oxide as KMS-1/r-GO aerogel to effectively remove Cs^+ and Sr^{2+} from aqueous solution. *Chem. Eng. J.* **2019**, *369*, 803–812. [CrossRef]
40. Yin, L.; Kong, X.; Shao, X.; Ji, Y. Synthesis of DtBuCH18C6-coated magnetic metal–organic framework Fe3O4@UiO-66-NH2 for strontium adsorption. *J. Environ. Chem. Eng.* **2019**, *7*, 103073. [CrossRef]
41. Ma, J.; Zhang, Y.; Ouyang, J.; Wu, X.; Luo, J.; Liu, S.; Gong, X. A facile preparation of dicyclohexano-18-crown-6 ether impregnated titanate nanotubes for strontium removal from acidic solution. *Solid State Sci.* **2019**, *90*, 49–55. [CrossRef]
42. Kudo, T.; Ito, T.; Kim, S.-Y. Adsorption behavior of Sr (II) from high-level liquid waste using crown ether with ionic liquid impregnated silica adsorbent. *Energy Procedia* **2017**, *131*, 189–194. [CrossRef]
43. Zhang, A.; Xiao, C.; Liu, Y.; Hu, Q.; Chen, C.; Kuraoka, E. Preparation of macroporous silica-based crown ether materials for strontium separation. *J. Porous Mater.* **2010**, *17*, 153–161. [CrossRef]

Disclaimer/Publisher's Note: The statements, opinions and data contained in all publications are solely those of the individual author(s) and contributor(s) and not of MDPI and/or the editor(s). MDPI and/or the editor(s) disclaim responsibility for any injury to people or property resulting from any ideas, methods, instructions or products referred to in the content.

Review

Metal Recovery from Wastewater Using Electrodialysis Separation

Maria del Mar Cerrillo-Gonzalez, Maria Villen-Guzman *, Jose Miguel Rodriguez-Maroto and Juan Manuel Paz-Garcia

Department of Chemical Engineering, University of Malaga, 29010 Malaga, Spain; mcerrillog@uma.es (M.d.M.C.-G.); maroto@uma.es (J.M.R.-M.); juanma.paz@uma.es (J.M.P.-G.)
* Correspondence: mvillen@uma.es

Abstract: Electrodialysis is classified as a membrane separation process in which ions are transferred through selective ion-exchange membranes from one solution to another using an electric field as the driving force. Electrodialysis is a mature technology in the field of brackish water desalination, but in recent decades the development of new membranes has made it possible to extend their application in the food, drug, and chemical process industries, including wastewater treatment. This work describes the state of the art in the use of electrodialysis (ED) for metal removal from water and wastewater. The fundamentals of the technique are introduced based on the working principle, operational features, and transport mechanisms of the membranes. An overview of the key factors (i.e., the membrane properties, the cell configuration, and the operational conditions) in the ED performance is presented. This review highlights the importance of studying the inter-relation of parameters affecting the transport mechanism to design and optimize metal recovery through ED. The conventional applications of ED for the desalination of brackish water and demineralization of industrial process water and wastewater are discussed to better understand the key role of this technology in the separation, concentration, and purification of aqueous effluents. The recovery and concentration of metals from industrial effluents are evaluated based on a review of the literature dealing with effluents from different sources. The most relevant results of these experimental studies highlight the key role of ED in the challenge of selective recovery of metals from aqueous effluents. This review addresses the potential application of ED not only for polluted water treatment but also as a promising tool for the recovery of critical metals to avoid natural resource depletion, promoting a circular economy.

Keywords: metal separation; membrane; selectivity; purification; industrial effluent

Citation: Cerrillo-Gonzalez, M.d.M.; Villen-Guzman, M.; Rodriguez-Maroto, J.M.; Paz-Garcia, J.M. Metal Recovery from Wastewater Using Electrodialysis Separation. *Metals* **2024**, *14*, 38. https://doi.org/10.3390/met14010038

Academic Editor: Antonije Onjia

Received: 30 October 2023
Revised: 23 December 2023
Accepted: 26 December 2023
Published: 28 December 2023

Copyright: © 2023 by the authors. Licensee MDPI, Basel, Switzerland. This article is an open access article distributed under the terms and conditions of the Creative Commons Attribution (CC BY) license (https://creativecommons.org/licenses/by/4.0/).

1. Introduction

During recent decades, the rising population and industrialization have increased demand for the Earth's natural resources and the occurrence of subsequent environmental problems. One of these problems is the generation of large amounts of wastewater containing metals that are a risk to the population and ecosystems due to their toxicity [1]. Some metals affect biological function and can accumulate in different organs, causing hazardous effects [2]. The most prevalent metals found in wastewater are Al, As Cd, Cr, Cu, Hg, Pb, Ni, Zn, Co, Fe, and Mn, and their composition in wastewater differs according to the type of industry (see Table 1). The main sectors that contribute to this problem are mining, smelting, foundries, and other chemistries such as textiles and refineries [2].

The negative effect of some metals in wastewater makes it necessary to treat these industrial effluents before discharging to reduce the pollution. The conventional method to remove metals from industrial effluents has been chemical precipitation, adjusting the pH of the effluents, and removing the precipitated particles through sedimentation or filtration [3]. Other chemical-based separation methods include coagulation/flocculation and flotation. These methods can be applied to treat industrial effluents with high metal concentrations. Although these methods are simple and require inexpensive equipment, they involve a large amount of chemicals to reduce metal concentrations and a large-volume sludge formation which requires post-treatment [4]. As an alternative method, adsorption is widely selected to remove metal from aqueous effluents. Several studies have focused on developing cheaper and more efficient adsorbent materials in the last decade, for example, natural materials, industrial by-products, or biological and agricultural waste [5]. However, biosorption (i.e., adsorption with biological materials) has some shortcomings, such as a lack of specificity in metal binding, large amounts of biomass required if the biosorption capacity is low, and a limited reusability of biomass after desorption for real applications in industrial effluents [6]. Another method applied to remove metals from aqueous effluents is ion exchange due to its higher ion selectivity and the reusability of ion-exchange material. The main drawbacks of this technology are high operational costs, which limit application at industrial scale, and the formation of fouling by solids and organic compounds contained in industrial effluents [6]. Electrochemical processes based on passing a direct current through an aqueous solution containing metal between electrodes have also been widely studied. Metal selectivity, no consumption of additional chemicals, high efficiency, and lower amounts of sludge produced are the main advantages of these processes. However, electrochemical technologies possess some drawbacks, such as a high dependence on the pH values of aqueous effluents and high operational costs related to electrical energy requirements and electrode replacement [3].

Table 1. Typical metals found in industrial wastewater, data from [3].

Industries	Al	As	Cd	Cr	Cu	Hg	Pb	Ni	Zn	Co	Fe	Mn
Paper mills				x	x	x	x	x	x			x
Organic chemistry	x	x	x	x		x	x		x			
Fertilizer	x	x	x	x	x	x	x	x	x			
Petroleum refinery	x	x	x	x	x			x	x			
Steel works		x	x	x	x	x	x	x	x	x		
Aircrafts	x		x	x	x	x		x			x	
Textile mills				x								
Power plants				x								
Pharmaceutical		x	x	x	x			x	x	x		
Engineering		x	x	x				x	x	x	x	
Metal smelters		x										
Electroplating				x				x	x			
Mining									x			
Ferromanganese production											x	x

Some alternative technologies to conventional methods are based on pressure-driven membrane processes, such as ultrafiltration, nanofiltration, and reverse osmosis [7]. Another membrane technology is electrodialysis (ED), which involves the migration of cations and anions through ion-exchange membranes under the effect of an electric field. This technique has been employed industrially to treat saline waters and brines. However, its selective separation of charged species offers excellent advantages, increasing its interest for treating water containing metals. Some advantages related to the selective separation of ions are the high separation efficiency, low operating pressure, small operating footprint, no need for adding chemicals, and reduced sludge formation. Moreover, this technology can treat effluents with low concentrations of metal ions. All these advantages make electrodialysis one of the most effective and promising technologies for treating industrial effluents [8].

The potential of electrodialysis is not only focused on treating wastewater effluent-containing metals but also on the revalorization of waste and promoting the circular economy. This work presents an overview of the state of the art of this technique, summarizing the current research and industrial applications of electrodialysis.

2. Methodology and Review Structure

This review was guided by the Kitchenham framework for literature reviews [9], including the three main stages: planning, conducting, and reporting results. Prior to searching the available literature related to the main topic, the need and novelty of this review must be defined. In this step, the research questions were defined, and the keywords were selected to filter the information that could answer the research questions (Table 2).

Table 2. Research questions and keywords that guided the review.

	Research Questions
Q1	What are the fundamentals of electrodialysis?
Q2	Can electrodialysis be used to treat wastewater?
Q3	Is electrodialysis used to separate metals from industrial wastewater?

Once the research questions and keywords were selected, the data extraction and article selection were performed. For that, we considered mainly works from the last fifteen years (2008–2023), which were published in three databases: Scopus, Web of Science, and Sciencedirect. The search terms were the same for all databases, "electrodialysis" and "metals", with additional strings of "wastewater" and "recovery", and they were searched in the title, abstract and keywords. Table 3 presents the number of works that matched with our research terms in each database.

Table 3. Research results in databases using the selected search terms.

Database (2008–2023)	Electrodialysis and Metals	Electrodialysis and Metals and Wastewater	Electrodialysis and Metals and Wastewater and Recovery
Scopus	676	239	111
Web of Science	648	185	111
Sciencedirect	227	80	44

Finally, the results obtained are presented and discussed. For that, the current work has been organized into five sections in total. First, a general introduction presents the environmental challenges posed by metal contamination in industrial wastewater, and the potential of electrodialysis to the circular economy as a promising and efficient treatment technology. The current section outlines the structure of this review, and the methodology followed to filter and select the information. In the third section, the operational principles and the mass transport mechanism of the process are presented to understand the fundamentals of the electrodialysis process. The fourth section presents the state of the art of the application of electrodialysis for water treatment. The section is divided into three subsections. In the first one, the general application of electrodialysis is presented, particularly focused on desalination and industrial wastewater treatment. The second and third subsections are focused on the use of electrodialysis not only to treat industrial wastewater but also to recover metals. The difference between these two subsections is the number of metals recovered from effluents (i.e., recovering only one or multiple metals from the mixture). Finally, the fifth section contains the conclusions reached from the discussion of the information presented in the current review work.

3. Fundamentals of Electrodialysis Processes

3.1. Operational Principle

The working principle of ED consists of the migration of cations and anions through cation-exchange membranes (CEMs) and anion-exchange membranes (AEMs), respectively, induced by an applied electric field set between a pair of electrodes. In a simple electrodialysis cell (Figure 1), a pair of membranes is used. Anions migrate towards the anode, and cations towards the cathode, resulting in an overall decrease in the feed stream salt concentration. The membranes are separated by channels ending with the electrode compartments.

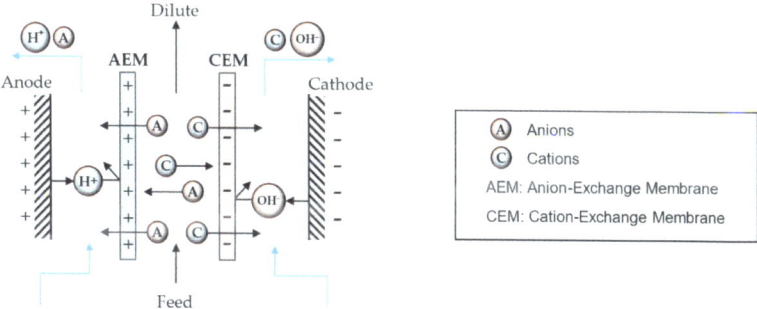

Figure 1. Schematic of a simple (two-membrane) electrodialysis cell.

The two-membrane ED cell results in the production of an acid and a base from the fed salt. The unit cell is formed by an AEM, a CEM, and a channel in-between where the concentrated salt solution is fed. The application of an electrical current between the electrodes promotes the dissociation of water at the electrodes [10].

Multichannel cells consist of a system of alternative membranes creating channels which produce diluted and concentrated outflows (Figure 2). The flow containing ions is fed in the channels between membranes. Anions migrate toward the anode crossing the AEM but they are retained in the compartment due to the presence of an adjacent CEM. In this channel, there are also cations coming from the opposite directions, migrating to the cathode through the CEM, but they are blocked by the AEM. This means the migrated anions and cations remain blocked in the same compartment, creating a concentrated channel. On the other hand, the adjacent channels to the concentrate are denoted as dilute channels due to the reduction in both anion and cation concentrations. The combination of an AEM, a concentrated channel, a CEM, and a dilute spacer is known as a cell pair.

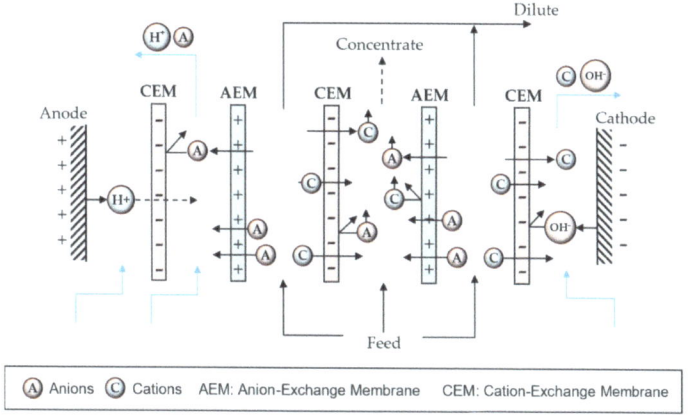

Figure 2. Depiction of a multichannel electrodialysis stack.

In aqueous solution, water electrolysis reactions are expected to take place: namely, water oxidation at the anode, Equation (1), and water reduction at the cathode, Equation (2).

$$\text{Anode}: 2H_2O \rightarrow 4H^+ + O_2 \uparrow + 4e^- \quad (1)$$

$$\text{Cathode}: 2H_2O + 2e^- \rightarrow H_2 \uparrow + 2OH^- \quad (2)$$

However, different electrochemical reactions may occur depending on the nature of the involved ions. For example, in the presence of chloride ions, chlorine gas can be formed at the anode. Similarly, certain metal ions may electrodeposit at the cathode surface. Different cell configurations can be used to prevent certain electrochemical reactions. Figure 2 shows a CEM at the anode end to prevent the anion, assumed to be Cl$^-$, from reaching the electrode surface and forming Cl$_2$ gas.

The development of membrane technologies has allowed new cell configurations, expanding the application of electrodialysis [11]. An interesting enhancement is the use of bipolar membranes in the cell stack (Figure 3). A bipolar membrane (BM) is composed of an anion-exchange layer (AEL) and a cation-exchange layer (CEL), and does not allow anions and cations through it. This phenomenon promotes the dissociation of water available within the thin layer between AEL and CEL into protons (H$^+$) and hydroxyl (OH$^-$). This cell configuration generates an acidic and alkaline compartment on both sides of the bipolar membrane due to H$^+$ and OH$^-$ crossing the CEL and AEL, respectively [12].

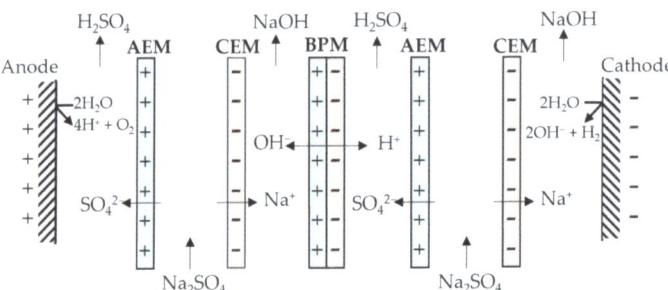

Figure 3. Depiction of a three-compartment electrodialytic cell with a bipolar membrane.

The use of monovalent anion-exchange membranes (MVAs) and monovalent cation-exchange membranes (MVCs) has been recently proposed to selectively separate ions in different ED compartments [9]. In this case, an MVA is used in the cell, which allows only the migration of monovalent anions, producing two compartments: one with multivalent anions retained and the other with monovalent ions (Figure 4). This process is classified as "selectrodialysis". In addition to using MVAs and MVCs, selectrodialysis can be achieved using different pH and chelating/complexation agents.

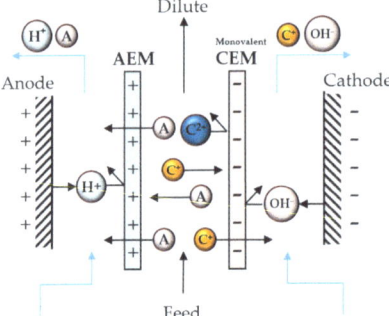

Figure 4. Scheme of a unit cell with a monovalent cation-exchange membrane.

One of the main drawbacks in the membrane processes is the fouling of their surfaces, caused by suspended particles or the precipitation of dissolved solids. Solid particles cause the deterioration of the membranes, increasing their electrical resistance with a subsequent lowering of the ED performance. In addition to performing pretreatments and cleaning procedures to prevent surface fouling, electrodialysis reversal (EDR) has been proposed to mitigate the fouling problem [13]. The technique consists of switching periodically the polarity of electrodes to force the migration of charged components to the opposite direction [14]. Several studies reported the success of EDR systems due to their self-cleaning mechanism.

3.2. Mass Transport Mechanism

Mass and charge transport in electrolyte solutions and ion-exchange membranes is described with the same set of equations. Under an electric field, ions are mainly transported by electromigration. Diffusion transport becomes significant at the boundary layers, near the membranes and the electrodes, due to high concentration gradients. The coupling of these transport mechanisms is known as electrodiffusion and can be described by the Nernst–Plank transport equation, which considers the diffusive, electromigration, and convective flux to calculate the total flux of ions, J_i (mol/m^2/s):

$$J_i = -D_i^* \nabla c_i - \frac{z_i F D_i^*}{RT} c_i \nabla \varphi + c_i u \quad ; \quad i = 1, 2, \ldots, N_i \tag{3}$$

where c_i (mM) is the concentration, D_i^* (m^2/s) is the effective diffusion coefficient, z_i is the ionic charge, F ($\approx$95,485 A/mol) is the Faraday constant, φ (V) is the electric potential, and u (m/s) is the velocity of water. The term $z_i F D_i / RT$ (m^2/s/V) is known as the ionic migration coefficient. The set of equations is completed with the following electroneutrality condition:

$$\sum_{i=1}^{N_i} c_i z_i = 0 \tag{4}$$

At the submicro scale, at the interface of the electrically charged membranes, the electroneutrality condition does not hold, and electrical double layers form. This can be described by means of the Poisson equation:

$$\nabla^2 \phi = -\frac{F}{\varepsilon} \sum_{i=1}^{N_i} c_i z_i \tag{5}$$

where ε (F/m) is the medium permittivity ($\varepsilon = \varepsilon_0 \varepsilon_r$; $\varepsilon_0 \approx 8.85 \times 10^{-12}$ F/m and $\varepsilon_r \approx 80.2$ for water at 20 °C).

The performance of an electrodialysis treatment depends on numerous variables, such as the membrane properties, the cell configuration, and the operational conditions. Regarding the transport mechanisms, the most important parameters are the electric current applied, the electrical conductivity of the electrolytes and the membranes, the concentration and pH of the electrolyte solutions, and the stream flow rate. All these parameters are inter-related between each other, and the study of their effects is essential to design and optimize the process.

Ion-exchange membranes (IEMs) are crucial elements of any ED process, and their characteristics play an important role in the performance of the process. Membrane properties can be divided into physicochemical (e.g., thickness) and electrochemical properties (e.g., permeability, selectivity, charge density, and area resistance) [15]. IEMs are dense polymeric membranes containing fixed charges in the polymer matrix, which can selectively enable the passage of oppositely charged ions (counter ions) while obstructing similarly charge ions (co-ions). The enhancement in the permselectivity between counter and co-ions as well as between counter ions with different (monovalent and multivalent) or equal valences (Cl$^-$ and NO$_3^-$) has allowed the expansion of IEMs to multiple applications [16]. However, improving the permselectivity involves an increase in cross-linking in the polymer matrix, which also results in an increase in the area resistance, which is not desirable from the point of view of energy efficiency. Thus, membrane designers must find a balance between permselectiviy and area resistance. Lu et al. [16] elaborated a comprehensive review of IEMs, focused on the progress of membrane manufacturing techniques, ion transport mechanisms, and experimental approaches to determine ion selectivity. In the same line, Tekinalp et al. [17] published a review on cation-exchange membranes and their properties for selective metal separation by electrodialysis. They discussed the counter ion selectivity based on the membrane properties and the operational conditions, emphasizing the implication of the boundary layer at the membrane surface in the transport ratio of competing counter ions.

The Intensity of the applied electric current is also one of the key factors in an electrical separation process. By default, operating with the highest possible current density is desirable to increase the driving force and achieve the maximum ion flows. However, high electric current values have drawbacks, such as increased energy requirements, possible membrane damage, and increased concentration polarization at the membrane surfaces. [8]. Concentration polarization appears due to the difference in transport numbers of the ions (i.e., the fraction of the current carried by each ionic species) in the solution and the membranes. In the solution electrolyte, the ionic current derives from the transport of both cations and anions. In the membrane, the ionic current is only transported by the counter ions inside the membranes, resulting in a higher transport number through the membranes than in the solution. This phenomenon leads to the formation of concentration gradients at the solution–membrane interface, decreasing the concentration with respect to the bulk in the dilute compartment and increasing it in the concentrate compartment.

Depending on the applied current density, the concentration of specific ions can deplete near the surface membrane of the dilute region, which means there may not be any available ions to transport the current in that region. The current density that generates the depletion of the ion concentration to zero is known as the limiting current density. Working over the limiting current density is not recommended due to the increased electrical resistance and voltage drop, negatively affecting the process efficiency [18]. Furthermore, when approaching the limiting current density, the lack of ions in the system promotes the migration of protons and hydroxides, which are replenished through the water self-ionization reaction. This phenomenon is referred to as water splitting [19].

The ion concentration in the feed solution plays an important role in the process's efficiency. A low concentration in the solutions involves low ionic conductivity, affecting ions' migration across the IEMs due to lower diffusion- and electromigration-driven forces. On the other hand, high concentrations of ions can cause some problems. First, the recovery rate of the target ions can be reduced because the residence time is insufficient to separate the desirable number of ions. Second, although electrical resistance decreases, applying a high value of electric current is necessary to mobilize the ions to the recovery compartment. Finally, in the case of metallic ions, a high concentration can cause their precipitation if the pH is not appropriate. Hence, one of the first parameters that should be decided in designing an ED stack is the feed flow, and then, the concentration of the dilute and concentrate output flows.

The pH is also an important parameter, especially when the feed flow contains metallic ions, in which the ED treatment is typically carried out at low pH to avoid metal precipitation. However, low pH affects ED variables such as concentration polarization, current efficiency, and energy consumption [20]. Protons are ions with the highest molar conductivity (350.1 Ω^{-1} cm^2 mol^{-1} at 25 °C [21]). So, when the pH is low, protons are the predominant current transporter in the systems, affecting the recovery efficiency because the current is used to mobilize the protons instead the other cations.

Controlling the pH and optimizing its value is crucial to the excellent performance of ED treatment. According to Abou-Shady et al. [20], who studied the effect of pH on the separation of Pb(II) and NO_3^- from aqueous solution by electrodialysis, the optimal pH range for metallic ion solutions is between 3 and 5. Operating at pH < 3 makes the technique noneconomical due to high energy consumption, while operating at pH > 5 has undesirable effects, such as metal precipitation.

Water electrolysis tends to generate an acidic medium and an alkaline medium in the anode and cathode compartments, respectively. The alkalization of the catholyte may produce the precipitation of salts and hydroxides, while the acidification of anolyte leads to an increase in the energy consumption due to the high ionic conductivity of protons. Furthermore, water electrolysis reactions involve the formation of O_2 and H_2 at the anode and cathode, respectively, which may produce polarization at the electrodes due to attached bubbles and overpressure in the compartments.

4. Application of the Technique

4.1. General Application

Electrodialysis has been applied for the desalination of brackish water. The first commercial ED equipment was sold by Ionics Inc. (Watertown, MA, USA) in 1954 to Arabian-American Oil Co. and installed in Saudi Arabia to produce demineralized water. Since then, the use of ED to remove salt from brackish water has extended to the United States and Europe. However, in the last decade, with the development of new membrane technologies, electrodialysis has become an alternative to reverse osmosis for seawater desalination [22]. An example of this is the Maspalomas desalination plant (Spain), built in 1986. Initially, the plant featured electrodialysis reversal technologies, which were replaced by reverse osmosis membranes in 2006 due to the need to increase its capacity (33,500 m^3 per day) [23].

Even so, electrodialysis has an economic advantage in a certain feed water salt composition range [24]. Electrodialysis is normally used in small- and medium-size plants (100–20,000 m^3/day) with a water salinity of 1–5 g/L of total dissolved solid (as higher salt contents entail higher energy requirements), while reverse osmosis (RO) is recommended for the desalination of water with dissolved solid concentrations higher than 10 g/L [22]. Moreover, some of the advantages of ED compared to RO are high water recovery rates, long lifetime of membranes, and less membrane fouling when using electrodialysis reversal. On the other hand, one of the disadvantages is that neutral toxic components, such as some viruses or bacteria, are not removed, and the produced water needs a post-treatment to be potable [18]. Another disadvantage of electrodialysis is the relatively high energy consumption, 2.6–5.5 kWh/m^3, compared to the energy requirement of reverse osmosis, 1.5–2.5 kWh/m^3 [25].

Although reverse osmosis leads in the water desalination market due to its advantages, ED combined with RO has a great potential [26–30]. RO generates large volumes of concentrate, typically 10–50% of the feed flow, that is usually discharged into the sea or treated by evaporation. With the aim of reducing the environmental and economic impact of the process, ED has been proposed to recover water and other valuable products from the RO concentrate, thereby minimizing the volume of discharge and contributing to the zero liquid discharge system [10]. In short, the role of ED is concentrating the RO waste even more to obtain an enriched stream of ions that favors the recovery of these ions in subsequent steps, and other streams of freshwater that can be mixed with the RO dilute, to increase the recovery of water in the process [27]. Brackish water is first treated with RO, and, subsequently, the concentrate stream is fed to an ED stack, increasing the concentration of total dissolved solids (TDs) from 1% to 10% and enhancing the overall water recovery from 82.5% (RO) to 92.1% (RO + ED).

In addition to brackish water desalination, other important applications of conventional electrodialysis are the demineralization of industrial process water and industrial wastewater treatment. In an industrial process, water is usually demineralized and neutralized before being used as process water to avoid equipment corrosion. Moreover, industrial wastewater is usually treated before being discharged to other water bodies due to its high salt concentration or toxic constituents that pose a risk to the environment. Because of the huge volume of water required in industrial processes, especially in cooling systems, a typical practice is reusing a fraction of treated wastewater to, on one hand, save costs related with the supply of fresh water, and, on the other hand, reduce wastewater discharge. Electrodialysis is particularly appropriate for the treatment of cooling water blowdown since high recovery rates and high brine concentrations can be achieved [31]. Furthermore, the temperature operation range of ion-exchange membranes is suitable for most cooling systems [22].

Regarding the treatment of wastewater, several applications of ED have been investigated to enhance the properties of municipal wastewater effluents. The reuse of tertiary treated effluents from wastewater treatment plants (WWTPs) is considered one of the solutions to the scarcity of water resources. However, conventional tertiary effluents are not always suitable for use, as some quality standards are not achieved. For example, the high salinity of effluents from WWTPs hinders the reuse of water for irrigation. Hence, ED can be used to reduce the salinity content of the effluent from WWTPs [32].

The economic and technical feasibility of applying ED to tertiary effluent has been reported in several studies [10]. Goodman et al. [33] investigated the capacity of a system based on EDR to remove salts from treated municipal wastewater. The pilot plant was constituted by a multimedia filtration unit and an EDR system with a capacity of 144 m^3/day. After prefiltration and coagulation–disinfection with $Fe_2(SO_4)_3$ and NaClO, the effluent was transferred to the multimedia filtration unit, then to the EDR system to remove the salt. The pilot plant reduced the total dissolved solid from 1104 mg/L to 328 mg/L, thus below the limit of 375 mg/L indicated by the Australian government and World Health Organization [34]. Moreover, the conductivity of the recycled water was reduced by 72%, demonstrating that the EDR-treated water is a viable alternative resource to provide quality water for agriculture. Gally et al. [32] evaluated the application of ED to turn sewage effluent into water with a proper quality to be reused as industrial water. The results confirmed the efficiency of ED in the reduction in electrical conductivity and the extraction of a high percentage of ions, especially, the removal of corrosive (Cl^-) and encrusting (Ca^{2+}, Mg^{2+}) ions.

In addition to reducing the salinity of wastewater for reclamation, electrodialysis has the potential to produce value-added streams. Llanos et al. [35] investigated the combined used of electrodialysis to, on one hand, reduce the conductivity of wastewater effluent, and, on the other hand, electrochemically (ECh) synthesize hypochlorite in the anolyte, which can be used as disinfectant. The integrated ED-ECh process achieved a final dilute and disinfect stream with a total electrical consumption of 1.03 Wh dm^{-3} and using a reduced volume of the anolyte (volumetric ratio anolyte/dilute 4:96).

In terms of revalorization, municipal wastewater (MWW) also has the potential to be a source for the recovery of nutrients, such as nitrates, phosphate, and potassium, which can be used as fertilizers. In this context, electrodialysis has been studied to be used as a pre-treatment step to concentrate the stream and enhance the subsequent precipitation/crystallization steps of nutrients [36]. Mohammadi et al. [37] studied the recovery of nitrate from MWW using single- and a two-stage electrodialysis processes. Under the optimized operational conditions (flow rate of 60 L h^{-1}; a four cell pairs; dilute-to-concentrated volume ratio of 2/0.5), the nitrate concentration in the dilute channel reached zero with a concentration ratio of 4.6 and energy consumption of 1.44 kWh/kg NO_3^-. The nitrate concentration ratio was enhanced by the two-stage process, and reached a ratio of 19.2 with an energy consumption of 4.34 kWh/kg NO_3^-. Cai et al. [38] developed an electrodialysis process with a magnesium anode to recover phosphate and ammonia as struvite from synthetic wastewater. The pilot-scale ED system removed 65% of phosphate from the wastewater stream, which had a phosphate concentration of 10 mg L^{-1}, while the phosphate concentration in the anode chamber was kept at 30 mg L^{-1} to promote the precipitation of phosphate with magnesium as struvite. In another study, carried out by Rota et al. [39], electrodialysis was proposed to treat solutions with a low phosphorus concentration with the aim of recovering it. The experiments were carried out in a five-compartment electrodialysis cell with two AEMs and two CEMs. Working under limiting current density conditions (0.6 mA cm^{-2}), a concentration factor of 9.7 was reached, obtaining a product stream with phosphate concentration of 0.120 g L^{-1}.

4.2. Applications to Wastewater Containing Metals

In recent years, electrodialysis has been expanded to the concentration and recovery of metals from industrial effluent. The development of new membrane materials has enhanced the techniques to treat acidic industrial effluents. The following section presents different approaches to concentrate and recover different metals typically contained in industrial wastewater.

4.2.1. Single Metal Recovery from Aqueous Effluents

The most relevant results of several studies dealing with the use of electrodialysis applied to remove metals from aqueous effluents are summarized in Table 4.

Liu et al. [40] proposed using a bipolar membrane electrodialysis system (BMED) to remove arsenic from copper slag obtained as waste during the hydrometallurgical process. They studied the effect of experimental parameters such as current density and particle size on the recovery and removal of metals. The procedure followed consisted of three steps: (a) use of a BMED system to leach metals from the copper slag and separate some metallic cations, (b) separation of solid and liquid phases, and (c) use of a BMED system to recover arsenic and remove metallic cations from the liquid phase. This method was found to be effective in the recovery of arsenic (more than 79% in the form of H_3AsO_4) and the removal of metal cations. The potential use of electrodialysis for arsenic removal from geothermal water has also been evaluated [41]. The operational parameters optimized were the pH, As concentration, and the discharged voltage. Results showed a reduction of more than 90% in arsenic within 60 min.

The combination of leaching and electrodialysis processes as a method to reduce the concentration of cadmium in phosphate ore was found to be an effective method. The influence of operational parameters such as reaction time, the chemical properties and concentration of the extracting agent, liquid-to-solid ratio, pH, temperature, and current density was evaluated. From the results, it was concluded that a relevant reduction in cadmium content in phosphate ore was not achieved by using simple batch leaching. However, results showed that the percentage of cadmium removed was up to 84% at optimum conditions (current density of 10 mA/cm^2) [42].

As a carcinogenic, mutagenic, and toxic metal, a lot of efforts have been made to reduce the amount of chromium. Liu et al. [43] evaluated the development of an electrokinetic system improved by a BM to recover Cr(VI) in the form of H_2CrO_4 from chromite ore processing residue. The electrolyte concentration in anode and cathode chambers was observed to have a direct effect on the cell voltage, resulting in optimal electrolyte concentrations: 0.6 mol/L HNO_3 in the anode chamber and 1.0 mol/L $NaNO_3$ in the cathode chamber. Regarding current density, the optimal value was 3.0 mA/cm^2. This study was carried out in cells equipped with different numbers of chromite ore processing residue chambers. Results showed that two- and three-chamber-equipped systems had higher current efficiency and lower specific energy consumption values. Under optimal experimental conditions, the Cr(VI) recovery efficiencies were higher than 80%. Similarly, the recovery of Cr(III) and Cr(VI) as Na_2CrO_4 (a relevant raw material in several areas of manufacturing) using a modified bipolar membrane electrodialysis system was also proposed as a promising method to treat industry waste containing heavy metals [44,45].

Liu et al. [46] studied bipolar electrodialysis (BMED) and electrodeposition processes to recover copper contained in hazardous sludge produced by electroplating processes. The authors evaluated the influence of the number of sludge compartments equipped in the system, concluding that the specific energy consumption decreased with the number of compartments. The removal of copper from sludge was up to 96%, and the recovery in the form of copper foil via electrodeposition was 57% under optimized conditions (current density of 50 mA/cm^2, pH solution less than or equal 0.5, and initial copper concentration less than or equal 4 g/L). Hernandez et al. [47] proposed the recovery of copper from pregnant leaching solutions using a reactive electrodialysis cell equipped with two anion-exchange membranes and a bipolar electrode. Under optimized conditions (temperature of 55 °C, current density of 80 A/m^2, and flow rate of 100 mL/s), the copper recovery was up to 99%. Both works confirmed that further work should be carried out to optimize the copper recovery rate and reduce the specific energy consumption.

Table 4. Summary of main results of studies focused on ED applied to treat aqueous effluents with recovery of a single metal.

Metal	Sources	Recovery (%)	Time	Max V	Current Density (mA cm^{-2})	Energy (kWh/g)	Feed	pH	Membrane	Ref.
As	Copper slag	96.50	45 h	9	3	-	-	-	BPMs	[40]
As	Geothermal water	91 As(III) 98 As(V)	60 min	25	-	-	5 mg/L As(III) 60 mg/L As(V)	8	CMB and AHA	[41]
Cd	Phosphate ore	84.30	24 h	8	10	-	50 mL 0.5 M Acetic acid + 2 g phosphate ore	4.5	CEM and AEM	[42]
Cr	Chromite ore	82	350 h	5	3	0.395	50 mL water NaNO$_3$ + 50 g solid	13.5	BPMs	[43]
Cr	Aqueous solution	87.8 Cr(III)	24 h	4	0.5	0.73	5 g/L Na$_2$SO$_4$	12	BPM, CEM and AEM	[44]
Cr	Chromium slag	70.6 Cr(VI)	300 h	-	3	-	50 mL water + 50 g solid	-	BPMs	[45]
Cu	Electroplating sludge	96.4	5 h	-	50	5.3	4 g/L Cu^{2+}	0.5	BPM, CEM, AEM	[46]
Cu	Pregnant leaching solutions	99.2	5 h	2.5	8	2.11	2.5 g/L Cu^{2+}	-	-	[47]
Ni	Electroless plating bath	82.34	3 h	-	3.5	0.0182	325 mg/L Ni^{2+}	3	AEM	[48]
Ni	Electroplating sludge	94	28 h	-	20	-	S/L ratio 1:15	-	BPM and CEM	[49]
Pb	Lead battery manufacture	75	4 h	40	-	7 kWh/m^3	5 mg/L Pb^{2+} 1000 mg/L SO$_4^-$	-	AEM: PC SA CEM: PC SK	[50]
Zn	Electroplating waste	86.6	60 min	-	2.5	-	0.748 M ZnSO$_4$·7H$_2$O + Citric acid	4	CMH-AMH Ralex membranes	[51]
Li	Spent LIB leachate	63.91	3 h	15	-	-	-	7	-	[52]

(BPM: bipolar membrane; CEM, CMB, CMH: types of cation-exchange membranes; AEM, AHA, AMH: types of anion-exchange membranes).

The recovery of Ni from spent electroless nickel-plating baths with high concentrations of nickel and phosphorus was evaluated. The authors designed a two-chamber cell equipped with an AEM to couple electrodialysis and electrodeposition processes. The results confirmed that the experimental system allows the efficient recovery of Ni and the removal of P without using additional chemical reagents, which is associated with less pollution [48]. Alternatively, Liu et al. [49] designed a bipolar membrane electrodialysis system to recover nickel from electroplating sludge. They concluded that the H^+ produced during the electrodialysis process has a relevant influence on the solubilization of Ni from the solid matrix. The $Ni(OH)_2$ obtained during the treatment was found to have the same purity and physiochemical properties as commercial products, which is a promising result from the point of view of the circular economy.

Voutetaki et al. [50] proved the use of electrodialysis to treat aqueous effluents produced in the lead–acid industry. They implemented an ED pilot plant to separate the sulfate and lead ions, typically contained in battery industrial wastewater with a concentration of 500–2000 mg/L and 5 mg/L, respectively. The pilot-scale ED setup was formed by 66 cell pairs of CEMs and AEMs of 5 m^2 effective membrane area, operated in batch mode. It was found that the high concentration ratio of sulfate-to-lead ions (400:1) had a negative effect on the removal of lead. To improve this, different experiments were carried out to optimize the operational conditions, resulting in 75% of Pb with an energy consumption of 7 kWh/m^3 water treated, after 4 h at 40 V, 300 L/h, and a 90% dilute-to-concentrate tank volume ratio.

Babilas et al. [51] evaluated the formation of complex as a method to improve the electrodialysis technology applied to the selective recovery of zinc from industrial wastewater. They studied the effect of several chelating agents (i.e., citric, malic, and lactic acids) and the ion-exchange membrane type on zinc removal from wastes contaminated with ferric ions. The combination of electrodialysis with the addition of citric acid was found to be a promising strategy to selectively recover metals from aqueous effluents. The possibility of using a suitable chelating agent for electrodialysis has also been applied to the recovery of lithium from aqueous effluents. It should be noted that the recovery of lithium from solid matrices is gaining interest due to the increasing consumption of batteries. Xing et al. [52] proposed the addition of several chelating agents to improve the performance of the electrodialysis approach applied to synthetic lithium-ion battery solutions. The role of the chelating agent was dependent on the transitional metal content of the aqueous effluent. The application of bipolar-membrane-assisted electrodialysis under optimum conditions resulted in the recovery of 64% of lithium with a purity of 99%. The use of a chelating agent, such as chloride ions, has also been proven to treat Hg(II) ions to avoid its easy reduction to elemental mercury inside the cell and to increase the ED removal rate of Hg [53]. However, it has been tested in solid matrices instead of wastewater. In this context, experimental research is still needed to further improve the removal of Hg from water using electrodialysis. Sun et al. [54] proposed a model to optimize the electrodialysis process for the recovery of Hg from seaweed extracts. The optimal conditions were achieved at 7.17 V and 72.54 L/h with a Hg removal rate of 76.45% from an initial solution with a concentration of 5.04 mg/L. From simulation results, it was concluded that electrodialysis could be a promising technique to recover Hg from aqueous solutions.

This section summarizes how effluents containing different amounts of metal can be concentrated by applying ED. As can be concluded from the results, the experimental setup and conditions should be selected according to the properties of the aqueous effluent to recover the target metal effectively.

4.2.2. Multiple Metal Recovery from Aqueous Effluents

Industrial wastewater frequently contains metals that are used in different concentrations in many industry sectors; so, they are contaminants that cause serious environmental problems when discharged and their removal and recovery constitute an environmental and economic challenge. Several studies focusing on the use of electrodialysis to recover multiple metals from aqueous effluents are summarized in Table 5.

Table 5. Summary of main results of studies focused on ED applied to treat aqueous effluents with multiple-metal recovery.

Metal	Sources	Recovery (%)	Time	Max V	Current Density (mA cm^{-2})	Energy	Feed	Stage	Membrane	Ref
Cr(VI) Ni	Industrial effluent	97.9 97.1	90 min	25	-	38.57 Wh/L	50 mg/L 50 mg/L	1	Ionac MC 3470 Ionac MA 3475	[55]
Fe Ni Cu	Printed circuit boards	-	50 min	30	50	-	-	1	PC Acid 60 (PCCell GmbH)CMH RALEX® (MEGA	[56]
Co Zn As(III)	Sulfuric acid solution	93 100 58	5 h	-	5	-	0.01 M 0.01 M	1	MK-40 CEM MK-40 AEM	[57]
Se(IV) Se(VI)	Brackish water	80 80	-	25	-	-	50–1000 µg/L	1	-	[58]
Cu Ni	Electroplating wastewater	90.7 90.2	25 min	12	22.5	-	22.3 mg/L 24.4 mg/L	1	AMX Astom CMX Astom	[59]
Ni Co Li Mn	Lithium-ion battery leaching solution	99.8 87.3 99 99	180 min	18	-	9.65 kWh/mol 15.3 kWh/mol	0.01 M 0.003 M 0.003 M 0.003 M	3	Neosepta AMXNeosepta CMXPCA PC 400D	[60]
Li Co	Lithium-ion battery leaching solution	66 33	144 h	-	1	-	1.75 g LiCoO$_2$ 350 mL 0.1 M HCl	1	Neosepta CMX	[61]

(CEM, CMX: types of cation-exchange membranes; AEM, AMX: types of anion-exchange membranes).

Kirmizi et al. [55] used a self-designed electrodialysis cell to separately remove chromium(VI) and nickel(II) ions from aqueous effluents. The electrodialytic cell was divided into a compartment containing the diluted solution and two electrolyte compartments separated by a pair of AEMs (Ionac MA 3475) and CEMs (MC 3470). The cathode and anode were made of carbon fiber and stainless steel, respectively. Distilled water with pH adjusted to 3 with H_2SO_4 was used as the electrolyte solution, and Na_2SO_4 was added at 7 mM. This study observed that the removal efficiency decreased with increasing metal concentration, although high concentrations of metals can reduce the concentration polarization phenomenon and increase efficiency. This has been attributed to the fact that a reduced membrane retention time can negatively affect the process. For both Cr(VI) and Ni(II), increasing the voltage and Na_2SO_4 concentration had an increasing effect on energy consumption, while increasing the pH had a decreasing effect. An optimal removal of 92.3 ± 1% was achieved in 90 min for a 40 mL/min feed rate and 50 mg/L of Ni(II) ions at pH = 3, voltage value of 20 V, and energy consumption of approximately 30 Wh/L. In the case of Cr(VI), a removal of 97.9 ± 1% was achieved in 90 min for optimal values of 25 V and a feeding rate of 40 mL/min and 50 mg/L of Cr(VI) ions at pH = 3, with an approximate energy consumption of 40 Wh/L. Finally, it was concluded that it is necessary to operate with a high concentration of metal, a low voltage, and a low addition of salt to the electrolytes to obtain a high current efficiency.

In another study, Shestakov et al. [56] proposed electrodialysis to recover iron, nickel, and copper from wastewater generated in the manufacture of printed circuit boards, for which they used an electrodialytic system with PC Acid 60 (PCCell GmbH) and CM (H) RALEX® (MEGA) ion-exchange membranes. To each metal solution, sodium ethylenediaminetetraacetate with a molar ratio of Na_2EDTA/metal = 1:1.2 and 2–3 drops of a 40% aqueous solution of HNO_3 was added. When separating solutions containing one and all three salts of the target metals, the retention coefficients of the Ni^{2+} and Cu^{2+} cations differedsignificantly. In contrast, those of Fe^{3+} cations did not vary by more than 2–3%. When separating a multicomponent solution of three salts, the retention coefficients of Ni^{2+} and Cu^{2+} cations decreased compared to the corresponding values for mono salt solutions. In any case, electrodialysis separation of multicomponent solutions is an efficient method for exhaustive metal recovery, as applied to Fe^{3+} and Cu^{2+} cations.

Sadyrbaeva et al. [57] proposed a new method to extract cobalt and zinc ions from sulfuric acid solutions, silver and lead ions from nitric acid solutions, and copper ions from hydrochloric acid solutions via electrodialysis with 1,2-dichloroethane liquid membranes with di-acid(2-ethylhexyl)phosphoric (D_2EHPA) and tri-n-octylamine (TOA) as a carrier. In this case, by using a current density of approximately 5 mA/cm^2 and times between 0.5 and 5.0 h, an almost complete extraction (93–100%) of the metals was achieved from aqueous solutions with an initial concentration of metallic salts of 0.01 M. The metal extraction rate decreased with the decrease in the pH of the starting solution and with the increase in the TOA concentration in the liquid membrane, establishing an optimum at 0.1 M, while the increase in the carrier concentration (D2EHPA) did not present significant effects, showing an optimum between 0.2 and 0.4 M.

The removal of arsenite As(III), selenite Se(IV), and selenate Se(VI) from brackish water using an electrodialysis system was studied by Aliaskari et al. [58]. Contaminant removal was investigated at different pH values (3–11). The results showed that the removal of As and Se was pH-dependent since the loading of Se(IV), Se(VI), and As(III) species is pH-dependent. The removal of As(III) increased at pH > 9, while the removal of As(V), Se(IV), and Se(VI) decreased. Two sources of groundwater contaminated with As and Se were also investigated. The effect of chloride ion competition was studied by testing different values of salinity (ranging from 1 to 10 g/L of total dissolved solids). The results showed that increasing salinity resulted in delayed removal of As and Se. Additionally, various feed concentrations of Se(IV), Se(VI), and As(III) were investigated, ranging from 50 to 1000 µg/L. It was observed that higher feed concentrations of As and Se led to a higher molar flux, but the removal rates were not affected by the ion concentration. Furthermore, increasing the electrical potential from 5 to 25 V led to a significant increase in contaminant removal. The removal rates for As(V), Se(IV), and Se(VI) increased from less than 5% at 5 V to over 80% at 25 V. The maximum removal observed for As(III) was 58% at 25 V.

Min et al. [59] studied electroplating wastewater treatment using electrodialysis. The electrodialysis cell was equipped with five pairs of ion-exchange membranes and one pair of platinum-plated titanium electrodes. The CEM was CMX-SB (NEOSEPTA) and the AEM was AMX-SB (NEOSEPTA). The wastewater contained mainly copper and nickel with concentrations of 22.4 and 24.4 mg/L, respectively; its conductivity was 6300 µS/cm, and the pH was 2.18. The electrolyte solution contained 4% (w/w) Na_2SO_4. The voltage used (6 to 18 V) in the process was a very important factor in the separation efficiency. As it increased from 6 to 12 V, the separation efficiency of Cu^{2+} and Ni^{2+} improved, but if a higher voltage was applied (>12 to 18 V), concentration polarization occured, and the efficiency decreased. The optimal applied voltage, with a water conductivity of 6300 µS/cm, was 12 V with Cu^{2+} and Ni^{2+} removal efficiencies >99% after 25 min, with final concentrations of Cu^{2+} and Ni^{2+} in the concentrate of 1000 and 1200 mg/L and recovery rates of 90.7% and 90.2%, respectively. From the results, it is concluded that electrodialysis could effectively treat metallic industrial wastewater and recover significant amounts of metals.

Another field of interest in metal recovery is spent lithium-ion batteries. For this purpose, electrodialysis must be considered an emerging green process capable of recovering valuable metals from solid matrices. The main challenge of electrodialysis is the difficulty of separating various similarly charged metal ions due to the low selectivity of ion-exchange membranes. In order to improve the process selectivity, several complexing agents, such as ethylenediaminetetraacetic acid, citric acid, malic acid, and lactic acid, are used to form negatively charged complex anions so that ions with different charges can be separated. However, there are very few studies on applying electrodialysis for metal separation from spent LIBs.

In a study conducted by Chan et al. [60], they investigated the separation and recovery of lithium, nickel, manganese, and cobalt from mixtures obtained from spent lithium-ion batteries. The process involved three stages of electrodialysis coupled with EDTA using an AEM (PCA PC 400D) and a CEM (Neosepta CMX). In the first stage, at a pH of approximately 2, 99.3% of nickel was successfully recovered. In the second stage, at a pH of around 3, 87.3% of cobalt was separated. Finally, in the third stage, electrodialysis with a monovalent CEM (Neosepta CMS) was used to separate 99% of lithium from manganese.

The later breakdown of the EDTA-metal complexes, Ni (from stage 1) and Co (from stage 2), was carried out by adding a 2.0 M H_2SO_4 solution until the pH reached a value less than 0.5. The solid acid (EDTA) was recovered after filtration and washing with water and could be reused. Likewise, the H_2SO_4 solution recovered in the anode compartment could be reused. All metals recovered were more than 99% pure.

A new technique that combines hydrometallurgical extraction with electrodialysis to selectively recover lithium (Li) and cobalt (Co) from lithium-ion battery waste has been presented by Cerrillo et al. in a recent article [61]. The combined method helps reduce the consumption of leaching solution by regenerating the acid through electrolysis. Several extractions were conducted on $LiCoO_2$ powder to investigate the dissolution process of this common cathode material in lithium-ion batteries. A 0.1 M HCl solution was used as the extraction agent with a liquid-to-solid ratio of 200. The purpose of using hydrochloric acid was to test the effectiveness of chloride ions in reducing Co^{3+} to Co^{2+}. To ensure consistency, two cells were used in series for the hydrometallurgical–electrodialytic experiments. The electrodialytic cells consisted of three compartments separated by two Neopsepta CMX-fd CEMs. The anodic, central, and cathodic compartments were kept apart using these membranes. A CEM was used to separate the anode from the central compartment to prevent chloride ions from reaching the anode, which could otherwise oxidize and produce chlorine gas. The anode was made of titanium coated with metal oxides, while the cathode was made of stainless steel. The electrical current was kept constant at 50 mA, which corresponds to a current density of 1 mA cm^{-2}. The incoming $LiCO_2$-HCl suspension was continuously pumped from an external container to the central compartment of the electrodialysis cell, passing through a separatory funnel and a glass fiber filter that prevented particles from reaching the interior of the electrodialysis cell. The experiments led to the recovery of 62% lithium and 33% cobalt in the catholyte; 80% of the cobalt was electrodeposited on the cathode.

Siekierka et al. [62] proposed a method for selectively recovering metal transition cations from leaching battery waste using a reverse electrodialytic process. This approach generates energy while recovering the desired metals. Typically, the reverse electrodialytic process is used for energy production by leveraging the salinity gradient between seawater and river water. In this study, highly concentrated spent battery acid leachate is applied to generate a potential difference in the reverse electrodialytic cell due to the transport of ionic species across the membrane to a dilute solution. The system utilizes two commercial AEMs (ASE S-5158) and one central selective CEM (PAN-5C8Q) developed by the authors. Both electrodes use a 0.2 M sodium carbonate solution. Based on the analysis, it has been estimated that the maximum amount of energy that can be extracted per square meter is 0.44 watts. Furthermore, the potential power generation for all alkali and transition metal cations increased with the salinity of the high-concentration solution. The energy efficiency of the presented method was 45.5%. This technique holds great potential for managing waste batteries by converting them into valuable products like cobalt salts and producing additional electrical energy.

The combined recovery of metal ions contained in a mixture is a challenge from the point of view of the selectivity and purity of the recovered component. In this context, electrodialysis is a promising technology for the recovery of multiple metals from aqueous effluents, as was concluded from previously described works. These studies demonstrate the effectiveness of electrodialysis in diverse applications for metal recovery from a diverse variety of industrial wastewaters. However, most of the studies reviewed were carried out at the lab scale. So, experimental research is still needed to further improve the process and optimize the work conditions, with the aim of scaling the process to an industrial level.

5. Conclusions

Electrodialytic separation is a promising technique that offers new possibilities for the selective separation of metals from wastewater effluents, particularly from industrial sources. The specific design of the cell, in terms of the number of compartments, the disposition of the ionic-exchange membranes, the pH control, and the use of special membranes such as monovalent membranes or bipolar membranes, allow not only desalination but also the selective separation of target contaminants, which facilitates circular economy strategies within the field of wastewater treatment and management. The results obtained from experimental studies dealing with the optimization of ED applied to aqueous effluents containing metals highlight the importance of evaluating factors such as membrane properties, cell configuration, and operational conditions. Advances in membrane technologies have been focused on expanding the applications of ED, optimizing the performance of the technology, and overcoming some limitations, such as fouling or high costs due to the energy consumption required. A new opportunity for ED was opened with the use of renewable energy sources. Moreover, the production of gaseous streams, such as H_2 from water electrolysis, can be used to generate electricity. Nevertheless, further investigations are required to implement this technology at an industrial scale, since most studies have been carried out at the lab scale. Furthermore, it would be necessary to carry out an economic evaluation of ED scaling and a comparison with other separation techniques to demonstrate the applicability of ED in the field of metal separation from aqueous solutions.

Author Contributions: Conceptualization, M.V.-G. and J.M.P.-G.; methodology, M.d.M.C.-G. and M.V.-G.; investigation, M.d.M.C.-G.; writing—original draft preparation, M.d.M.C.-G., M.V.-G. and J.M.R.-M.; writing—review and editing, M.V.-G. and J.M.P.-G.; visualization, J.M.R.-M.; supervision, J.M.R.-M. and M.V.-G.; project administration, M.V.-G. All authors have read and agreed to the published version of the manuscript.

Funding: The authors acknowledge the funding from the University of Malaga (B1-2021_35).

Data Availability Statement: No new data were created or analyzed in this study. Data sharing is not applicable to this article, since the information in the article has been obtained from literature revision (cited in reference section).

Acknowledgments: Cerrillo-Gonzalez acknowledges the postdoctoral grant obtained from the University of Malaga.

Conflicts of Interest: The authors declare no conflicts of interest.

References

1. Ahmed, J.; Thakur, A.; Goyal, A. Chapter 1: Industrial Wastewater and Its Toxic Effects. In *Biological Treatment of Industrial Wastewater*; The Royal Society of Chemistry: London, UK, 2021; pp. 1–14. [CrossRef]
2. Briffa, J.; Sinagra, E.; Blundell, R. Heavy Metal Pollution in the Environment and Their Toxicological Effects on Humans. *Heliyon* **2020**, *6*, e04691. [CrossRef]
3. Shrestha, R.; Ban, S.; Devkota, S.; Sharma, S.; Joshi, R.; Tiwari, A.P.; Kim, H.Y.; Joshi, M.K. Technological Trends in Heavy Metals Removal from Industrial Wastewater: A Review. *J. Environ. Chem. Eng.* **2021**, *9*, 105688. [CrossRef]
4. Barakat, M.A. New Trends in Removing Heavy Metals from Industrial Wastewater. *Arab. J. Chem.* **2011**, *4*, 361–377. [CrossRef]
5. Villen-Guzman, M.; Cerrillo-Gonzalez, M.M.; Paz-Garcia, J.M.; Rodriguez-Maroto, J.M.; Arhoun, B. Valorization of Lemon Peel Waste as Biosorbent for the Simultaneous Removal of Nickel and Cadmium from Industrial Effluents. *Environ. Technol. Innov.* **2021**, *21*, 101380. [CrossRef]
6. Doble, M.; Kumar, A. Treatment of Waste from Metal Processing and Electrochemical Industries. In *Biotreatment of Industrial Effluents*; Butterworth-Heinemann: Oxford, UK, 2005; pp. 145–155. [CrossRef]
7. Goh, P.S.; Wong, K.C.; Ismail, A.F. Membrane Technology: A Versatile Tool for Saline Wastewater Treatment and Resource Recovery. *Desalination* **2022**, *521*, 115377. [CrossRef]
8. Arana Juve, J.M.; Christensen, F.M.S.; Wang, Y.; Wei, Z. Electrodialysis for Metal Removal and Recovery: A Review. *Chem. Eng. J.* **2022**, *435*, 134857. [CrossRef]
9. Kitchenham, B. *Procedures for Performing Systematic Reviews*; Keele University: Keele, UK, 2004; Volume 33.
10. Gurreri, L.; Cipollina, A.; Tamburini, A.; Micale, G. Electrodialysis for Wastewater Treatment—Part I: Fundamentals and Municipal Effluents. In *Current Trends and Future Developments on (Bio-) Membranes: Membrane Technology for Water and Wastewater Treatment—Advances and Emerging Processes*; Elsevier: Amsterdam, The Netherlands, 2020; pp. 141–192, ISBN 9780128168240.
11. Van der Bruggen, B. Ion-Exchange Membrane Systems—Electrodialysis and Other Electromembrane Processes. In *Fundamental Modeling of Membrane Systems: Membrane and Process Performance*; Elsevier: Amsterdam, The Netherlands, 2018; pp. 251–300. [CrossRef]

12. Pärnamäe, R.; Mareev, S.; Nikonenko, V.; Melnikov, S.; Sheldeshov, N.; Zabolotskii, V.; Hamelers, H.V.M.; Tedesco, M. Bipolar Membranes: A Review on Principles, Latest Developments, and Applications. *J. Membr. Sci.* **2021**, *617*, 118538. [CrossRef]
13. Hansima, M.A.C.K.; Makehelwala, M.; Jinadasa, K.B.S.N.; Wei, Y.; Nanayakkara, K.G.N.; Herath, A.C.; Weerasooriya, R. Fouling of Ion Exchange Membranes Used in the Electrodialysis Reversal Advanced Water Treatment: A Review. *Chemosphere* **2021**, *263*, 127951. [CrossRef]
14. Tanaka, Y. Chapter 2: Electrodialysis Reversal. In *Membrane Science and Technology*; Elsevier: Amsterdam, The Netherlands, 2007; Volume 12, pp. 383–404. [CrossRef]
15. Güler, E.; Elizen, R.; Vermaas, D.A.; Saakes, M.; Nijmeijer, K. Performance-Determining Membrane Properties in Reverse Electrodialysis. *J. Membr. Sci.* **2013**, *446*, 266–276. [CrossRef]
16. Luo, T.; Abdu, S.; Wessling, M. Selectivity of Ion Exchange Membranes: A Review. *J. Membr. Sci.* **2018**, *555*, 429–454. [CrossRef]
17. Tekinalp, Ö.; Zimmermann, P.; Holdcroft, S.; Burheim, O.S.; Deng, L. Cation Exchange Membranes and Process Optimizations in Electrodialysis for Selective Metal Separation: A Review. *Membranes* **2023**, *13*, 566. [CrossRef]
18. Moura Bernardes, A.; Zoppas Ferreira, J.; Siqueira Rodrigues, M.A. *Electrodialysis and Water Reuse: Novel Approaches*; Springer: Berlin/Heidelberg, Germany, 2014; ISBN 9783642402494.
19. Ebbers, B.; Ottosen, L.M.; Jensen, P.E. Electrodialytic Treatment of Municipal Wastewater and Sludge for the Removal of Heavy Metals and Recovery of Phosphorus. *Electrochim. Acta* **2015**, *181*, 90–99. [CrossRef]
20. Abou-Shady, A.; Peng, C.; Almeria, O.J.; Xu, H. Effect of PH on Separation of Pb (II) and NO_3^- from Aqueous Solutions Using Electrodialysis. *Desalination* **2012**, *285*, 46–53. [CrossRef]
21. Dean, J.A. *Lange's Handbook of Chemistry*, 15th ed.; McGraw-Hill Professional Publishing: New York, NY, USA, 1999; ISBN 0-07-016384-7.
22. Strathmann, H. Electrodialysis, a Mature Technology with a Multitude of New Applications. *Desalination* **2010**, *264*, 268–288. [CrossRef]
23. Dow Water & Process Solutions Aumenta La Eficiencia de La Planta de Agua Municipal MasPalomas I | IAgua. Available online: https://www.iagua.es/noticias/espana/dow-water-and-process-solutions/17/02/22/dow-water-process-solutions-aumenta (accessed on 26 May 2022).
24. Generous, M.M.; Qasem, N.A.A.; Akbar, U.A.; Zubair, S.M. Techno-Economic Assessment of Electrodialysis and Reverse Osmosis Desalination Plants. *Sep. Purif. Technol.* **2021**, *272*, 118875. [CrossRef]
25. Soliman, M.N.; Guen, F.Z.; Ahmed, S.A.; Saleem, H.; Khalil, M.J.; Zaidi, S.J. Energy Consumption and Environmental Impact Assessment of Desalination Plants and Brine Disposal Strategies. *Process Saf. Environ. Prot.* **2021**, *147*, 589–608. [CrossRef]
26. Jiang, C.; Wang, Y.; Zhang, Z.; Xu, T. Electrodialysis of Concentrated Brine from RO Plant to Produce Coarse Salt and Freshwater. *J. Membr. Sci.* **2014**, *450*, 323–330. [CrossRef]
27. Xu, X.; Lin, L.; Ma, G.; Wang, H.; Jiang, W.; He, Q.; Nirmalakhandan, N.; Xu, P. Study of Polyethyleneimine Coating on Membrane Permselectivity and Desalination Performance during Pilot-Scale Electrodialysis of Reverse Osmosis Concentrate. *Sep. Purif. Technol.* **2018**, *207*, 396–405. [CrossRef]
28. Yan, H.; Wang, Y.; Wu, L.; Shehzad, M.A.; Jiang, C.; Fu, R.; Liu, Z.; Xu, T. Multistage-Batch Electrodialysis to Concentrate High-Salinity Solutions: Process Optimisation, Water Transport, and Energy Consumption. *J. Membr. Sci.* **2019**, *570–571*, 245–257. [CrossRef]
29. Korngold, E.; Aronov, L.; Daltrophe, N. Electrodialysis of Brine Solutions Discharged from an RO Plant. *Desalination* **2009**, *242*, 215–227. [CrossRef]
30. McGovern, R.K.; Zubair, S.M.; Lienhard, V.J.H. The Benefits of Hybridising Electrodialysis with Reverse Osmosis. *J. Membr. Sci.* **2014**, *469*, 326–335. [CrossRef]
31. Soliman, M.; Eljack, F.; Kazi, M.K.; Almomani, F.; Ahmed, E.; El Jack, Z. Treatment Technologies for Cooling Water Blowdown: A Critical Review. *Sustainability* **2021**, *14*, 376. [CrossRef]
32. Gally, C.R.; Benvenuti, T.; Da Trindade, C.D.M.; Rodrigues, M.A.S.; Zoppas-Ferreira, J.; Pérez-Herranz, V.; Bernardes, A.M. Electrodialysis for the Tertiary Treatment of Municipal Wastewater: Efficiency of Ion Removal and Ageing of Ion Exchange Membranes. *J. Environ. Chem. Eng.* **2018**, *6*, 5855–5869. [CrossRef]
33. Goodman, N.B.; Taylor, R.J.; Xie, Z.; Gozukara, Y.; Clements, A. A Feasibility Study of Municipal Wastewater Desalination Using Electrodialysis Reversal to Provide Recycled Water for Horticultural Irrigation. *Desalination* **2013**, *317*, 77–83. [CrossRef]
34. WHO. *Guidelines for Drinking-Water Quality: Fourth Edition Incorporating the First Addendum*; WHO: Geneva, Switzerland, 2017.
35. Llanos, J.; Cotillas, S.; Cañizares, P.; Rodrigo, M.A. Novel Electrodialysis–Electrochlorination Integrated Process for the Reclamation of Treated Wastewaters. *Sep. Purif. Technol.* **2014**, *132*, 362–369. [CrossRef]
36. Mohammadi, R.; Tang, W.; Sillanpää, M. A Systematic Review and Statistical Analysis of Nutrient Recovery from Municipal Wastewater by Electrodialysis. *Desalination* **2021**, *498*, 114626. [CrossRef]
37. Mohammadi, R.; Ramasamy, D.L.; Sillanpää, M. Enhancement of Nitrate Removal and Recovery from Municipal Wastewater through Single- and Multi-Batch Electrodialysis: Process Optimisation and Energy Consumption. *Desalination* **2021**, *498*, 114726. [CrossRef]
38. Cai, Y.; Han, Z.; Lin, X.; Duan, Y.; Du, J.; Ye, Z.; Zhu, J. Study on Removal of Phosphorus as Struvite from Synthetic Wastewater Using a Pilot-Scale Electrodialysis System with Magnesium Anode. *Sci. Total Environ.* **2020**, *726*, 138221. [CrossRef]
39. Rotta, E.H.; Bitencourt, C.S.; Marder, L.; Bernardes, A.M. Phosphorus Recovery from Low Phosphate-Containing Solution by Electrodialysis. *J. Membr. Sci.* **2019**, *573*, 293–300. [CrossRef]

40. Liu, Y.; Dai, L.; Ke, X.; Ding, J.; Wu, X.; Chen, R.; Ding, R.; Van der Bruggen, B. Arsenic and Cation Metal Removal from Copper Slag Using a Bipolar Membrane Electrodialysis System. *J. Clean. Prod.* **2022**, *338*, 130662. [CrossRef]
41. Pham, M.T.; Nishihama, S.; Yoshizuka, K. Effect of Operational Conditions on Arsenic Removal from Aqueous Solution Using Electrodialysis. *Solvent Extr. Ion. Exch.* **2021**, *39*, 655–667. [CrossRef]
42. Benredjem, Z.; Delimi, R.; Khelalfa, A.; Saaidia, S.; Mehellou, A. Coupling of Electrodialysis and Leaching Processes for Removing of Cadmium from Phosphate Ore. *Sep. Sci. Technol.* **2016**, *51*, 718–726. [CrossRef]
43. Liu, Y.; Zhu, H.; Zhang, M.; Chen, R.; Chen, X.; Zheng, X.; Jin, Y. Cr(VI) Recovery from Chromite Ore Processing Residual Using an Enhanced Electrokinetic Process by Bipolar Membranes. *J. Membr. Sci.* **2018**, *566*, 190–196. [CrossRef]
44. Wu, X.; Zhu, H.; Liu, Y.; Chen, R.; Qian, Q.; Van der Bruggen, B. Cr(III) Recovery in Form of Na_2CrO_4 from Aqueous Solution Using Improved Bipolar Membrane Electrodialysis. *J. Membr. Sci.* **2020**, *604*, 118097. [CrossRef]
45. Dai, L.; Ding, J.; Liu, Y.; Wu, X.; Chen, L.; Chen, R.; Van der Bruggen, B. Recovery of Cr(VI) and Removal of Cationic Metals from Chromium Slag Using a Modified Bipolar Membrane System. *J. Membr. Sci.* **2021**, *639*, 119772. [CrossRef]
46. Liu, Y.; Lv, M.; Wu, X.; Ding, J.; Dai, L.; Xue, H.; Ye, X.; Chen, R.; Ding, R.; Liu, J.; et al. Recovery of Copper from Electroplating Sludge Using Integrated Bipolar Membrane Electrodialysis and Electrodeposition. *J. Colloid Interface Sci.* **2023**, *642*, 29–40. [CrossRef]
47. Hernández, J.; Tapia, J. Direct Copper Recovery from Pregnant Leaching Solutions (PLS), Using a Custom Electrolytic Cell, Based on Reactive Electrodialysis (RED). *Miner. Process. Extr. Metall. Trans. Inst. Min. Metall.* **2023**, *132*, 110–116. [CrossRef]
48. Yan, K.; Huang, P.; Xia, M.; Xie, X.; Sun, L.; Lei, W.; Wang, F. An Efficient Two-Chamber Electrodeposition-Electrodialysis Combination Craft for Nickel Recovery and Phosphorus Removal from Spent Electroless Nickel Plating Bath. *Sep. Purif. Technol.* **2022**, *295*, 121283. [CrossRef]
49. Liu, Y.; Lian, R.; Wu, X.; Dai, L.; Ding, J.; Wu, X.; Ye, X.; Chen, R.; Ding, R.; Liu, J.; et al. Nickel Recovery from Electroplating Sludge via Bipolar Membrane Electrodialysis. *J. Colloid Interface Sci.* **2023**, *637*, 431–440. [CrossRef]
50. Voutetaki, A.; Plakas, K.V.; Papadopoulos, A.I.; Bollas, D.; Parcharidis, S.; Seferlis, P. Pilot-Scale Separation of Lead and Sulfate Ions from Aqueous Solutions Using Electrodialysis: Application and Parameter Optimization for the Battery Industry. *J. Clean. Prod.* **2023**, *410*, 137200. [CrossRef]
51. Babilas, D.; Dydo, P. Selective Zinc Recovery from Electroplating Wastewaters by Electrodialysis Enhanced with Complex Formation. *Sep. Purif. Technol.* **2018**, *192*, 419–428. [CrossRef]
52. Xing, Z.; Srinivasan, M. Lithium Recovery from Spent Lithium-Ion Batteries Leachate by Chelating Agents Facilitated Electrodialysis. *Chem. Eng. J.* **2023**, *474*, 145306. [CrossRef]
53. Hansen, H.K.; Ottosen, L.M.; Kliem, B.K.; Villumsen, A. Electrodialytic Remediation of Soils Polluted with Cu, Cr, Hg, Pb and Zn. *J. Chem. Technol. Biotechnol.* **1997**, *70*, 67–73. [CrossRef]
54. Sun, J.; Su, X.; Liu, Z.; Liu, J.; Ma, Z.; Sun, Y.; Gao, X.; Gao, J. Removal of Mercury (Hg(II)) from Seaweed Extracts by Electrodialysis and Process Optimization Using Response Surface Methodology. *J. Ocean Univ. China* **2020**, *19*, 135–142. [CrossRef]
55. Kırmızı, S.; Karabacakoğlu, B. Performance of Electrodialysis for Ni(II) and Cr(VI) Removal from Effluents: Effect of Process Parameters on Removal Efficiency, Energy Consumption and Current Efficiency. *J. Appl. Electrochem.* **2023**, *53*, 2039–2055. [CrossRef]
56. Shestakov, K.V.; Lazarev, S.I.; Polyanskii, K.K.; Ignatov, N.N. Recovery of Iron, Nickel, and Copper in Waste Water from Printed Circuit Board Manufacture by Electrodialysis Method. *Russ. J. Appl. Chem.* **2021**, *94*, 555–559. [CrossRef]
57. Sadyrbaeva, T.Z. Membrane Extraction of Ag(I), Co(II), Cu(II), Pb(II), and Zn(II) Ions with Di(2-Ethylhexyl)Phosphoric Acid under Conditions of Electrodialysis with Metal Electrodeposition. *Theor. Found. Chem. Eng.* **2021**, *55*, 1204–1220. [CrossRef]
58. Aliaskari, M.; Ramos, R.L.; Schäfer, A.I. Removal of Arsenic and Selenium from Brackish Water Using Electrodialysis for Drinking Water Production. *Desalination* **2023**, *548*, 116298. [CrossRef]
59. Min, K.J.; Choi, S.Y.; Jang, D.; Lee, J.; Park, K.Y. Separation of Metals from Electroplating Wastewater Using Electrodialysis. *Energy Sources Part A Recovery Util. Environ. Eff.* **2019**, *41*, 2471–2480. [CrossRef]
60. Chan, K.H.; Malik, M.; Azimi, G. Separation of Lithium, Nickel, Manganese, and Cobalt from Waste Lithium-Ion Batteries Using Electrodialysis. *Resour. Conserv. Recycl.* **2022**, *178*, 106076. [CrossRef]
61. Cerrillo-Gonzalez, M.M.; Villen-Guzman, M.; Vereda-Alonso, C.; Gomez-Lahoz, C.; Rodriguez-Maroto, J.M.; Paz-Garcia, J.M. Recovery of Li and Co from $LiCoO_2$ via Hydrometallurgical–Electrodialytic Treatment. *Appl. Sci.* **2020**, *10*, 2367. [CrossRef]
62. Siekierka, A.; Yalcinkaya, F.; Bryjak, M. Recovery of Transition Metal Ions with Simultaneous Power Generation by Reverse Electrodialysis. *J. Environ. Chem. Eng.* **2023**, *11*, 110145. [CrossRef]

Disclaimer/Publisher's Note: The statements, opinions and data contained in all publications are solely those of the individual author(s) and contributor(s) and not of MDPI and/or the editor(s). MDPI and/or the editor(s) disclaim responsibility for any injury to people or property resulting from any ideas, methods, instructions or products referred to in the content.

Article

Performance Assessment of Wood Ash and Bone Char for Manganese Treatment in Acid Mine Drainage

Ivana Smičiklas [1,*], Bojan Janković [1], Mihajlo Jović [1], Jelena Maletaškić [1], Nebojša Manić [2] and Snežana Dragović [1]

[1] "VINČA" Institute of Nuclear Sciences—National Institute of the Republic of Serbia, University of Belgrade, Mike Petrovića Alasa 12-14, P.O. Box 522, 11001 Belgrade, Serbia; bojan.jankovic@vin.bg.ac.rs (B.J.); mjovic@vin.bg.ac.rs (M.J.); jelena.pantic@vin.bg.ac.rs (J.M.); sdragovic@vin.bg.ac.rs (S.D.)

[2] Faculty of Mechanical Engineering, University of Belgrade, Kraljice Marije 16, P.O. Box 35, 11120 Belgrade, Serbia; nmanic@mas.bg.ac.rs

* Correspondence: ivanat@vin.bg.ac.rs

Abstract: Developing efficient methods for Mn separation is the most challenging in exploring innovative and sustainable acid mine drainage (AMD) treatments. The availability and capacity of certain waste materials for Mn removal warrant further exploration of their performance regarding the effect of process factors. This study addressed the influence of AMD chemistry (initial pH and concentrations of Mn, sulfate, and Fe), the solid/solution ratio, and the contact time on Mn separation by wood ash (WA) and bone char (BC). At an equivalent dose, WA displayed higher neutralization and Mn removal capacity over the initial pH range of 2.5–6.0 due to lime, dicalcium silicate, and fairchildite dissolution. On the other hand, at optimal doses, Mn separation by BC was faster, it was less affected by coexisting sulfate and Fe(II) species, and the carbonated hydroxyapatite structure of BC remained preserved. Efficient removal of Mn was feasible only at final pH values ≥ 9.0 in all systems with WA and at pH 6.0–6.4 using BC. These conclusions were confirmed by treating actual AMD with variable doses of both materials. The water-leaching potential of toxic elements from the AMD/BC treatment residue complied with the limits for inert waste. In contrast, the residue of AMD/WA treatment leached non-toxic quantities of Cr and substantial amounts of Al due to high residual alkalinity. To minimize the amount of secondary waste generated by BC application, its use emerges particularly beneficial after AMD neutralization in the finishing step intended for Mn removal.

Keywords: acid mine drainage; Mn separation; waste valorization; wood ash; bone char

1. Introduction

Acid drainage resulting from mining activities poses a substantial worldwide problem for protecting and managing water resources, sediments, and soil [1–3]. The adjustment of pH and elimination of metal ions, such as manganese (Mn), iron (Fe), copper (Cu), zinc (Zn), arsenic (As), cadmium (Cd), and lead (Pb), are fundamental objectives of acid mine drainage (AMD) remediation processes (physical, chemical, and biological), which implementation depends on reckoning environmental, technical, and economic factors [4]. The neutralization of AMD by conventional chemicals (lime, pebble quicklime, caustic soda, soda ash briquettes, ammonia, magnesium hydroxide, and magna lime) is by far the most utilized [5]. Nevertheless, the neutralization is often combined with other technologies (aeration, flocculation, coagulation, filtration, etc.) to meet effluent quality requirements [6].

The actual cost of the AMD treatment approach includes the price of the chemical reagent and the cost of energy needed to transport the materials [7], thus, the efficiency of alternative low-cost and locally available materials has been progressively studied in recent years to ensure the sustainability of AMD treatment. A special impetus to the research of waste materials applicability in AMD remediation is given through the European Union's ambition to reduce its consumption footprint and double its circular material use rate in the coming decade [8]. Alkaline industrial wastes and byproducts, for instance, the bauxite residue, coal fly ash, byproducts of industrial smelting of metals (slags), and quicklime manufacturing, are being proposed for AMD remediation [9,10]. Likewise, abundant organic waste materials, such as rice husks and dairy manure compost, exhibit high metal removal efficiency, while chitin and chitosan additionally act as neutralizing and sulfate removal agents [11].

Compared to other toxic metals in AMD, manganese (Mn) treatment is the most difficult due to its complex chemistry and high pH values necessary for precipitation. Given various Mn oxidation states, the precipitation occurs at a pH of 9.0–9.5, but in some cases, even a pH of 10.5 is necessary for complete removal [5]. Concentrations of Mn in surface waters are frequently less than 200 µg/L and rarely exceed 1 mg/L [12], whereas, in mining water, they can be orders of magnitude higher, reaching several hundred mg per liter in highly contaminated mine drainage [13]. These effluents' impacts on natural freshwaters encompass the temporal changes in dissolved Mn concentrations, Mn behavior in the suspended load and bottom sediments, and the geochemical fate of other trace elements with a high affinity for Mn-oxyhydroxides [13,14]. The two fundamental mechanisms in control for Mn removal from AMD are precipitation and sorption, both of which are dependent upon Eh-pH conditions, the concentration of Mn, and competing ions. The weaknesses of Mn chemical treatment include the high costs of pH increase and the high pH of the sludge, thus, effective removal of Mn continues to be a challenge.

Various affordable and waste-derived materials have been investigated so far for Mn separation [15,16]; however, their applicability in actual AMD is more challenging due to the strong acidity and competitive species [15]. The usability of materials generated by wood combustion (wood ash) and thermal treatment of animal bones (bone char) for treating Mn in AMD was seldom explored, yet with encouraging results. The intensive use of wood as an energy-supply source in some regions urged a search for methods of wood ash utilization to minimize the environmental risks caused by disposal. Its chemical composition fluctuates among species of timber, combustion temperature, type, and hydrodynamics of the furnace [17]. Nevertheless, the major components, i.e., lime (CaO), portlandite ($Ca(OH)_2$), and calcite ($CaCO_3$), make it broadly applicable in pH control of AMD. A comparison of $Ca(OH)_2$ and wood ash in AMD treatment (initial Mn concentration 6.2 mg/L), revealed that dosing of both materials increased the solution pH from 3.5 to 8.3, causing a more effective drop in Mn concentration by wood ash (2.68 mg/L) than by $Ca(OH)_2$ (4.35 mg/L) [18]. Wood ash was found to act by precipitation, co-precipitation, and adsorption of trace metals from AMD, including Fe, As, Co, Cu, Ni, Zn, Mg, Al, and Mo. Moreover, the sludge from wood ash application had better settling capacities than the $Ca(OH)_2$ generated sludge.

The bone char obtained by treating ground animal bones at moderate temperatures is composed of calcium phosphate (57–80%), calcium carbonate (6–10%), and carbon (7–10%) [19]. The bone mineral component is commonly referred to as carbonated hydroxyapatite [20]. Manganese removal study from AMD by a commercial bone char suggested the intraparticle diffusion as the main rate-limiting step with contribution from boundary layer diffusion when smaller sizes of bone char particles are used [21]. The bone char not only removed metal species from the effluent but also increased its pH value. At the equal solid/liquid ratio, equilibrium sorbed amounts of Mn were higher at initial pH 6.5 (maximum sorption capacity, q_m = 22 mg/g) than at initial pH 5.0 (q_m = 14 mg/g). In continuous fixed bed column runs, no significant change in the breakthrough volume was detected with different flow rates, but the breakthrough volume increased by increasing the initial pH from 2.9 to 5.5 [22]. The maximum loading capacity of bone char in continuous tests with AMD effluents was 6.03 mg/g. The outcomes of studies using synthetic hydroxyapatite [23], bioapatite obtained from fish bone (Apatite II^{TM}) [24], and natural fluorapatite [25], also indicate the potential for efficient Mn removal at pH values considerably lower than those required for Mn-hydroxide precipitation. Moreover, apatite materials are suitable for the separation of not only Mn but diverse pollutants that may occur in AMD, controlling trace element mobility to surrounding soils and surface water [26–28].

While the integration of wood ash and thermally treated animal bones in Mn treatment aligns with circular economy principles and holds promise, a significant gap in understanding their efficacy still exists due to inadequate recognition of the impact of AMD chemistry and other process variables on their performance. Notably, the levels of coexisting Fe ions and the relevance of the Fe to Mn ratio are important issues requiring more extensive investigation [13]. Thus, the main objective of this study was to address these knowledge gaps and provide deeper insights into the performance of wood ash and bone char in Mn separation.

The influence of process variables, specifically reagent dose, reaction time, initial pH, Mn concentration, and the concentrations of sulfate and coexisting Fe ions, on the efficiency of Mn removal by wood ash and bone char was addressed. Owing to distinct physicochemical properties, neutralization capacities, and mechanisms for Mn removal, these materials are anticipated to manifest divergent responses to variations in process factors. The impacts of factors were ascertained through two approaches: (i) a classical methodology involving the alteration of one independent factor across multiple levels while keeping others constant, and (ii) an experimental design approach encompassing the simultaneous variation of all factors to evaluate the intensity of their effects and glean insights into potential interactions. Furthermore, the study assessed the practical viability of wood ash and bone char in treating actual AMD sourced from the lead and zinc mine "Sase" in Srebrenica, Bosnia and Herzegovina. Additionally, an examination was conducted regarding the leaching potential of hazardous elements from the resultant residues and prospects of Mn recovery.

2. Materials and Method

2.1. Wood Ash and Bone Char Preparation and Characterization

The wood ash used in the study (WA) represents the bottom ash from the combustion process in the conventional 250 kW boiler fueled by wood chips. During the combustion of the wood chips, mainly pine wood, the average temperature in the combustion chamber was 800 °C. The WA sample containing bigger particles (0.1–10 mm) was collected from the ash pane at the bottom of the boiler, crushed in a ball mill, and sieved through a 1 mm screen to make representative material for further testing and analysis.

The bone char (BC) was prepared by treating crushed bovine bones in an electrical furnace (ELEKTRON, Banja Koviljača, Serbia) at 400 °C in ambient air for 4 h, according to the procedure previously described [29]. The sample mass loss of approximately 33% is associated with the loss of water and decomposition of the organic phase, which was found advantageous for obtaining a high-capacity sorbent for trace metals [29,30]. The sample was homogenized and ground to a particle size < 0.2 mm.

The pH values and electrical conductivity (EC) of WA and BC were determined in deionized water at a solid/liquid ratio of 1:5, using the inoLab pH Level 1 and inoLab Cond 7110 (WTW, Weilheim, Germany), respectively. The point of zero charge (pH_{PZC}) was determined in an inert electrolyte by adjusting its initial pH values and monitoring the equilibrium pH values [31]. The investigated materials (0.1 g) were agitated in an overhead shaker (10 rpm) with 20 mL of 0.1 mol/L and 0.01 mol/L KNO_3 solutions with the initial pH in the range 1–12 (adjusted using 0.1 M HNO_3 and 0.1 M KOH). The suspensions were centrifugated and filtrated, and the final pH values were measured in clear supernatants. The pH shift upon reaction with the BC and WA ($\Delta pH = pH_{final} - pH_{initial}$) was calculated, and the pH_{PZC} was identified at $\Delta pH = 0$.

Pseudo-total element concentrations were measured by Inductively Coupled Plasma Optical Emission Spectrometry (ICP-OES) on the Avio 200 instrument (Perkin Elmer, Shelton, CT, USA). The samples were prepared by precise mass measurement and digestion with heating in a mixture of HNO$_3$ and HCl (3:1) for 4 h under reflux. The results of two replicated analyses are reported in % or mg/kg dry material.

Identification of crystalline phases was made at room temperature by X-ray powder diffraction (XRPD) using Ultima IV (Rigaku, Tokyo, Japan) diffractometer equipped with Cu Kα1,2 radiation using a generator voltage (40.0 kV) and a generator current (40.0 mA). The range of 5–70°2θ was used for all powders in a continuous scan mode with a scanning step size of 0.02° and at a scan rate of 5°/min. Phase analysis was performed by the PDXL2 software (version 2.0.3.0, Rigaku, Tokyo, Japan) [32], with reference to the patterns of the International Centre for Diffraction Data database (ICDD) version 2012 [33]. All the structure information was taken from the American Mineralogist Crystal Data Structure Base (AMCDSB) [34].

2.2. Effects of Independent Process Factors on Mn Removal by BC and WA

Manganese removal experiments were performed in the batch mode by mixing the reagents and solutions in 50 mL polypropylene tubes on the Heidolph Reax overhead shaker (10 rpm), centrifugation for 5 min at 9000 rpm, and filtering through syringe filters with a 0.45 μm pore size. Synthetic AMD solutions were prepared from p.a. purity salts (MnCl$_2$ × 4 H$_2$O, Na$_2$SO$_4$, and FeCl$_2$ × 4 H$_2$O, Merck KGaA, Darmstadt, Germany).

The initial pH effect on Mn removal was studied at a fixed dose of BC and WA (1 g/L), Mn concentration (100 mg/L), and contact time (24 h) while adjusting the pH in the range of 1.5–6.0 by adding dropwise 0.1 mol/L HCl or 0.01 mol/L NaOH.

The effect of BC and WA dose variation (0.3–10 g/L) was examined for three initial concentrations of Mn (5 mg/L, 50 mg/L, and 100 mg/L) at the fixed initial pH of 2.5 and a contact time of 24 h.

Based on the above-described experiments, the doses for treating 100 mg/L Mn solution at initial pH 2.5 were fixed for investigated waste materials (1.2 g/L WA and 10 g/L BC), and the influence of contact time (5 min–24 h), sulfate concentration (500–10,000 mg/L), and Fe concentrations (5–500 mg/L) were investigated.

The experiments were conducted in duplicate. Manganese and Fe concentrations were measured using the ICP-OES technique, and the efficiency of each treatment was calculated.

2.3. Assessment of the Effects of Process Variables and Their Interactions—Experimental Design

Previous experiments showed that all investigated factors affect the efficiency of Mn removal from synthetic AMD at the constant levels of other independent process variables. To better understand the system response under the influence of several experimental factors, the experimental design (DOE) approach was applied. Table 1 denotes selected independent factors and their lower and higher values.

Table 1. Selected independent factors and their levels.

Level	Factor						
	A	B	C	D	E	F	G
	Reagent Type	Mn Conc. (mg/L)	Sulfate Conc. (mg/L)	Fe Conc. (mg/L)	Time (h)	Initial pH	Reagent Dose (g/L)
−1	BC	5	100	5	0.5	2.0	0.5
+1	WA	100	10,000	400	24	4.5	5.0

The fractional factorial design matrix in 16 runs was chosen as a screening technique. In the two-level DOE, one treatment setup refers to a combination of levels of all selected independent factors, and it is helpful to assess their contribution and identify essential interactions between the variables. The fractional factorial design matrix was created, and the results were analyzed using the statistical software (Minitab, demo version, Minitab LLC., State College, PA, USA). As the primary output variable, the efficiency of Mn removal was determined for each experimental run. Furthermore, final pH values and Fe removal efficiency were monitored.

2.4. AMD Treatment

The actual AMD used in the study was a sample of pit water collected at the lead and zinc mine "Sase" Srebrenica (Bosnia and Herzegovina). Treatments have been conducted at constant contact time (24 h) and mixing speed (10 rpm) by varying the dose of WA (1–3 g/L) and BC (1–30 g/L). The pH values of AMD were determined before and after the treatments, and the ICP-OES was used to determine the concentrations of Mn and other toxic metals.

2.5. Stability of the AMD Treatment Residues

The selected residues obtained after AMD treatments were further investigated regarding the leaching potential and metal recovery. Two replicated experiments were conducted for each residue. Solid phases following the AMD reaction with 2.5 g/L WA and 30 g/L BC were separated by filtration (qualitative filter paper), dried at 100 °C for 24 h, and exposed to leaching by deionized water at a 1:10 solid-to-solution ratio for 20 h, using an overhead laboratory shaker at 20 rpm [35]. After filtering the leachates (0.45 μm pore size syringe filters), the concentration of elements was determined using the ICP-OES. Furthermore, the possible reuse of the materials and recovery of AMD major cations from WA/AMD and BC/AMD residues was investigated using HCl solutions with concentrations of 0.01 mol/L, 0.1 mol/L, and 1 mol/L. The solid-to-solution ratio was maintained at 5 g/L. The suspensions were agitated for 24 h, and the metal concentrations and final pH values were measured in the supernatants after centrifugation and filtration.

3. Results

3.1. Chemical and Mineralogical Characteristics of WA and BC

According to the results in Table 2, the primary metals in WA were Ca (31.0%), K (15.5%), and Mg (5.3%), as reported for a variety of wood and woody biomass ashes [36,37], followed by Al (2.36%), Mn (1.22%), Na (0.844%) and Fe (0.398%).

High Mn concentrations are characteristic for wood ash, i.e., a mean concentration of 13,160 mg/kg (min. 775 mg/kg, max. 35,740 mg/kg) was reported considering the ashes of 28 types of wood and woody biomass [36]. Among the trace metals, copper (163 mg/kg) and Cr (40.6 mg/kg) were dominant, whereas Pb, Cd, and As concentrations in WA were below detection limits. The content of P was 2.06%.

The complex mineralogy of pristine WA is shown in Figure 1a. The shape of the XRD baseline indicates the presence of amorphous solids, while the principal crystalline phases were dicalcium silicate (Sy: Ca_2SiO_4) and fairchildite (F: $K_2Ca(CO_3)_2$). Although calcite is a common component in the bottom biomass ashes [18,38], ash mineralogy is controlled by the relative ratios of the dominant ash-forming cations and reaction temperature. The formation of fairchildite instead of calcite relies on the wood's initial K:Ca ratio [39] and can be associated with a high K content in the WA sample (Table 2). The interaction between Mg and Ca or K is insignificant during ash formation, usually leading to the periclase (P: MgO) formation [39], as identified in the WA. The relatively high level of CaO (L) in WA is typical for wood sources. Furthermore, the minor shares of potassium dithionate (Ks: $K_2S_2O_6$), anorthite (An: $CaAl_2Si_2O_8$), and potassium-iron-phosphate phase (Kp: $KFePO_4$) were present in the sample.

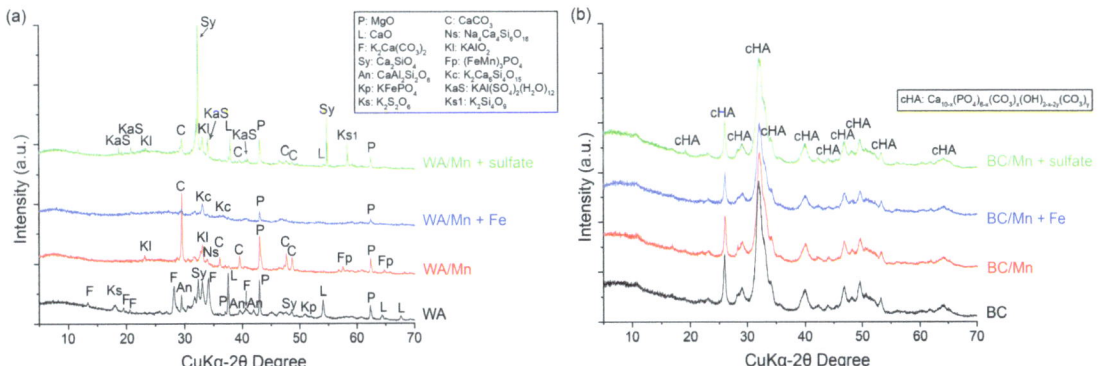

Figure 1. XRD patterns of pristine samples (**a**) WA and (**b**) BC, and the solid residues of their interaction with solutions having 100 mg/L Mn (red line); 100 mg/L Mn and 500 mg/L Fe (blue line), 100 mg/L Mn + 10,000 mg/L SO$_4$ (green line). Initial pH 2.5, dose 1.2 g/L WA and 10 g/L BC, contact time 24 h.

Table 2. pH, pH$_{PZC}$, electrical conductivity, and pseudo-total concentrations of elements in WA and BC.

Parameter	WA		BC	
pH	12.6 ± 0.1		7.7 ± 0.1	
EC (mS/cm)	34.8 ± 0.5		0.887 ± 0.035	
pH$_{PZC}$	12.0 ± 0.1		6.9 ± 0.2	
Element	Concentration	Units	Concentration	Units
P	2.06 ± 0.08	%	18.6 ± 0.6	%
Ca	31.0 ± 2.8	%	37.7 ± 1.7	%
Mg	5.30 ± 0.15	%	0.796 ± 0.032	%
K	15.5 ± 0.4	%	104 ± 5	mg/kg
Na	0.844 ± 0.031	%	0.475 ± 0.018	%
Al	2.36 ± 0.13	%	<LOD	-
Mn	1.22 ± 0.06	%	0.953 ± 0.031	mg/kg
Fe	0.398 ± 0.012	mg/kg	31.6 ± 1.1	mg/kg
Cu	163 ± 6	mg/kg	0.897 ± 0.033	mg/kg
Cr	40.6 ± 0.7	mg/kg	3.51 ± 0.17	mg/kg
Ni	14.8 ± 0.1	mg/kg	<LOD	-
Zn	6.59 ± 1.17	mg/kg	147 ± 8	mg/kg
Co	5.45 ± 0.22	mg/kg	<LOD	-
V	4.27 ± 0.34	mg/kg	<LOD	-
Se	0.295 ± 0.015	mg/kg	<LOD	-
Pb	<LOD	-	0.599 ± 0.030	mg/kg
As	<LOD	-	<LOD	-
Cd	<LOD	-	<LOD	-

LOD (Limit of Detection, mg/L)—Pb: 0.007, As: 0.005, Cd: 0.001, Al: 0.005, Co: 0.003, V: 0.001, Se: 0.002, Ni: 0.002.

The structure of the bone mineral phase was maintained after the treatment, and the XRD spectrum of pristine BC shows only peaks corresponding to carbonated hydroxyapatite (cHA: Ca$_{10-x}$(PO$_4$)$_{6-x}$(CO$_3$)$_x$(OH)$_{2-x-2y}$(CO$_3$)$_y$) (Figure 1b). The chemical composition of the BC (Table 1) supports high Ca and P content (37.7% and 18.6%). Magnesium (0.796%) and Na (0.475%) are typical bone constituents as well [40]. The potassium and trace elements content in BC was significantly lower than in WA, except for Zn (147 mg/kg).

While BC exhibited a mildly alkaline reaction (pH 7.7), WA acts as a strong base in water, bringing pH to 12.6. Wood bottom ash samples generally exhibit higher pH and better neutralization capacity than wood fly ash [41]. Consequently, the water solubility of WA constituents provoked EC of 34.8 mS/cm, much higher compared to BC (0.887 mS/cm).

The pH$_{PZC}$ of BC was found to be at pH 6.9 (Table 2, Supplementary Material Figure S1), in good agreement with near-neutral values determined for hydroxyapatite samples obtained by different procedures and from different precursors [29,42–44]. In line with the literature data [45], the pH$_{PZC}$ of WA was in the alkaline region (12 ± 0.1).

3.2. The Influence of Initial pH on Mn Removal

Removal of Mn from 100 mg/L solutions varied significantly with the change in the initial pH and the reagent type, maintaining the experiment's other conditions constant (Figure 2a,b).

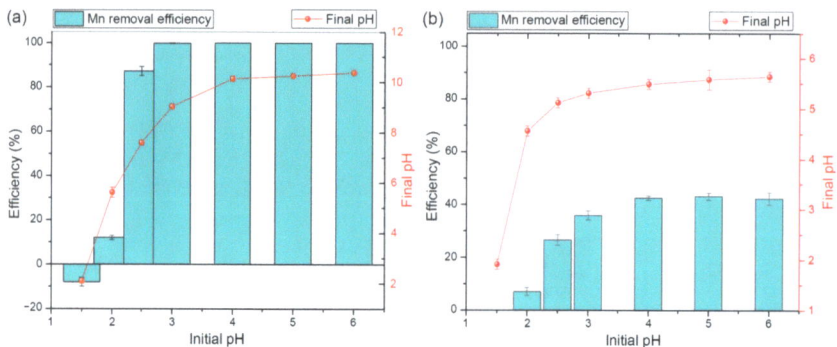

Figure 2. Relations between the initial pH, final pH, and Mn removal efficiency in the system with WA (**a**) and BC (**b**). Dose 1 g/L, Mn concentration 100 mg/L, 24 h contact time.

Manganese concentration after the experiment conducted at an initial pH of 1.5 remained virtually unchanged using BC, whereas the net increase was detected following WA application. At the 1 g/L dose, the WA and BC neutralization capacity was exhausted, marginally raising the pH to 2.1 and 1.9, respectively. Under such pH conditions, the leaching of Mn from the WA composition (Table 2) occurred. The reaction of WA and BC with the solution of initial pH 2.0 resulted in a 12.1% and 7.1% decrease in Mn concentration. WA application triggered a drastic increase in process efficiency (to 87.2%) already at an initial pH of 2.5, and with a further initial pH increase (3–6), Mn was removed completely (99.9–100%). The efficiency of BC improved to 42.5% with increasing initial pH to 4.0 and remained at a constant level up to the initial pH of 6.0.

Excellent WA performance was related to the abrupt increase in the final pH (by 3.6–6.1 pH units, depending on the initial pH), and the Mn was eliminated in all systems with final pH ≥ 9.0 (Figure 2a). Final pH values increased as well after BC addition, reaching a maximum of 5.6 (Figure 2b). The most significant changes in final pH were observed in the initial pH range 2.0–3.0 (by 2.5–2.3 pH units) while retarded with a further increase in initial pH values. As demonstrated for mineral and biogenic samples, the apatite dissolution rate decreases by increasing pH [24,25].

According to the mine drainage pollution classes suggested by Hill [1], Class I is described as acidic with a typical pH range of 2.0–4.5, and Class II denotes partially oxidized or neutralized drainage (pH 3.5–6.6). By applying an equivalent dose of materials, WA showed a higher capacity to separate Mn over the initial pH range corresponding to these categories. The amount of Mn removed per unit mass of reagent varied between 12.1 mg/g and 100 mg/g for WA and between 7.1 mg/g and 42.5 mg/g for BC (Supplementary Material Figure S2a).

3.3. The Influence of Reagent Dose and Mn Concentration

The effect of the solid/solution ratio on the removal efficiency of Mn was examined at a constant initial pH value of 2.5. As displayed in Figure 3a,c, an increase in the dose of reagents affects the decrease in Mn concentration in the solution, but the absolute values of these changes depend on the type of added material and the initial Mn concentration.

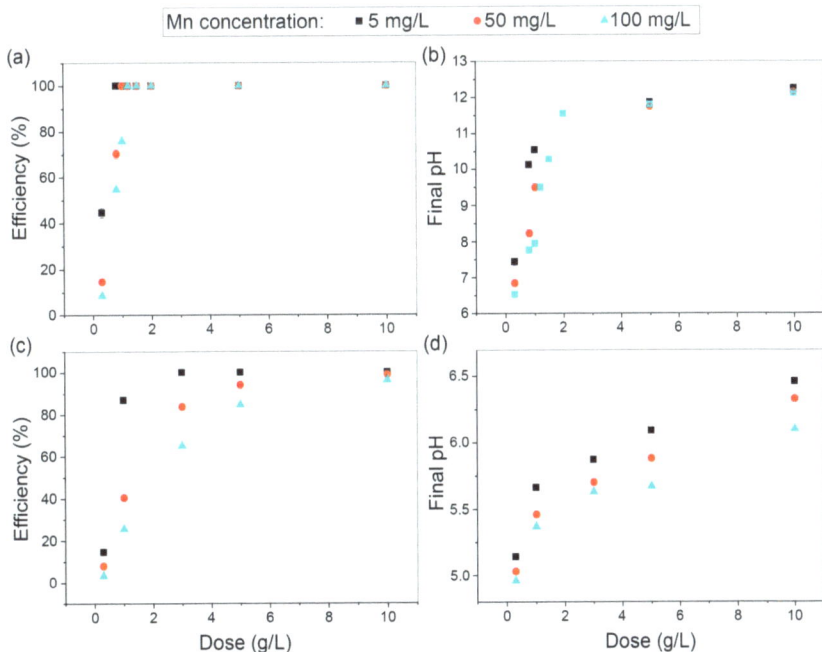

Figure 3. Effect of reagent dose on Mn removal efficiency and final pH values in the systems with WA (**a,b**) and BC (**c,d**). Initial pH 2.5, contact time 24 h.

Using WA, Mn was removed entirely from the 5 mg/L solution at a dose of 0.8 g/L. For the complete removal of 50 mg/L Mn, the dose of 1 g/L was necessary, while adding 1.2 g/L WA was successful for the treatment of 100 mg/L Mn solution. The amount of Mn removed per unit mass of WA commonly increased with the increase in its initial concentration but varied broadly with the change in the reagent dose (Supplementary Material Figure S2b).

Increasing Mn removal efficiency was also observed with increasing BC dose (Figure 3c), but effective treatment of acidic wastewater required considerably higher BC amounts than WA. Adding 3 g/L BC, Mn was separated from the 5 mg/L solution below the detection limit. Process efficiency continuously increased in solutions with 50 and 100 mg/L Mn with the increase in BC dose, reaching 99.0 and 96.2%, respectively, for 10 g/L BC. The highest Mn removal from 5 mg/L solution per unit mass of BC was 1.6 mg/g at the dose of 3 g/L (Supplementary Material Figure S2c). Starting with more concentrated Mn solutions (50 and 100 mg/L), maximally 20.0 mg/g and 25.5 mg/g BC were respectively removed with the BC dose of 1 g/L. With a dose rise to 10 g/L, the amount of Mn removed per gram of BC declined to 5.0 mg/g and 9.6 mg/g.

As the Mn concentration in the solution increased, raising the WA dose from 0.8 to 1.2 g/L was necessary to reach a final pH value of ~9 (Figure 3b). Given the pH critical for Mn's complete removal by wood ash, it was associated with hydroxide/oxyhydroxide precipitation [38]. An additional increase in the amount of WA can be considered unnecessary and unfavorable from the treated water quality point of view, i.e., high alkalinity.

Compared to pristine WA, in the representative solid residue of WA/synthetic AMD interaction (initial pH 2.5, 1.2 g/L WA, 100 mg/L Mn), the MgO phase was preserved, whereas diffraction maximums of CaO and Ca_2SiO_4 disappeared, signifying their role in bringing the solution pH to an alkaline region (Figure 1a). In addition to lime, calcium silicate is known for neutralizing the active acidity in AMD [3] by capturing free H^+ ions and forming neutral H_4SiO_4, which remains in the bulk solution. The fairchildite (F: $K_2Ca(CO_3)_2$) was also dissolved, while calcite (C: $CaCO_3$) precipitated. Additional precipitates were identified as tetrasodium tetracalcium cyclo-hexasilicate (Ns: $Na_4Ca_4Si_6O_{18}$), potassium aluminum oxide (Kl: $KAlO_2$), and graftonite (Fp: $(FeMn)_3PO_4$).

Quite the opposite, Mn removal in the presence of BC occurred at final pH values of 5.0–6.5 (Figure 3d), far from the Mn precipitation threshold. The final pH increased with the increase in BC dose at all tested Mn concentrations, similarly as in the work of Sicupira et al. [21], displaying that the BC/AMD ratio is a vital operating variable in Mn removal by bone char.

The mineral composition of BC remained unaffected after the interaction of 10 g/L BC with 100 mg/L Mn solution at initial pH 2.5, but a decrease in the intensity of the carbonated hydroxyapatite peaks is associated with the BC dissolution in contact with the acidic medium (Figure 1b). Manganese sorption with apatite phase can be, thus, anticipated either by the exchange with Ca through solid-state diffusion or a dissolution-precipitation mechanism. It was indicated that the dissolution reaction of biogenic hydroxyapatite is controlled by fast adsorption of protons on specific surface sites, followed by a slow removal of Mn by forming phosphate precipitate [$Mn_3(PO_4)_2 \cdot 7H_2O$] on the Apatite II^{TM} substrate [24]. Higher solubility of calcium fluorapatite compared to a manganese phosphate phase favored the dissolution-precipitation mechanism, in line with the evidenced mole-per-mole exchange between Mn and Ca in natural apatite [25]. Although the Mn-phosphate phase could not be identified in the BC residue due to the small concentration of Mn in the solid phase (9.62 mg/g) and the detection limits of the XRD technique, its formation cannot be dismissed.

Moreover, at the same BC dose, the Mn sorption increase decreases the final pH (Figure 3d). This phenomenon is characteristic of the specific cation sorption (chemisorption) that may also contribute to Mn removal by BC. The BC pH_{PZC} 6.9 (Table 2) signifies a solution pH value where the number of positively and negatively charged functional groups is balanced and a surface charge is neutral. The final pH values of interest for Mn sorption were lower than the pH_{PZC}, thus, BC exhibited an overall positive surface charge. In this case, the electrostatic attraction forces between the BC surface and the Mn(II) cannot contribute to the sorption [27]. Prevailing groups on hydroxyapatite surface in such conditions are positively charged $\equiv CaOH_2^+$ and neutral $\equiv POH^0$ species [21], and they may react with aqueous Mn through the surface complex formation:

$$\equiv CaOH_2^+ + Mn^{2+} <=> \equiv CaO\text{-}Mn^+ + 2H^+$$

$$2 \equiv CaOH_2^+ + Mn^{2+} <=> (\equiv CaO)_2Mn + 4H^+$$

$$\equiv POH + Mn^{2+} <=> \equiv POMn^+ + H^+$$

$$2 \equiv POH + Mn^{2+} <=> (\equiv PO)_2Mn + 2H^+$$

3.4. The Influence of Contact Time on Mn Removal

Based on the previous experiment, the manganese removal rate from the synthetic AMD (initial pH 2.5, 100 mg/L Mn) was investigated at 1.2 g/L WA and 10 g/L BC doses. Already after 5 min of contact, 70.3% of Mn was removed by BC and 20.7% by WA (Figure 4a,b). The efficiency of the treatment changed markedly up to 6 h, which was sufficient for the BC system to reach equilibrium (96%). The equilibrium time of Mn removal from 100 mg/L solution at the initial pH of 5.8 was found to vary from several hours to a few days using a commercial bone char, depending on particle size and bone char/solution ratio [21]. On the other hand, WA removed 95.6% of Mn in 6 h, and the performance continued to increase up to 100% within 17 h of contact.

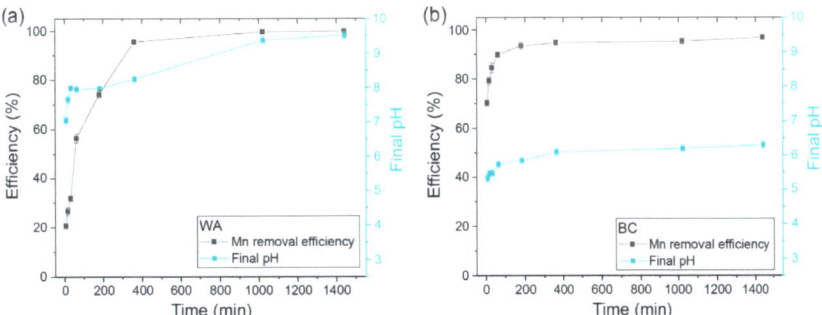

Figure 4. The effect of contact time on Mn removal by WA (**a**) and BC (**b**). Initial Mn concentration 100 mg/L, dose 1.2 g/L WA and 10 g/L BC, initial pH 2.5.

The amount of Mn removed per unit mass of reagents increased with time, reaching 83 mg/g of WA and 9.7 mg/g of BC at equilibrium conditions (Supplementary Material Figure S2d).

The final solution pH also increased with time (Figure 4a,b). WA required more time to achieve equilibrium pH (~17 h). The curve of pH change with time was complex, demonstrating the fast increase during the first 30 min, a quasi-plateau between 30 min and 3 h, and, subsequently, a slow increase until the equilibrium pH of 9.4. This trend results from heterogenous WA composition (Figure 1a) and different dissolution rates of mineral phases that contribute to the pH rise [46].

The Mn(II) is the prevalent state in natural waters [13]. The stability fields of manganese species in aqueous solution predicted by the Eh-pH diagrams depend on Mn activity, and for 100 mg/L Mn in solution, oxidation to Mn(III) and Mn(IV) and Mn_2O_3 and/or Mn_3O_4 precipitation in the absence of other species occur at pH $\geq$ 8 [47]. With WA, the final pH increased to the neutral range in the first minutes of contact (Figure 4a), in which the Mn(II) oxidation rate is very slow. Thus, initially, the uptake of Mn corresponds to Mn(II) removal by chemisorption at an overall positively charged WA surface (final pH < pH_{PZC}, Table 2). At pH > 8, particularly at pH 9, Mn(II) oxidation becomes faster and can be additionally accelerated if catalyzed by a solid surface or autocatalytic effect of the formed solid Mn oxide minerals [47]. Therefore, precipitation of Mn oxides occurred, as indicated by the complete disappearance of Mn from the solution.

The solution pH increased from 2.5 to 5.3 within 5 min reaction with BC. Fast carbonated hydroxyapatite dissolution in an acidic medium [24] increases pH and phosphate release significantly. In this period, as much as 70.3% of the Mn was removed, which strongly supports the dissolution and precipitation mechanism. Additionally, 26% Mn was removed with further pH increase from 5.3 to 6.2 up to 6 h of contact, and afterward, pH remained stable, as did Mn sorption.

3.5. The Influence of Sulfate Concentration

The concentration of sulfate anions in mine drainage effluents takes a wide range of values, typically from around five hundred to several thousand mg per liter [1]. Figure 5 shows variations in Mn removal efficiency with increased sulfate concentration for preselected conditions (initial pH 2.5, 100 mg/L Mn, doses of 1.2 g/L WA, and 10 g/L BC).

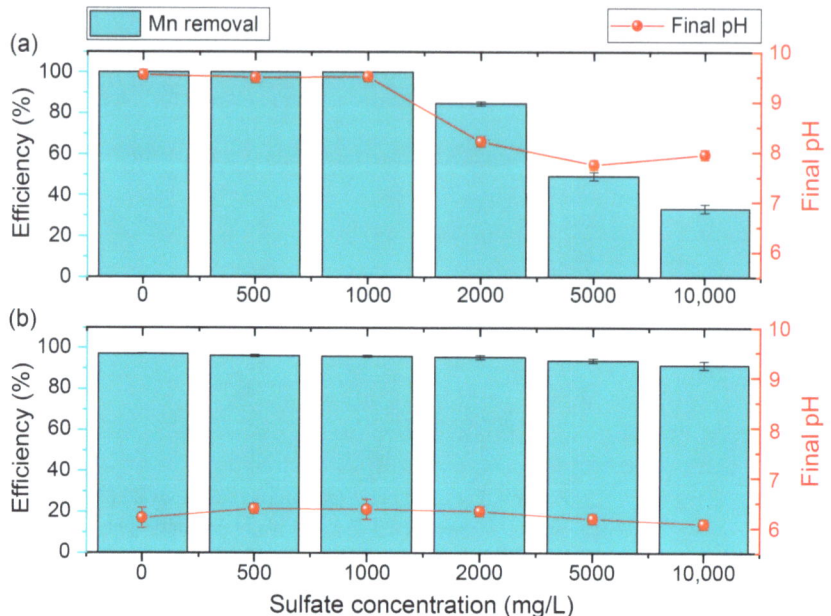

Figure 5. The effect of sulfate concentration on Mn removal by WA (**a**) and BC (**b**). Initial Mn concentration 100 mg/L, dose 1.2 g/L WA and 10 g/L BC, initial pH 2.5, contact time 24 h.

Up to 1000 mg/L, sulfate anions did not interfere with Mn removal. With WA, the final pH values of ~9.5 provided conditions for total Mn removal (Figure 5a). The presence of sulfate in the solution slows down the Mn oxidation rates even at pH > 9, at least in part due to the presence of the $MnSO_4$ (aq) complex [47]. However, the reaction time of 24 h was sufficient to accomplish complete Mn removal by oxidation/precipitation at such a pH. A decrease in WA efficiency to 84.7% occurred at 2000 mg/L sulfate and continued with increasing sulfate concentration so that at 10,000 mg/L, only 33.4% Mn was separated. In parallel, the solution pH gradually decreased, reaching a final pH of 8.0 at the highest sulfate concentration.

The variations in process efficacy and final pH values were considerably less pronounced when BC was added to synthetic AMD (Figure 5b). Over the entire range of sulfate concentration, the final pH was virtually constant (6.1 ± 0.1). Sulfate anions affected the process efficiency only at concentrations ≥ 2000 mg/L, causing the decrease to 91.5% at a concentration of 10,000 mg/L.

The amount of Mn removed per unit mass of reagents declined with the increase in sulfate content from 83.3 mg/g to 27.8 mg/g using WA and from 9.7 mg/g to 9.1 mg/g using the BC (Supplementary Material Figure S2e).

The residue of WA interaction with synthetic AMD having 10,000 mg/L sulfate shows high-intensity peaks of Ca_2SiO_4 and preserved MgO phase (Figure 1a). Also, the intensity of CaO peaks was reduced compared to pristine WA, the fairchildite ($K_2Ca(CO_3)_2$) dissolved completely, and the peaks of calcite appeared. The diffraction maximums of other newly formed phases are best matched with dipotassium silicate (Ks1: $K_2Si_4O_9$) and potassium aluminum sulfate hydrate (KaS: $KAl(SO_4)_2 \times 12H_2O$). The presence of extra low-intensity diffraction peaks in the XRD spectrum indicates the presence of other crystalline phases, which could not be identified reliably due to their low content and sample complexity. Precipitated sulfates may explain the limited solubility of primary pH regulating minerals, in line with the recorded final solution pH of 8.0, and the consequent adverse effect on Mn removal efficiency by WA.

On the other hand, the corresponding BC residue's crystal structure remained unchanged in relation to the starting BC (Figure 1b). Similarly, Fourier Transform Infrared (FTIR) spectra of the products from the natural fluorapatite reaction with Mn sulfate solution in the wide range of pH (2.0–7.0) did not differ from the FTIR spectrum of the starting material [25].

3.6. The Influence of Fe Concentration

Figure 6 displays the impact of coexisting Fe(II) ions on Mn removal by WA and BC and concurrent pH and Fe concentration changes. At a constant initial Mn concentration (100 mg/L) and preselected doses of reagents, an increase in Fe concentration negatively affected Mn ions removal efficiency. This effect was more pronounced with WA, causing a substantial decrease in Mn separation to 47.0% at Fe concentration of 100 mg/L and continuing steep fall, entirely blocking this process at 500 mg/L (Figure 6a). The removal of Fe ions from the solution was complete up to the highest initial Fe concentration of 500 mg/L, where a drop in Fe removal efficiency to 41.1% was also recorded.

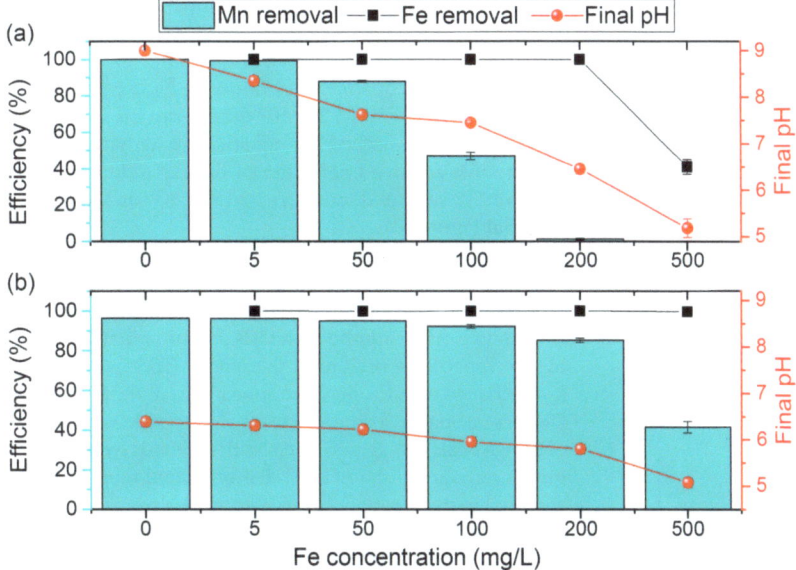

Figure 6. The effect of Fe concentration on Mn removal by WA (**a**) and BC (**b**). Initial Mn concentration 100 mg/L, dose 1.2 g/L WA and 10 g/L BC, initial pH 2.5, contact time 24 h.

These changes coincide with the significant pH decrease from 8.3 at 5 mg/L Fe to 5.2 at 500 mg/L Fe) (Figure 6a). The XRD spectra of the residue obtained at the concurrent Fe ions concentration of 500 mg/L demonstrate the exhaustion of the WA neutralization capacity (Figure 1a), with crystalline phases identified as periclase ($Mg_{0.99}Fe_{0.01}O$) and dipotassium hexacalcium pentadecaoxotetrasilicate (Kc: $K_2Ca_6Si_4O_{15}$).

Over the range of oxidation-reduction potentials characteristic for normal water environments, Fe (II) oxidizes to Fe (III), forming amorphous $Fe(OH)_3$ precipitates at pH > 5 [48]. Not only are lower pH and Eh conditions essential for the oxidation of Fe(II) compared to Mn(II), but the kinetics of the redox reactions of Fe(II) is also significantly faster [48]. Seeing the final pH values (Figure 6a), the co-removal of Mn and Fe by WA that took place in parallel up to their 1:2 ratio was governed by Fe oxidation and $Fe(OH)_3$ precipitation, whereas Mn was primarily removed by sorption. Sorption of Mn was likely enhanced by oxidation/precipitation at final pH > 8 (with 5 mg/L and 50 mg/L Fe) and declined rapidly with the final pH decrease. When the Fe concentration reached 500 mg/L, even Fe removal was incomplete as the final pH was ~5.

The negative influence of Fe on Mn oxidation and precipitation at Fe/Mn ratios >4 was previously reported [13]. In the work of Calugaru et al. [49], half-calcined dolomite used at a dose of 7.5 g/L (0.3 g / 40 mL) was able to separate both metals efficiently at Mn concentrations up to ~100 mg/L and the Fe/Mn ratio of 10:1 as long as the equilibrium pH values were 9.7–9.8. However, keeping the same Fe/Mn ratio and increasing the initial Mn and Fe concentrations, a decrease in Fe and particularly Mn removal occurred due to a decrease in equilibrium pH to 6.8–6.9. The results of both our studies indicate the importance of the final pH in the system.

By applying BC, Mn removal of 96.2–85.2% was achieved from 5 mg/L Fe up to 200 mg/L Fe, while it declined to 41.5% at 500 mg/L Fe (Figure 6b). In parallel, Fe ions were entirely removed in all experimental runs, and the pH value of the solution decreased from 6.3 to 5.1 (Figure 6b). At initial pH 2.5, partial dissolution of a preselected dose of BC has provided a sufficient pH increase, phosphate anions release, and an abundance of active centers for effective Mn(II) removal by phosphate precipitation and chemisorption even at twice the concentration of Fe ions.

At initial Fe concentrations of 200 mg/L and 500 mg/L, the amount of Mn removed per unit mass of BC (8.5 mg/g and 4.1 mg/g) was higher compared to WA (1.0 mg/g and 0 mg/g) (Supplementary Material Figure S2f). Quantities of Fe removed by WA increased sharply from 4.2 to 167 mg/g with the increase in its concentration from 5 mg/L to 200 mg/L and reached a virtual plateau with further Fe concentration increase, whereas Fe amounts removed per gram of BC expanded from 0.5 mg/g to 49.9 mg/g, and the apparent linear trend indicates that the BC capacity was not exhausted (Supplementary Material Figure S3).

Apart from the characteristic carbonated hydroxyapatite peaks being of lower intensity, no other changes in the XRD spectrum of BC after interaction with 100 mg/L Mn and 500 mg/L Fe solution were detected (Figure 1b). In some previous investigations, Fe(II) sorption by apatite materials was anticipated. The study conducted with synthetic hydroxyapatite showed its high capacity (55.2 mg/g) for Fe, and it was attributed to Fe(II) ion exchange with Ca since the appearance of the Fe characteristic peaks in the EDS (Energy Dispersive Spectroscopy) spectrum of the residue coincided with a reduction in the intensity of the Ca peak [50]. Experiments with synthetic hydroxyapatite for the remediation of acidic mine drainage (initial pH 2.9) demonstrated more rapid Fe removal than Mn [51]. It was assigned to the sorption due to detected Ca release. However, the final pH values were not reported. Our study's final pH values of ≥ 5 imply possible Fe(II) oxidation/precipitation. On the other hand, at 500 mg/L Fe, BC removed Fe more efficiently than WA at the same final pH value of ~5, emphasizing that these two materials have different mechanisms of action. Additional experiments and analyses are necessary to address BC's mechanism of Fe removal.

3.7. Comparison of the Effects of Independent Process Variables

The selected responses (Mn and Fe removal efficiency, final pH) were monitored as a result of the simultaneous change in the levels of all input variables according to the matrix presented in Table 3. The combined influence of the process factors provoked Mn removal efficiency variation from 0% to 100%, and the applied design allowed the comparison of the effects of factors and identification of the most significant ones.

Table 3. Fractional factorial design matrix with a combination of factors levels in each experimental run and corresponding system responses.

Run	Independent Process Variables							Responses		
	Reagent	Mn (mg/L)	Sulfate (mg/L)	Fe (mg/L)	Time (h)	Initial pH	Dose (g/L)	Mn Removal(%)	Fe Removal (%)	Final pH
1	BC	100	10,000	5	0.5	2.0	5.0	26.6	93.4	4.9
2	WA	100	10,000	400	24	4.5	5.0	99.9	99.9	10.4
3	WA	5	1000	5	24	2.0	5.0	100	100	11.3
4	BC	5	1000	5	0.5	2.0	0.5	0	0	2.6
5	BC	100	1000	400	24	2.0	5.0	10.8	67.2	4.0
6	WA	5	1000	400	24	4.5	0.5	0	48.2	4.7
7	BC	100	10,000	400	0.5	4.5	0.5	2.4	2.9	4.8
8	BC	5	10,000	400	24	2.0	0.5	1.5	0	2.3
9	BC	5	10,000	5	24	4.5	5.0	99.4	100	6.9
10	WA	100	1000	5	0.5	4.5	5.0	99.9	100	11.6
11	WA	100	1000	400	0.5	2.0	0.5	0	1.2	2.4
12	WA	5	10,000	5	0.5	4.5	0.5	100	100	10.7
13	BC	5	1000	400	0.5	4.5	5.0	35.7	48.8	5.2
14	WA	100	10,000	5	24	2.0	0.5	0	0	2.4
15	WA	5	10,000	400	0.5	2.0	5.0	0	33.2	6.0
16	BC	100	1000	5	24	4.5	0.5	20.9	100	5.9

The main effects plot for Mn removal shows the calculated average response of each factor level and their distance from the overall average response for all variables presented by the horizontal line (Figure 7a). The critical impact on Mn removal can be ascribed to the variation of reagent dose, initial pH, and concentration of coexisting Fe ions. The changes in initial pH and solid/solution ratio from a lower to a higher level were beneficial for the process in contrast to increasing Fe concentration. A change of the reagent type from BC to WA displayed a slight positive effect on Mn removal, considering the average response of the system at all levels of all other factors. Finally, the effects of initial Mn and sulfate concentration and time were least pronounced.

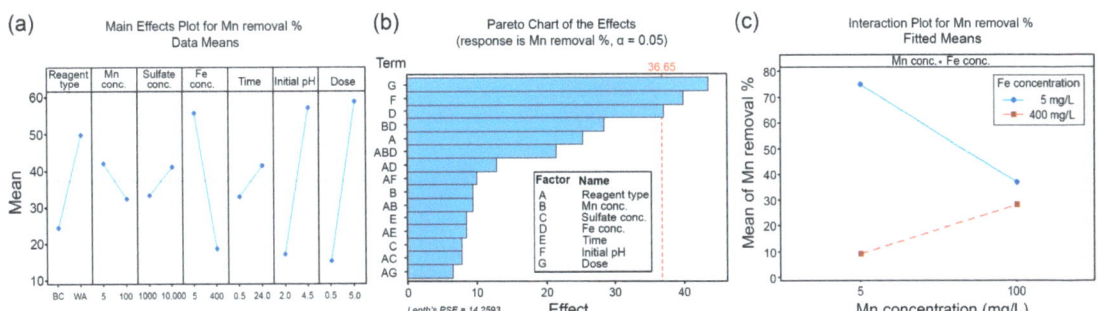

Figure 7. Main effect plots for Mn removal (**a**), Pareto chart of the effects of variables (**b**), and interaction plot for the effects of Mn and Fe concentrations (**c**).

The Pareto chart confirmed the significance (at $\alpha = 0.05$) of three main factors in the order: reagent dose > initial pH > Fe concentration (Figure 7b). None of the interaction effects (Supplementary Material Figure S4) displayed a statistical significance (Figure 7b), however, it is interesting to note that the effect of the Fe and Mn concentration (BD) interaction was most pronounced and higher compared to the main effects of the reagent type, Mn and sulfate concentration and contact time. As shown in Figure 7c, a higher Fe level hinders Mn removal by both reagents more significantly at lower than at higher Mn concentrations in the solution, confirming the importance of the Mn/Fe ratio.

Main effect plots for final pH and Fe removal efficiency as system responses (Supplementary Material Figure S5a,b) revealed common positive impacts of increasing reagent dose and initial pH and the negative impact of increasing initial Fe concentration. A change in the reagent type from BC to WA positively influenced the final pH. Nevertheless, none of the factors or their interactions were statistically significant at $\alpha = 0.05$ (Supplementary Material Figure S6a,b).

3.8. Relationship between Process Responses: Final pH and Mn Removal Efficiency

The change in the levels of individual factors and all factors simultaneously displayed the effect on Mn separation and final pH in WA and BC systems. To establish general relationships between these two system responses, the results (Figures 2–6, Table 3) are plotted as Mn removal efficiency against the final pH (Figure 8).

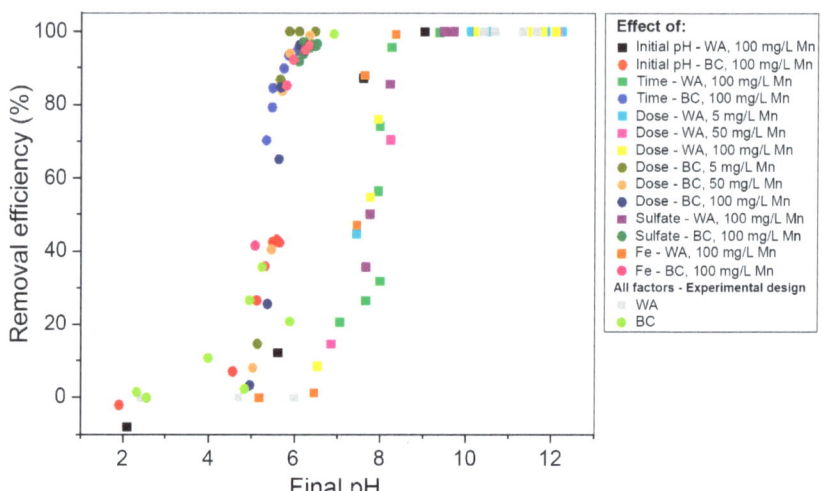

Figure 8. Relationships between the final pH and Mn removal efficiency by WA and BC under the influence of process factors.

Experimental data for different solid/solution ratios, initial pH, initial Mn, sulfate, Fe concentrations, and reaction times indicate a sharp Mn removal increase in narrow pH ranges, essentially separated according to the agents used. For BC, process efficiency was negligible below the final pH 5.0, while >95% for the final pH > 6.0. While previous studies outlined the importance of solution pH in Mn removal from AMD by apatite materials [22,25], the results from the present work link the influence of all examined factors on the process efficiency via the final pH values.

A pH of 8.4–9.0 was essential for Mn to precipitate as oxide and hydroxide during AMD treatment by various biomass ashes [38]. This study supports these findings and shows that Mn removal by WA shifts from insignificant at pH < 6.5 to complete at pH > 9.0 and takes place via Mn(II) sorption and oxidation/precipitation, depending on the pH in the system. The influence of high concentrations of coexisting species, insufficient time to dissolve alkaline WA minerals or a small dose of WA is reflected in less than 100% Mn removed because a critical final pH of 9 was not reached to make Mn(II) oxidation to higher oxidation states and deposition of Mn oxides thermodynamically favored.

3.9. Actual AMD Treatment with WA and BC

The actual AMD (Table 4) sample was characterized by a pH value of 3.4 and 116 mg/L of Mn. Additionally, AMD contained a high concentration of Zn (137 mg/L), significantly lower concentrations of Pb (1.92 mg/L), Ni (0.174 mg/L), Fe (0.140 mg/L), Cu (0.163 mg/L), Cd (0.264 mg/L), Co (0.072 mg/L), and Cr (0.008 mg/L).

Table 4. AMD composition before and after the treatment with WA and BC.

	AMD	WA Treatment 2.5 g/L	BC Treatment 30 g/L		AMD	WA Treatment 2.5 g/L	BC Treatment 30 g/L
pH	3.4 ± 0.1	9.0 ± 0.1	6.5 ± 0.1				
				Element (mg/L)			
As	<0.005	<0.005	<0.005	Fe	0.140 ± 0.005	0.029 ± 0.001	0.015 ± 0.001
Al	2.89 ± 0.13	<0.005	<0.005	Mn	116 ± 4	0.061 ± 0.003	0.402 ± 0.020
Ba	<0.001	<0.001	<0.001	Ni	0.174 ± 0.09	<0.002	<0.002
Cd	0.264 ± 0.015	<0.001	<0.001	Pb	1.92 ± 0.10	<0.007	<0.007
Co	0.072 ± 0.003	<0.003	<0.003	Se	<0.002	<0.002	<0.002
Cr	0.008 ± 0.001	<0.001	<0.001	Zn	137 ± 9	0.013 ± 0.001	0.147 ± 0.008
Cu	0.163 ± 0.008	0.006 ± 0.001	0.024	V	<0.001	<0.001	<0.001

The variations in Mn removal efficiency and final pH values of AMD treated with WA and BC are presented in Figure 9. The increase in WA dose from 1.0 g/L to 3.0 g/L provoked a substantial pH rise from 7.0 to 9.4 and a consequential increase in Mn removal from 2.5% to 100%. At 2.5 g/L, 99.9% of Mn was removed at pH 9.0. Likewise, a rapid progression in Mn removal (4.2–90.1%) was detected as the quantity of applied BC raised from 1.0 g/L to 15 g/L. Process efficiency continued to increase with a further increase in the BC dose up to 30 g/L, gradually reaching 99.9%. In the entire range of BC doses, the final pH of AMD increased from 5.1 to 6.5.

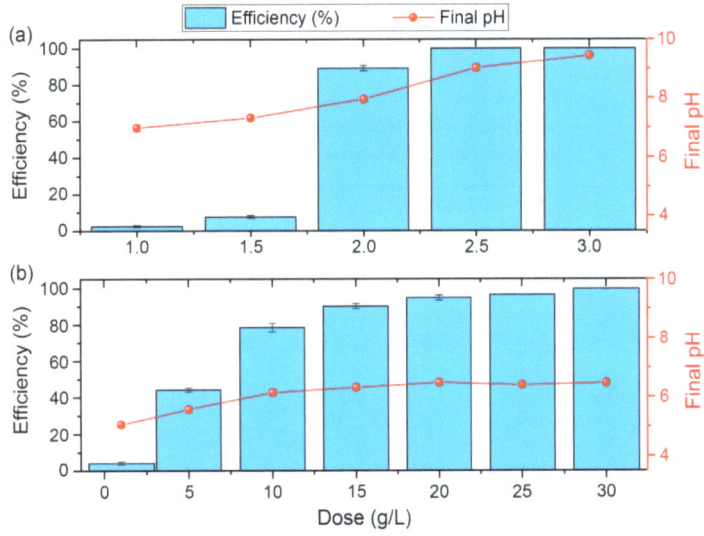

Figure 9. Mn removal efficiency and final pH values of actual AMD following the treatments by different doses of WA (a) and BC (b). Initial Mn concentration 116 mg/L, initial pH 3.4, contact time 24 h.

At AMD treatment conditions that can be considered optimal for Mn removal (2.5 g/L WA or 30 g/L BC), all coexisting metals were also effectively removed (Table 4). In agreement with the results obtained using synthetic AMD samples (Figure 8), this experiment confirms that successful removal of Mn from actual AMD is achievable by adjusting the doses of tested waste materials such as to provide a pH value of ≥9 after reaction with the WA, and a pH ~ 6.5, using the BC. However, the amount of BC to achieve this goal was twelve times the amount of WA, resulting in higher amounts of treatment residue.

WA and BC appear to be more practical agents for the remediation of Mn in AMD compared to limestone, which is a main precipitant due to its low cost and availability. By treatment of mine water at its natural pH (3.3) with 20.8 g/L of limestone, only 13% of the Mn initially present (16.5 mg/L) was removed, whereas the efficiency increased to 97%

when the initial pH of AMD was adjusted to 8.0 [52]. The application of limestone could not remove Mn effectively at high concentrations (140 mg/L), and even from synthetic solution with pH 6.5 and at high limestone doses (25 g/L), the Mn concentration was reduced to 120 mg/L [53]. However, 12.5 g/L of limestone mixed with 0.67 g/L sodium carbonate removed Mn completely from actual mine water with a pH of 6.5 as the final pH reached 9.6. For comparison, precipitation of 140 mg/L Mn with NaOH at pH 10 resulted in a residual metal concentration of 6.5 mg/L [53]. The actual AMD (pH 3.04; 5.94 mg/L Mn, 83.24 mg/L Fe) was also treated with waste biomaterials (shrimp shell (SS) and mussel byssus (MB)) [54]. The best removal efficiency of 96% for Fe and 78% for Mn was obtained with a mixture of waste materials (18 g/L SS and 10 g/L MB) when the final pH was 8.42 through the combined action of the sorption mechanisms on the biomaterials and precipitation as hydroxides. On the other hand, synthetic materials may show supremacy in Mn removal from AMD at much lower doses. For example, 0.5 g/L of hydroxyapatite (Hap) modified graphitic carbon nitride powders (Hap/gC$_3$N$_4$) applied in AMD with initial 3.5 and Mn concentration of 132.4 mg/L, removed 69% Mn [55].

3.10. *The Stability of AMD Treatment Resides and Prospects for Metal Recovery*

The stability of AMD treatment residues raises concerns about environmental and social impacts related to storage and disposal. Therefore, it was assessed by a batch test with deionized water. The concentrations of leachable elements measured in the liquid phase are expressed in mg/kg dry waste materials and compared with the threshold values set for the inert, nonhazardous, and hazardous wastes (Table 5).

Table 5. Water leachable concentrations of elements (mg/kg of dry waste material) in actual AMD treatment residues (EN 12457-2 test), and leaching limit values set by Council Decision 2003/33/EC for different waste categories.

	Actual AMD Treatment Residues		Waste Categories		
Element	WA	BC	Inert	Non-Hazardous	Hazardous
Al	173 ± 7	<LOD	-	-	-
As	<LOD	<LOD	0.5	2	25
Ba	0.518 ± 0.029	0.284 ± 0.012	20	100	300
Cd	<LOD	<LOD	0.04	1	5
Co	0.012 ± 0.005	0.023 ± 0.001	-	-	-
Cr	0.526 ± 0.022	0.005 ± 0.001	0.5	10	70
Cu	0.050 ± 0.002	0.050 ± 0.002	2	50	100
Fe	0.143 ± 0.006	0.129 ± 0.007	-	-	-
Mn	0.013 ± 0.002	5.940 ± 0.321	-	-	-
Mo	<LOD	<LOD	0.5	10	30
Ni	<LOD	<LOD	0.4	10	40
Pb	<LOD	<LOD	0.5	10	50
Sb	<LOD	<LOD	0.06	0.7	5
Se	<LOD	0.039 ± 0.002	0.1	0.5	7
V	<LOD	<LOD	-	-	-
Zn	0.488 ± 0.026	0.346 ± 0.018	4	50	200
pH	11.3 ± 0.1	7.1 ± 0.1	-	-	-

The concentrations of investigated elements were below the thresholds for inert waste, except for Cr in WA-residue (0.526 mg/kg vs. 0.5 mg/kg limit). Although not subject to classification, the leachable content of Al in WA-residue was substantial (173.3 mg/kg). The pH value of 11.3 was measured after resuspending this residue in deionized water, demonstrating high residual alkalinity. The enhanced leaching of Cr and Al over other metals can be attributed to the alkaline environment due to their amphoteric leaching behavior [56,57]. Given the neutral pH of the BC-residue (Table 5) and low concentrations of toxic and potentially toxic elements in pristine BC (Table 2), the leachability of all tested elements was lower compared to WA, except for Mn (5.940 mg/kg), indicating higher stability.

The prospects for recovery of AMD major metals from the treatment residues are presented in Figure 10. With increasing HCl concentration, Mn and Zn release rates from WA residue significantly increased, from 0.7% to 127% and from 0% to 96%, respectively, as the final pH decreased from 8.5 to 0.34. In 1 mol/L HCl, WA residue was virtually completely dissolved, thus releasing the Mn from the AMD treatment process and the WA composition (Table 2). Considering the BC residue (Figure 10b), Mn recovery with 0.01 mol/L HCl at final pH 5.3 was 7.2%, while Zn detachment was marginal (0.13% and 0.68%). The Mn and Zn recovery reached maximally 93% and 95% in 1 mol/L HCl due to almost complete BC residue dissolution at the final pH of 0.2.

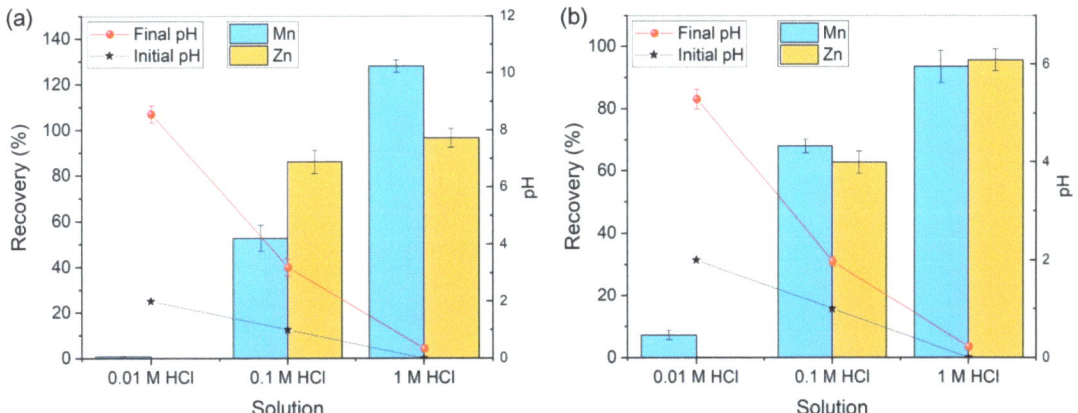

Figure 10. Recovery of AMD major cations (Mn and Zn) from WA (**a**) and BC (**b**) treatment residues using 0.1 mol/L CaCl$_2$, 0.01 mol/L HCl, 0.1 mol/L HCl, and 1 mol/L HCl. Solid to solution ratio 5 g/L, contact time 24 h.

Since the pH levels profoundly influence the removal of Mn, it is plausible to anticipate that the release of Mn ions could be facilitated by controlling the pH in the system. Notably, mineral acids, particularly HCl, exhibited remarkable efficacy in regenerating materials such as adsorptive membranes [58] and nanocarbon hybrid [59], which had previously sequestered Mn from solution via ion exchange or physical adsorption. The successful reuse of the sorbent underscores the preservation of its physical and chemical properties over several sorption-desorption cycles. However, results depicted in Figure 10 confirm that precipitated and chemisorbed forms of Mn were released when the residues were subjected to high acid concentrations, resulting in the dissolution of the materials employed. While recyclability is paramount for costly synthetic and modified materials [60], the regeneration of WA and BC, as waste and waste-derived materials, would present a modest economic rationale. Consequently, utilizing these residues in alternative technologies and for metal recovery emerges as a more pragmatic option.

4. Conclusions

Insight into the performance of unconventional reagents concerning AMD properties and other process factors can accelerate their practical application. This study's results show that wood ash and bone char can be valuable materials for Mn treatment in AMD and emphasize their use's comparative advantages and disadvantages. The summary key points and recommendations are following:

- Complete separation of Mn in AMD occurs only at final pH > 9.0 using WA. The WA is a cost-effective alternative to conventional alkaline reagents such as lime, and its consumption is attractive from environmental protection and circular economy

- standpoints. However, the issue of the high alkalinity of the treated water and the generated residue remains.
- The BC application is advantageous compared to WA for its ability to separate Mn at close to neutral conditions and faster establishment of equilibrium conditions. Furthermore, BC is a more useful reagent for treating sulfate-rich AMD, showing a 5% decrease in Mn removal efficiency up to sulfate concentrations of 10,000 mg/L. The influence of coexisting Fe ions up to a Fe/Mn ratio of 5:1 is significantly lower using BC.
- The impact of all investigated process factors and their interactions on Mn removal efficiency can be seen indirectly through their ability to bring the pH to or further from the optimal pH values for BC or WA action. Such insight helps optimize the conditions of the Mn removal process by adjusting the pH value in the system, which can be done successfully by optimizing the reagent dose.
- Depending on the AMD pH and composition, 3–12 times smaller quantities of WA than BC are needed, generating less residue. Still, the alkalinity of the WA-treatment residue impacts the enhanced leaching of elements with amphoteric leaching behavior, posing a potential risk to the environment.
- Further studies on potential hazards due to metal release from the residues should consider the conditions of their disposal or reuse. The reuse options for the residues of AMD treatment with WA and BC are particularly interesting, for instance, in metals recovery, prevention and control of AMD, or stabilization of contaminated soil. Given the high BC/AMD ratio needed to effectively treat high Mn concentrations at low pH, previous pH adjustment to a near-neutral region using a low-cost alkalizer would significantly reduce the required BC dose. In that sense, applying wood ash and BC in two consecutive steps is worth exploring.

Supplementary Materials: The following supporting information can be downloaded at: https://www.mdpi.com/article/10.3390/met13101665/s1, Figure S1. Determination of the point of zero charge (pH_{PZC}) of the BC and WA samples; Figure S2. The amounts of Mn removed per unit mass of reagent (mg/g) as affected by (a) initial pH (dose 1 g/L, initial Mn concentration 100 mg/L, contact time 24 h), (b) WA dose and Mn concentration (initial pH 2.5, contact time 24 h), (c) BC dose and Mn concentration (initial pH 2.5, contact time 24 h), (d) contact time (initial Mn concentration 100 mg/L, dose 1.2 g/L WA and 10 g/L BC, initial pH 2.5), (e) sulfate concentration (initial Mn concentration 100 mg/L, dose 1.2 g/L WA and 10 g/L BC, initial pH 2.5, contact time 24 h), (f) Fe concentration (initial Mn concentration 100 mg/L, dose 1.2 g/L WA and 10 g/L BC, initial pH 2.5, contact time 24 h); Figure S3: The amounts of Fe removed per unit mass of reagent (mg/g) as affected by the concentration of Fe. Initial Mn concentration 100 mg/L, dose 1.2 g/L WA and 10 g/L BC, initial pH 2.5, contact time 24 h; Figure S4. Interaction plot for Mn removal; Figure S5. Main effect plots for (a) Fe removal, and (b) final pH; Figure S6. Pareto chart of the effects of variables on (a) Fe removal, and (b) final pH.

Author Contributions: Conceptualization, I.S.; methodology, I.S., B.J., S.D. and M.J.; validation J.M. and M.J.; formal analysis, I.S., J.M. and M.J.; investigation, I.S., M.J., J.M. and N.M.; resources, I.S. and N.M.; writing—original draft preparation, I.S.; writing—review and editing, I.S., B.J. and S.D.; visualization—M.J.; supervision, S.D.; funding acquisition, I.S., B.J., J.M., J.M. and S.D. All authors have read and agreed to the published version of the manuscript.

Funding: This research was funded by the Ministry of Science, Technological Development, and Innovation of the Republic of Serbia (Contract No. 451-03-47/2023-01/200017).

Data Availability Statement: The datasets generated during the current study are available upon request.

Acknowledgments: The authors thank Slavko Smiljanić at the University of East Sarajevo, Faculty of Technology Zvornik, for providing the acid mine drainage sample.

Conflicts of Interest: The authors declare no conflict of interest.

References

1. Hill, R.D. *Mine Drainage Treatment: State of the Art and Research Needs*; U.S. Department of the Interior, Mine Drainage Control Activities, Federal Water Pollution Control Administration: Cincinnati, OH, USA, 1968.
2. Park, I.; Tabelin, C.B.; Jeon, S.; Li, X.; Seno, K.; Ito, M.; Hiroyoshi, N. A review of recent strategies for acid mine drainage prevention and mine tailings recycling. *Chemosphere* **2019**, *219*, 588–606. [CrossRef] [PubMed]
3. Skousen, J.G.; Ziemkiewicz, P.F.; McDonald, L.M. Acid mine drainage formation, control and treatment: Approaches and strategies. *Extr. Ind. Soc.* **2019**, *6*, 241–249. [CrossRef]
4. Kefeni, K.K.; Msagati, T.A.M.; Mamba, B.B. Acid mine drainage: Prevention, treatment options, and resource recovery: A review. *J. Clean. Prod.* **2017**, *151*, 475–493. [CrossRef]
5. Skousen, J. Overview of acid mine drainage treatment with chemicals. In *Acid Mine Drainage, Rock Drainage, and Acid Sulfate Soils*; John Wiley & Sons, Inc.: Hoboken, NJ, USA, 2014; pp. 325–337.
6. Jiao, Y.; Zhang, C.; Su, P.; Tang, Y.; Huang, Z.; Ma, T. A review of acid mine drainage: Formation mechanism, treatment technology, typical engineering cases and resource utilization. *Process Saf. Environ. Prot.* **2023**, *170*, 1240–1260. [CrossRef]
7. Johnson, D.B.; Hallberg, K.B. Acid mine drainage remediation options: A review. *Sci. Total Environ.* **2005**, *338*, 3–14. [CrossRef] [PubMed]
8. European Commission. *A New Circular Economy Action Plan For a Cleaner and More Competitive Europe*; COM(2020) 98 Final. Communication from the Commission to the European Parliament, the Council, the European Economic and Social Committee and the Committee of the Regions; EC: Brussels, Belgium, 2020; pp. 1–19.
9. Li, Y.; Li, W.; Xiao, Q.; Song, S.; Liu, Y.; Naidu, R. Acid mine drainage remediation strategies: A review on migration and source controls. *Miner. Metall. Process.* **2018**, *35*, 148–158. [CrossRef]
10. Lucas, H.; Stopic, S.; Xakalashe, B.; Ndlovu, S.; Friedrich, B. Synergism red mud-acid mine drainage as a sustainable solution for neutralizing and immobilizing hazardous elements. *Metals* **2021**, *11*, 620. [CrossRef]
11. Westholm, L.J.; Repo, E.; Sillanpää, M. Filter materials for metal removal from mine drainage—A review. *Environ. Sci. Pollut. Res.* **2014**, *21*, 9109–9128. [CrossRef]
12. Howe, P.D.; Malcolm, H.M.; Dobson, S.; World Health Organization & International Programme on Chemical Safety. *Manganese and Its Compounds: Environmental Aspects*; World Health Organization: Geneva, Switzerland, 2004; pp. 1–63.
13. Neculita, C.M.; Rosa, E. A review of the implications and challenges of manganese removal from mine drainage. *Chemosphere* **2019**, *214*, 491–510. [CrossRef]
14. Matveeva, V.A.; Alekseenko, A.V.; Karthe, D.; Puzanov, A.V. Manganese pollution in mining-influenced rivers and lakes: Current state and forecast under climate change in the Russian Arctic. *Water* **2022**, *14*, 1091. [CrossRef]
15. Li, Y.; Xu, Z.; Ma, H.; Hursthouse, S.A. Removal of manganese(II) from acid mine wastewater: A review of the challenges and opportunities with special emphasis on Mn-oxidizing bacteria and microalgae. *Water* **2019**, *11*, 2493. [CrossRef]
16. Rudi, N.N.; Muhamad, M.S.; Te Chuan, L.; Alipal, J.; Omar, S.; Hamidon, N.; Abdul Hamid, N.H.; Mohamed Sunar, N.; Ali, R.; Harun, H. Evolution of adsorption process for manganese removal in water via agricultural waste adsorbents. *Heliyon* **2020**, *6*, e05049. [CrossRef] [PubMed]
17. Chowdhury, S.; Mishra, M.; Suganya, O. The incorporation of wood waste ash as a partial cement replacement material for making structural grade concrete: An overview. *Ain Shams Eng. J.* **2015**, *6*, 429–437. [CrossRef]
18. Heviánková, S.; Bestová, I.; Kyncl, M. The application of wood ash as a reagent in acid mine drainage treatment. *Miner. Eng.* **2014**, *56*, 109–111. [CrossRef]
19. Fawell, J.; Bailey, K.; Chilton, J.; Dahi, E.; Fewtrell, L.; Magara, Y. *Fluoride in Drinking-Water*, 1st ed.; World Health Organization: Geneva, Switzerland, 2006; pp. 1–134.
20. Kono, T.; Sakae, T.; Nakada, H.; Kaneda, T.; Okada, H. Confusion between carbonate apatite and biological apatite (carbonated hydroxyapatite) in bone and teeth. *Minerals* **2022**, *12*, 170. [CrossRef]
21. Sicupira, D.C.; Silva, T.T.; Leão, V.A.; Mansur, M.B. Batch removal of manganese from acid mine drainage using bone char. *Brazilian J. Chem. Eng.* **2014**, *31*, 195–204. [CrossRef]
22. Sicupira, D.C.; Tolentino Silva, T.; Ladeira, A.C.Q.; Mansur, M.B. Adsorption of manganese from acid mine drainage effluents using bone char: Continuous fixed bed column and batch desorption studies. *Brazilian J. Chem. Eng.* **2015**, *32*, 577–584. [CrossRef]
23. Carrillo-González, R.; González-Chávez, M.C.A.; Cazares, G.O.; Luna, J.L. Trace element adsorption from acid mine drainage and mine residues on nanometric hydroxyapatite. *Environ. Monit. Assess.* **2022**, *194*, 280. [CrossRef]
24. Oliva, J.; Cama, J.; Cortina, J.L.; Ayora, C.; De Pablo, J. Biogenic hydroxyapatite (Apatite II^{TM}) dissolution kinetics and metal removal from acid mine drainage. *J. Hazard. Mater.* **2012**, *213–214*, 7–18. [CrossRef]
25. Vandeginste, V.; Cowan, C.; Gomes, R.L.; Hassan, T.; Titman, J. Natural fluorapatite dissolution kinetics and Mn^{2+} and Cr^{3+} metal removal from sulfate fluids at 35 °C. *J. Hazard. Mater.* **2020**, *389*, 122150. [CrossRef]
26. Nayak, A.; Bhushan, B. Hydroxyapatite as an advanced adsorbent for removal of heavy metal ions from water: Focus on its applications and limitations. *Mater. Today Proc.* **2021**, *46*, 11029–11034. [CrossRef]
27. Brazdis, R.I.; Fierascu, I.; Avramescu, S.M.; Fierascu, R.C. Recent progress in the application of hydroxyapatite for the adsorption of heavy metals from water matrices. *Materials* **2021**, *14*, 6898. [CrossRef]
28. Amenaghawon, A.N.; Anyalewechi, C.L.; Darmokoesoemo, H.; Kusuma, H.S. Hydroxyapatite-based adsorbents: Applications in sequestering heavy metals and dyes. *J. Environ. Manag.* **2022**, *302*, 113989. [CrossRef]

29. Dimović, S.; Smičiklas, I.; Plećaš, I.; Antonović, D.; Mitrić, M. Comparative study of differently treated animal bones for Co^{2+} removal. *J. Hazard. Mater.* **2009**, *164*, 279–287. [CrossRef]
30. Smičiklas, I.; Dimović, S.; Šljivić, M.; Plećaš, I.; Lončar, B.; Mitrić, M. Resource recovery of animal bones: Study on sorptive properties and mechanism for Sr^{2+} ions. *J. Nucl. Mater.* **2010**, *400*, 15–24. [CrossRef]
31. Fiol, N.; Villaescusa, I. Determination of sorbent point zero charge: Usefulness in sorption studies. *Environ. Chem. Lett.* **2009**, *7*, 79–84. [CrossRef]
32. *PDXL: Integrated X-ray Powder Diffraction Software*, Version 2.0.3.0; Rigaku Corporation: Tokyo, Japan, 2011; 196–8666.
33. Gates-Rector, S.; Blanton, T. The powder diffraction file: A quality materials characterization database. *Powder Diffr.* **2019**, *34*, 352–360. [CrossRef]
34. Downs, R.T.; Hall-Wallace, M. The American mineralogist crystal structure database. *Am. Mineral.* **2003**, *88*, 247–250.
35. BS EN 12457-2; Characterisation of Waste. Leaching. Compliance Test for Leaching of Granular Waste Materials and Sludges. Part 2: One Stage Batch Test at a Liquid to Solid Ratio of 10 L/kg for Materials with Particle Size below 4 mm (without or with Size Reduction). British Standards Institue: London, UK, 2002.
36. Vassilev, S.V.; Baxter, D.; Andersen, L.K.; Vassileva, C.G. An overview of the chemical composition of biomass. *Fuel* **2010**, *89*, 913–933. [CrossRef]
37. Campbell, A.G. Recycling and disposing of wood ash. *Tappi J.* **1990**, *73*, 141–146.
38. Bogush, A.A.; Dabu, C.; Tikhova, V.D.; Kim, J.K.; Campos, L.C. Biomass ashes for acid mine drainage remediation. *Waste Biomass Valor.* **2020**, *11*, 4977–4989. [CrossRef]
39. Garvie, L.A.J. Mineralogy of paloverde (*Parkinsonia microphylla*) tree ash from the Sonoran Desert: A combined field and laboratory study. *Am. Mineral.* **2016**, *101*, 1584–1595. [CrossRef]
40. Von Euw, S.; Wang, Y.; Laurent, G.; Drouet, C.; Babonneau, F.; Nassif, N.; Azaïs, T. Bone mineral: New insights into its chemical composition. *Sci. Rep.* **2019**, *9*, 8456. [CrossRef] [PubMed]
41. Barbu, C.H.; Pavel, P.B.; Moise, C.M.; Sand, C.; Pop, M.R. Neutralization of acid mine drainage with wood ash. *Rev. Chim.* **2018**, *68*, 2768–2770. [CrossRef]
42. Labaali, Z.; Kholtei, S.; Naja, J. Co^{2+} removal from wastewater using apatite prepared through phosphate waste rocks valorization: Equilibrium, kinetics and thermodynamics studies. *Sci. Afr.* **2020**, *8*, e00350. [CrossRef]
43. Bin Mobarak, M.; Pinky, N.S.; Chowdhury, F.; Hossain, M.S.; Mahmud, M.; Quddus, M.S.; Jahan, S.A.; Ahmed, S. Environmental remediation by hydroxyapatite: Solid state synthesis utilizing waste chicken eggshell and adsorption experiment with Congo red dye. *J. Saudi Chem. Soc.* **2023**, *27*, 101690. [CrossRef]
44. Ganta, D.D.; Hirpaye, B.Y.; Raghavanpillai, S.K.; Menber, S.Y. Green synthesis of hydroxyapatite nanoparticles using *Monoon longifolium* leaf extract for removal of fluoride from aqueous solution. *J. Chem.* **2022**, *2022*, 4917604. [CrossRef]
45. Kalembkiewicz, J.; Galas, D.; Sitarz-Palczak, E. The physicochemical properties and composition of biomass ash and evaluating directions of its applications. *Polish J. Environ. Stud.* **2018**, *27*, 2593–2603. [CrossRef]
46. Petronijević, N.; Radovanović, D.; Štulović, M.; Sokić, M.; Jovanović, G.; Kamberović, Ž.; Stanković, S.; Stopic, S.; Onjia, A. Analysis of the mechanism of acid mine drainage neutralization using fly ash as an alternative material: A case study of the extremely acidic Lake Robule in eastern Serbia. *Water* **2022**, *14*, 3244. [CrossRef]
47. Hem, J.D. *Chemical Equilibria and Rates of Manganese Oxidation*; Water Supply Paper 1667-A; U.S. G.P.O.: Washington, DC, USA, 1963; pp. A1–A64. [CrossRef]
48. Liu, J.; Chen, Q.; Yang, Y.; Wei, H.; Laipan, M.; Zhu, R.; He, H.; Hochella, M.F. Coupled redox cycling of Fe and Mn in the environment: The complex interplay of solution species with Fe- and Mn-(oxyhydr)oxide crystallization and transformation. *Earth-Science Rev.* **2022**, *232*, 104105. [CrossRef]
49. Calugaru, I.L.; Genty, T.; Neculita, C.M. Treatment of manganese, in the presence or absence of iron, in acid and neutral mine drainage using raw vs half-calcined dolomite. *Miner. Eng.* **2021**, *160*, 106666. [CrossRef]
50. Brundavanam, S.; Poinern, G.E.J.; Fawcett, D. Kinetic and adsorption behaviour of aqueous Fe^{2+}, Cu^{2+} and Zn^{2+} using a 30 nm hydroxyapatite based powder synthesized via a combined ultrasound and microwave based technique. *Am. J. Mater. Sci.* **2015**, *5*, 31–40.
51. Popa, M.; Bostan, R.; Varvara, S.; Moldovan, M.; Rosu, C. Removal of Fe, Zn and Mn ions from acidic mine drainage using hydroxyapatite. *J. Environ. Prot. Ecol.* **2016**, *17*, 1472–1480.
52. Silva, A.M.; Cruz, F.L.S.; Lima, R.M.F.; Teixeira, M.C.; Leão, V.A. Manganese and limestone interactions during mine water treatment. *J. Hazard. Mater.* **2010**, *181*, 514–520. [CrossRef] [PubMed]
53. Silva, A.M.; Cunha, E.C.; Silva, F.D.R.; Leão, V.A. Treatment of high-manganese mine water with limestone and sodium carbonate. *J. Clean. Prod.* **2012**, *29–30*, 11–19. [CrossRef]
54. Núñez-Gómez, D.; Lapolli, F.R.; Nagel-Hassemer, M.E.; Lobo-Recio, M.Á. Optimization of Fe and Mn removal from coal acid mine drainage (AMD) with waste biomaterials: Statistical modeling and kinetic study. *Waste Biomass Valor.* **2020**, *11*, 1143–1157. [CrossRef]
55. Agha Beygli, R.; Mohaghegh, N.; Rahimi, E. Metal ion adsorption from wastewater by g-C_3N_4 modified with hydroxyapatite: A case study from Sarcheshmeh acid mine drainage. *Res. Chem. Intermed.* **2019**, *45*, 2255–2268. [CrossRef]
56. Cui, Y.; Chen, J.; Zhang, Y.; Peng, D.; Huang, T.; Sun, C. pH-dependent leaching characteristics of major and toxic elements from red mud. *Int. J. Environ. Res. Public Health* **2019**, *16*, 2046. [CrossRef]

57. Smičiklas, I.; Jović, M.; Janković, M.; Smiljanić, S.; Onjia, A. Environmental safety aspects of solid residues resulting from acid mine drainage neutralization with fresh and aged red mud. *Water Air Soil Pollut.* **2021**, *232*, 490. [CrossRef]
58. Vo, T.S.; Hossain, M.M.; Jeong, H.M.; Kim, K. Heavy metal removal applications using adsorptive membranes. *Nano Converg.* **2020**, *7*, 36. [CrossRef]
59. Embaby, M.A.; Abdel Moniem, S.M.; Fathy, N.A.; El-kady, A.A. Nanocarbon hybrid for simultaneous removal of arsenic, iron and manganese ions from aqueous solutions. *Heliyon* **2021**, *7*, e08218. [CrossRef] [PubMed]
60. Mokgehle, T.M.; Tavengwa, N.T. Recent developments in materials used for the removal of metal ions from acid mine drainage. *Appl. Water Sci.* **2021**, *11*, 42. [CrossRef]

Disclaimer/Publisher's Note: The statements, opinions and data contained in all publications are solely those of the individual author(s) and contributor(s) and not of MDPI and/or the editor(s). MDPI and/or the editor(s) disclaim responsibility for any injury to people or property resulting from any ideas, methods, instructions or products referred to in the content.

Article

Selection of Operation Conditions for a Batch Brown Seaweed Biosorption System for Removal of Copper from Aqueous Solutions

Henrik K. Hansen [1,*], Claudia Gutiérrez [1], Natalia Valencia [1], Claudia Gotschlich [1], Andrea Lazo [1], Pamela Lazo [2] and Rodrigo Ortiz-Soto [3]

[1] Departamento de Ingeniería Química y Ambiental, Universidad Técnica Federico Santa María, Avenida España 1680, Valparaíso 2390123, Chile; claudia.gutierrez@usm.cl (C.G.); natalia.valencia@alumnos.usm.cl (N.V.); claudia.gotschlich@gmail.com (C.G.); andrea.lazo@usm.cl (A.L.)
[2] Instituto de Química y Bioquímica, Facultad de Ciencias, Universidad de Valparaíso, Valparaíso 2362735, Chile; pamela.lazo@uv.cl
[3] Escuela de Ingeniería Química, Pontificia Universidad Católica de Valparaíso, Valparaíso 2340025, Chile; rodrigo.ortiz@pucv.cl
* Correspondence: henrik.hansen@usm.cl

Abstract: Heavy metal exposure from wastewater is an important environmental issue worldwide. In the search for more efficient treatment technologies, biosorption has been presented as an alternative for contaminant removal from wastewaters. The aim of this work is to determine the operation parameters of copper adsorption followed by biosorbent regeneration. The algae *Durvillaea antarctica* and *Lessonia trabeculata* were used as biosorbents in batch experiments. These biosorbents were exposed to different conditions, such as pH, copper concentration, exposure time, mass-to-volume ratios and regeneration reagents. Batch sorption tests revealed an adequate pH of 4.5–5.0. The selected biosorbent was *D. antarctica* due to a considerably higher copper retention capacity. As a regenerating reagent, sulfuric acid was more efficient. For diluted copper solutions (10 to 100 mg L^{-1}), a biosorbent particle size of between 1.70 and 3.36 mm showed better retention capacity than larger particles and a biosorbent mass-to-volume ratio of 10 g L^{-1} was desirable for these metal concentrations.

Keywords: copper retention; biomass particle size; sorption isotherms; sorption kinetics

1. Introduction

Water resources, which are vital for life, have been reduced partly due to contamination mainly caused by industrial processes, agriculture and urbanization. In Chile, mining is the most relevant productive industrial activity, so focus has to be turned onto the waste generated by this kind of activity [1–4]. Mining, besides being the largest production activity of the country, is also the principal cause of heavy metal contamination.

Within great mineral reservoirs in Chile, the production of copper, iron, molybdenum, lead, zinc, gold and silver is considered, with copper and molybdenum (a byproduct of copper production) being the most interesting ones. To obtain these metals, large water consumption is required for use in different operations, such as extraction, grinding, concentration and refinement. Once used, the wastewater contains a high amount of heavy metals.

The toxicity of heavy metals and their effect on the environment has created a need to reduce their concentration in industrial effluents below the levels required by environmental legislation. This has initiated a search for alternative methods for the elimination of these elements from aqueous solutions.

Normally, heavy metals are removed by physicochemical treatment methods such as chemical precipitation, reverse osmosis, adsorption in activated carbon, electrodialysis, and ion exchange, but it has to be noted that these processes are expensive and could be inefficient. Because of this, biosorption is being studied as a heavy metal removal technique, which is a promising technology for wastewater treatment with low concentrations (1–100 mg L^{-1}). Biosorbent material can be regenerated several times. It is found in abundance in the environment and is an inexpensive resource that is easy to access; often biomass waste is used [5–8].

A biosorption process refers to the ability of materials of biological origin to retain heavy metals from diluted aqueous solutions in their structures [9]. It is considered a clean environmental remediation process for metal recovery and decontamination of wastewater by heavy metals and metalloids, such as copper, lead, cadmium, nickel and arsenic [10]. Biomass traditionally used in biosorption processes belongs to three groups: bacteria, fungi or algae. In relation to its source, it can be obtained directly from nature or as a waste from productive processes [11] (Franco et al., 2021).

Seaweeds (or marine algae) have previously been reported to be efficient in inorganic contaminant removal from wastewater [12,13]. In brown seaweeds, fibers are mainly cellulose and insoluble alginates [14]. These alginates are Ca, Mg, or Na salts of alginic acid (1,4-linked polymer of β-D-mannuronic acid and α-L-guluronic acid). Alginates are known for their high divalent metal cation uptake capacities [15] and are therefore a suitable adsorbent for copper removal. Two brown seaweed species that are abundant along the entire coastline of Chile contain significant amounts of alginates. In D. antarctica, typically between 10–20% and sometimes even around 50% of the total dry weight are alginates [16,17], whereas L. trabeculata was reported to contain similar amounts of alginates [18]. Therefore, both seaweed species would be possible metal cation accumulaters when treating aqueuos solutions with adsorption processes using these biosorbents. Furthermore, these seaweeds are actually so abundant in Chile that they appear as solid waste when cleaning the beaches and coast—meaning that they provide very low-cost sorbent material.

In Chile, both mineral processing wastewater and acidic mine drainage contain copper in concentrations that would be favorable for biosorption [19]. Until now, no copper uptake data have been published with regard to L. trabeculata, and only Cid et al. (2015) [20] have investigated this the behavior of D. antarctica under some specific conditions with respect to pH, biosorbent mass-to-solution volume, copper concentration and particle size, so it would be interesting to compare these biosorbents with previously reported copper retentions for other sorbents, in particular seaweeds, that would be more difficult to have access to in Chile. Table 1 shows a summary of research studies on the copper retention of a variety of brown seaweeds. Furthermore, to optimize the applicablity of the sorbents in a real treatment process, it would be necessary to evaluate the possiblity to regenerate the biosorbents, so that (1) biosorbent disposal would be minimized and (2) copper could be recovered.

Table 1. Comparison of maximum copper biosorption capacities (q_{max}) of different brown seaweeds.

Brown Seaweed	Cu conc. (mg L^{-1})	pH	Particle Size (mm)	Biomass/ Volumen (g/L)	Time to Equilibrium (hours)	Temperature (°C)	q_{max} (mg g^{-1})	Reference
Lessonia nigrescens	7.5–300	5	0.5–1	1	2	20	60.4	[21]
Cystoseira sp.	10–30	6	<0.5	0.1	2	28	180.4	[22]
Lessonia nigrescens blades	200–1000	3.2	5–20	1 / 4	168	25	56.2 / 47.3	[7]
Lessonia nigrescens stipes	200–1000	3.2	10–15	1 / 4	168	25	78.8 / 218.7	[7]
Sargassum tenerrimum	10–50	5	0.2–0.5	10	24	28	39.8	[23]
Iyengaria stellata	10–50	5	0.2–0.5	10	24	28	46.3	[23]
Lobophora variegata	10–50	5	0.2–0.5	10	24	28	38.0	[23]
Cystoseira indica	10–50	5	0.2–0.5	10	24	28	30.9	[23]
Sargassum cinereum	10–50	5	0.2–0.5	10	24	28	34.0	[23]
Durvillaea antarctica	7.5–300	5	0.5–1	1	2	20	91.5	[20]

Table 1. Cont.

Brown Seaweed	Cu conc. (mg L^{-1})	pH	Particle Size (mm)	Biomass/ Volumen (g/L)	Time to Equilibrium (hours)	Temperature (°C)	q_{max} (mg g^{-1})	Reference
Sargassum filipendula	19–265	4.5	0.855	-	-	20	84.1	[24]
Sargassum sinicola	2–256	-	0.2–0.5	10	24	-	116.6	[25]
Fucus serratus	0.6–25	5.5	0.355–0.5	0.09	8	20	101.8	[26]
Fucus vesiculosus	10–150	5	<0.5	0.5	2	23	105.5	[27]
Sargassum sp.	20–500	5.5	0.5	1	3	22	72.5	[28]
Sargassum sp.	-	6	<0.325	2	4	22	84.0–86.9	[29]
Fucus spiralis	10–150	4	<0.5	0.5	2	-	70.9	[30]
Ascophyllum nodosum	10–150	4	<0.5	0.5	2	-	58.8	[30]
Sargassum sp.	-	5	0.5–0.8	1	6	22	62.9	[31]
Padina sp.	-	5	0.5–0.8	1	6	22	72.4	[31]
Sargassum vulgare	10–250	4.5	1–4	2	6	22	59.1	[32]
Sargassum fluitans	10–150	4.5	1–4	2	6	22	50.8	[32]
Sargassum filipendula	10–250	4.5	1–4	2	6	22	56.6	[32]

In this part of the research work, a batch copper removal process is developed, using the algae *D. antarctica* and *L. trabeculata* as biosorbents. These biosorbents are available along the coast of Chile. The specific objectives of the study are (i) to choose an adequate operating pH, (ii) to choose the appropriate biosorbent for a continuous process, (iii) to choose particle size, (iv) to choose an adequate mass of biosorbent/volume of solution ratio, (v) to study adsorption kinetics, (vi) to study adsorption isotherm and (vii) to choose the regeneration reagent that recovers the most copper from the biosorbent.

2. Materials and Methods

Based on biosorption experiments, it is possible to quantitatively assess the retention capacity of a biosorbent by using a solution with a specific contaminant. For the evaluation of the retention capacity, a simple metal mass balance is used, which follows the logical assumption that the metal ion loss in the solution is the metal retained by the biosorbent, as shown in Equation (1).

$$q = \frac{V \cdot (C_i - C_{eq})}{M} \quad (1)$$

where C_i (mg L^{-1}) is the initial concentration of the element in the solution, V (L) is the initial solution volume, C_{eq} (mg L^{-1}) is the equilibrium concentration of the element in solution, M (g) is the biosorbent mass, and q (mg g^{-1}) is the retention capacity of the element by the biosorbent.

The biosorption phenomena are time dependent, thus it is necessary to obtain the adsorption rate for the design and evaluation of a potential biosorbent. Furthermore, the fitting of both biosorption kinetic and equilibrium data with conventional mathematical models would enlighten the efficiency of the metal uptake. Table 2 summarizes the models used in this work.

2.1. Reagents

The copper solutions were prepared by dissolving $CuSO_4 \cdot 5H_2O$ 99.5% (analytical grade) in distilled water. pH was adjusted by the addition of hydrochloric acid 37% GR for analysis (Merck, Rahway, NJ, USA) or by the addition of sodium hydroxide NaOH (5 M) prepared by dissolving 98% extra pure sodium hydroxide pellets (Loba Chemie, Mumbai, India).

Table 2. Mathematical biosorption models for kinetic and isotherm data.

Model Type	Equation	Parameter Description
Kinetic model		
Pseudo first order Lagergren	$q_t = q_{eq}(1 - e^{-k_{ad}t})$	q_t (mg g^{-1}) is the adsorbate retention in time t, q_{eq} (mg g^{-1}) is the adsorbate retention in equilibrium, k_{ad} (min^{-1}) is the adsorption first order constant and t (min) is the time.
Pseudo second order Ho & McKay	$q_t = \dfrac{t}{\dfrac{1}{k \cdot q_{eq}^2} + \dfrac{t}{q_{eq}}}$	k (g mg^{-1} min^{-1}) is the second order adsorption constant.
Isotherm model		
Freundlich	$q_{eq} = k \cdot C_{eq}^{1/n}$	k is the Freundlich capacity parameter and 1/n is the Freundlich intensity parameter.
Langmuir	$q_{eq} = \dfrac{q_m \cdot b \cdot C_{eq}}{1 + b \cdot C_{eq}}$	q_m is the maximum concentration of the metal on the biomass (mg metal g^{-1} dry biosorbent), b is a coefficient related to the affinity between the biosorbent and the metal, high values of b indicate a high affinity for the biosorbent and show a steep initial slope in the isotherm plot (L mg^{-1}).
Sips	$q_{eq} = \dfrac{q_m (K_S \cdot C_{eq})^{1/n_s}}{1 + (K_S \cdot C_{eq})^{1/n_s}}$	K_S (L mg^{-1}) is the equilibrium constant and n_s (-) is the model exponent.
Brunauer, Emmett and Teller	$q_e = \dfrac{q_m k_1 C_e \left[1 - (n+1)(k_2 C_e)^n + n(k_2 C_e)^{n+1}\right]}{(1 - k_2 C_e)\left[1 + \left(\dfrac{k_1}{k_2} - 1\right)k_2 C_e - \left(\dfrac{k_1}{k_2}\right)(k_2 C_e)^{n+1}\right]}$	q_m is the maximum adsorbate retention in the monolayer (mg g^{-1}), k_1 is the equilibrium constant of adsorption in the first layer (L mg^{-1}), k_2 is the equilibrium constant of adsorption in upper layers (L mg^{-1}) and n is the number of adsorption layers estimated.

2.2. Analytical

Each liquid sample was filtered through a N° 131 grade filter paper (Advantec, Dublin, CA, USA) by a vacuum pump (Welch, Ilmenau, Germany—model 2522). The copper concentration in the filtrate was determined by atomic absorption spectrophotometry in flame (Varian, Palo Alto, CA, USA—model SpectrAA 55) according to Chilean standard NCh 2313/10 Of. 96. pH was measured using an Orion (Thermo Scientific, Beverly, MA, USA) PerpHect logR model 370 pH meter with a combined pH electrode.

2.3. Preparation of Adsorbent

D. antarctica and *L. trabeculata* samples were collected in the bay of Valparaíso, Chile. After sampling, the algae were washed in tap water and then in distilled water to remove any salt present. The algae were dried at 50 °C until they obtained a constant weight. The dry biosorbents were first cut with a knife into regular-shaped pieces and then a jaw crusher was used to obtain the smaller-sized particles. A RO-TAP Sieve Shaker with test sieves from W.S. Tyler, model RX-29-10, was used to obtain different size fractions. The particle size ranges separated by sieving and chosen for the experiments were: 0.43–1.70 mm, 1.70–3.36 mm, 3.36–4.00 mm and 4.00–5.66 mm.

2.4. Experimental Plan

The conditions of every experimental run are summarized in Table 3. The analyzed parameters were: (a) operation pH, (b) algae used as biosorbent, (c) regenerating reagent, (d) biosorbent particle size, (e) biosorbent mass-to-solution volume ratio (M/V), (f) time sorption and (g) adsorption isotherm.

Table 3. Summary of experimental details.

Experimental Run	Cu conc. mg L^{-1}	M/V Ratio g L^{-1}	Particle Size mm	Time min	Biosorbent	pH
pH determination	2	10	4.00–5.66	30 60 120	D. Antarctica L. trabeculata	3.0–3.5 4.5–5.0
Biosorbent determination	30 100	10	4.00–5.66	30 60 120 300 1440	D. Antarctica L. trabeculata	4.5–5.0
Regenerating reagent determination	100	10	4.00–5.66	10 20 30 120	D. antarctica	4.5–5.0
Biosorbent particle size	10 100	20	4.00–5.66 3.36–4.00 1.70–3.36 0.43–1.70	360	D. antarctica	4.5–5.0
M/V Ratio	10 100	10 20 40	1.70–3.36	1440	D. antarctica	4.5–5.0
Biosorption kinetics	10 100	10	1.70–3.36	5 10 20 30 60 120 360 720	D. antarctica	4.5–5.0
Adsorption isotherm	10 25 50 75 100	10	1.70–3.36	360	D. antarctica	4.5–5.0

The pH in the solution was kept constant by adding drops of either 0.5 M HCl or 0.5 M NaOH solutions, assuring that the total liquid volume was not affected severely. The experiments were carried out in duplicate, without stirring and at an ambient temperature (20–25 °C). Relative standard deviations were lower than 5% in every experiment. The standard deviations and error margins are given in the tables and figures representing the experimental results.

3. Results

3.1. Determination of Sorption pH

One of the most relevant factors in metal ion retention with seaweeds is pH [33–35]; therefore, experiments were carried out at two pH intervals, 3.0–3.5 and 4.5–5.0, in the solutions. Lower pH values were not analyzed because it is generally known that the sorption is worse [31,36,37], and higher pH values were not chosen to avoid the effect of chemical precipitation.

For these experiments, 2 mg L^{-1} copper solutions were used, from which 500 mL was poured into every beaker and 5 g of dried *L. trabeculata* or *D. antarctica* was added in each case. After the treatment time was reached, the solution was filtered and the copper content in the liquid was measured. Experimental results for the copper retention capacity of *L. trabeculata* and *D. antarctica* at different times for pH intervals of 3.0–3.5 and 4.5–5.0, respectively, are displayed in Figure 1.

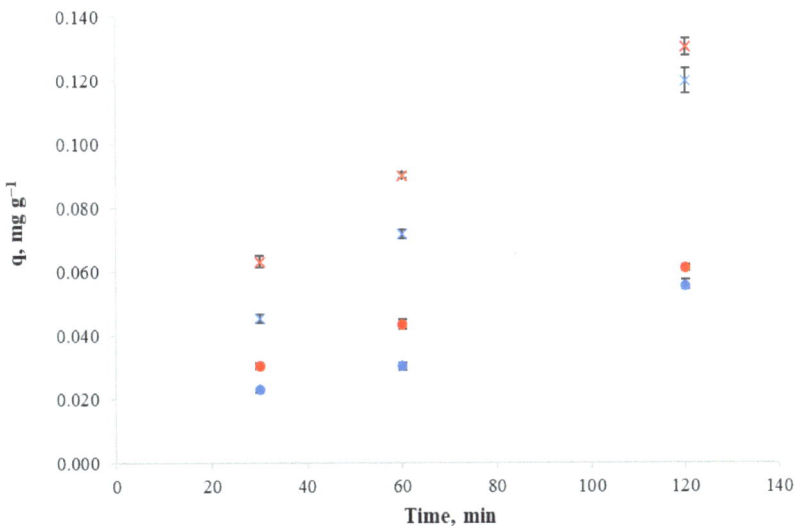

Figure 1. Copper retention capacity in time of different algae and pH. (●): *Lessonia trabeculata* at pH of 3.5–4.0. (●): *Lessonia trabeculata* at pH of 4.5–5.0. (×): *Durvillaea antarctica* at pH of 3.5–4.0. (×): *Durvillaea antarctica* at pH of 4.5–5.0.

In Figure 1, the increase in retention capacity of *L. trabeculata* and *D. antarctica* by time can be observed for both pH intervals. Furthermore, at the highest pH, a higher retention capacity is obtained at every time. Thus, the operation pH for the following experiments is 4.5–5.0.

3.2. Biosorbent Determination

In order to maximize the biosorption process, the highest copper retention has to be achieved in the minimum contact time, when focusing on continuous systems. For this reason, experiments at different times were carried out with *L. trabeculata* and *D. antarctica* as biosorbents, with the objective of choosing the most effective biosorbent.

For these experiments, 100 mg L^{-1} and 30 mg L^{-1} copper solutions were prepared, from which 500 mL was poured into every beaker and 5 g of dried *L. trabeculata* or *D. antárctica*, as it corresponds, was added in each case. After the treatment time was reached, the solution was filtered and the copper content in the liquid phase was measured. Experimental results for the copper retention capacity of both biosorbents at different times are shown in Figure 2.

In Figure 2, it can be noticed that from the beginning, *D. antarctica* has a considerably higher copper retention than *L. trabeculata* for both initial metal concentrations. This can be assumed because at 100 mg L^{-1} of copper initial concentration, the metal mass retention capacity of *D. antarctica* is, on average, 82% higher than that of *L. trabeculata*. For the initial copper concentration of 30 mg L^{-1}, the metal mass retention capacity of *D. antarctica* is, on average, 46% higher than *L. trabeculata*.

This can be explained by the alginate content of the algae, which in *D. antarctica* could be as high as in the range of between 30 and 55% d.wt., and in *L. trabeculata*, it is in the range of 15–21% d.wt. [17,38,39]. Alginates are responsible for the strong affinity that the algae show for heavy metals such as copper [40,41].

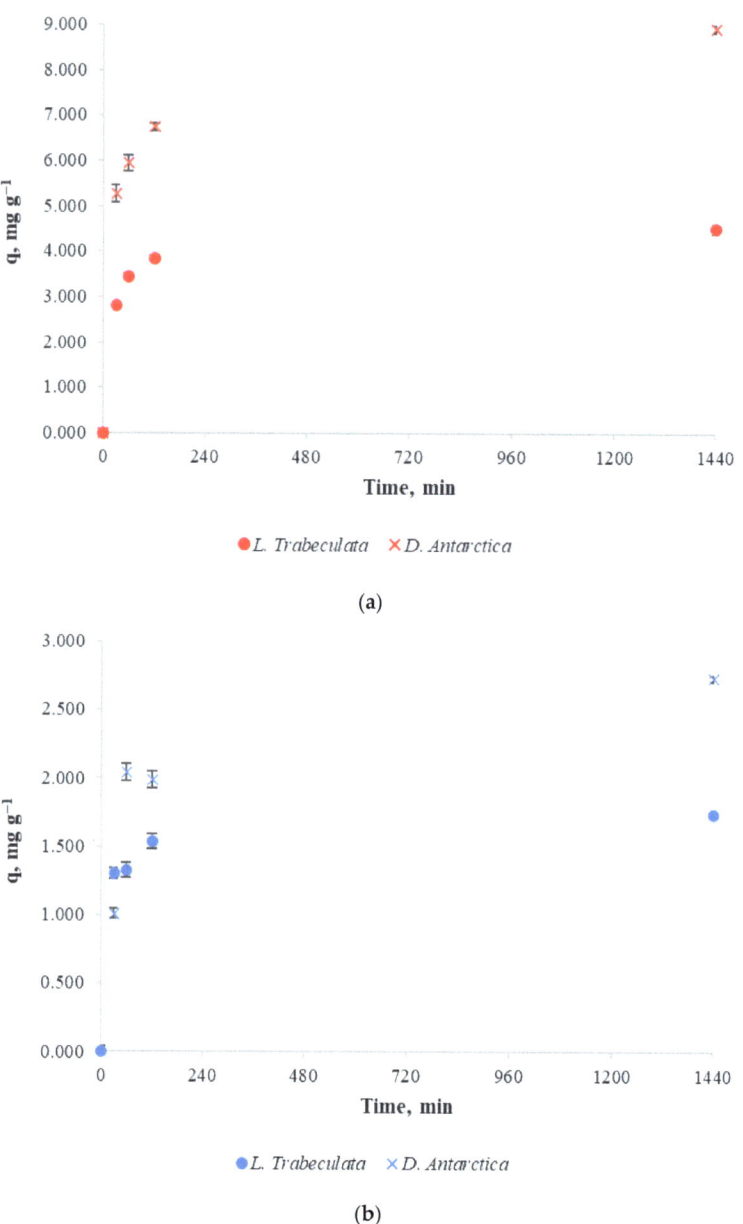

Figure 2. Copper retention capacity of different algae and copper initial concentrations. (**a**): (●): *Lessonia trabeculata* and copper initial concentration of 100 mg L^{-1}. (×): *Durvillaea antarctica* and copper initial concentration of 100 mg L^{-1}. (**b**): (×): *Durvillaea antarctica* and copper initial concentration of 30 mg L^{-1}. (●): *Lessonia trabeculata* and copper initial concentration of 30 mg L^{-1}.

3.3. Biosorbent Particle Size Determination

Experiments for determining the effect of biosorbent particle size in copper removal from the solution were carried out with the aim of obtaining the most copper removed, using *D. Antarctica* as the biosorbent with copper solutions of 10 and 100 mg L^{-1}, respectively, and a contact time of 360 min. Experimental results for copper retention capacity of every particle size for different initial copper concentrations are shown in Table 4.

Table 4. Biosorbent particle size determination experimental runs.

Experimental Run	Particle Size Range mm	Initial Cu Concentration mg L^{-1}	
		10	100
		Retention Capacity mg g^{-1}	
T_4	4.00–5.66	0.371 ± 0.014	2.372 ± 0.031
T_3	3.36–4.00	0.325 ± 0.013	2.531 ± 0.044
T_2	1.70–3.36	0.358 ± 0.013	2.681 ± 0.051
T_1	0.43–1.70	0.312 ± 0.015	2.657 ± 0.055

It can be noticed that for the initial copper concentration of 10 mg L^{-1}, the particle size corresponding to T_4 and T_2 experimental runs presented the best retention capacity. For an initial copper concentration of 100 mg L^{-1}, the best results were in the T_1 and T_2 experimental runs. Thus, a biosorbent particle size of 1.70 to 3.36 mm was selected for further experiments.

3.4. Mass/Volume Ratio Determination

In order to determine the biosorbent mass-to-solution volume ratio, 10 and 100 mg L^{-1} of initial copper concentrations and a contact time of 24 h were used so that the influence of biosorbent mass/volume ratio in retention capacity and metal removal could be observed by using 10, 20 and 40 g of biosorbent per liter of solution.

Experimental results for copper retention capacity and copper removal against different mass/volume ratios for both initial copper concentrations are shown in Figure 3. The decrease in copper retention capacity of the biosorbent as the mass/volume ratio increases can be observed for both initial concentrations. This indicates that for those concentrations, it is not beneficial to increase the biosorbent concentration. Thus, the mass/volume ratio of 10 g L^{-1} achieves high retention capacity.

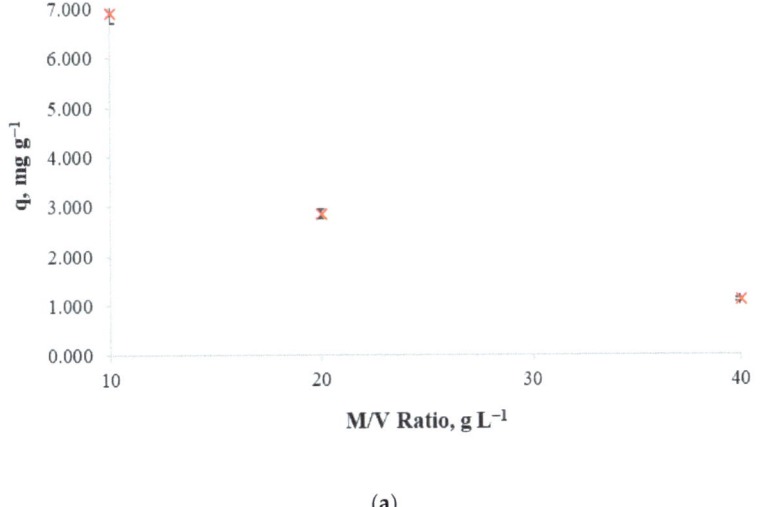

(a)

Figure 3. Cont.

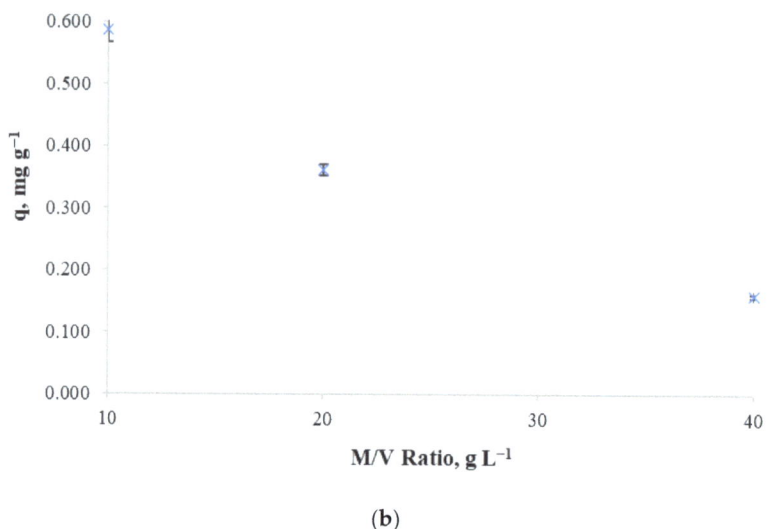

(b)

Figure 3. Copper retention capacity at different mass/volume ratio for (**a**): an initial copper concentration of 100 mg L^{-1}, (**b**) an initial copper concentration of 10 mg L^{-1}.

3.5. Biosorption Kinetics

In order to determine biosorption kinetics, a 500 mL copper solution with an initial concentration of 10 and 100 mg L^{-1} and a pH of 4.5–5.0 was used with 5 g of biosorbent mass. Experiments were based on a change in contact time, which was between 5 and 720 min. After that time, the biosorbent was withdrawn from the solution and the copper concentration in the solution was analyzed.

Experimental results for copper retention capacity against contact time and best fits for the Lagergren and Ho & Mckay models are shown in Figure 4. According to the figure, it can be noticed that the copper retention by the biosorbent increases considerably during the first minutes of contact until equilibrium is achieved at 360 min for both cases. The retention capacity does not increase significantly after that time due to process stabilization.

Experimental data was fitted to both the Lagergren and Ho & Mckay models and the obtained parameters for each model are presented in Table 5. It can be concluded from the determination coefficient R^2 that the Ho & Mckay model fits the experimental data better than the Lagergren model when a mass/volume ratio of 10 g L^{-1} and initial copper concentration of 100 mg L^{-1} is used.

Table 5. Lagergren and Ho & Mckay model parameter values.

Model	Initial Cu Concentration mg L^{-1}		Model	Initial Cu Concentration mg L^{-1}	
Ho & Mckay	10	100	Lagergren	10	100
q_{eq} mg g^{-1}	0.585 ± 0.009	6.513 ± 0.077	q_{eq} mg g^{-1}	0.589 ± 0.005	6.202 ± 0.041
k g mg^{-1} min^{-1}	0.076 ± 0.002	0.035 ± 0.001	k_{ad} min^{-1}	0.024 ± 0.001	0.145 ± 0.002
R^2	90.6%	95.8%	R^2	81.6%	89.9%

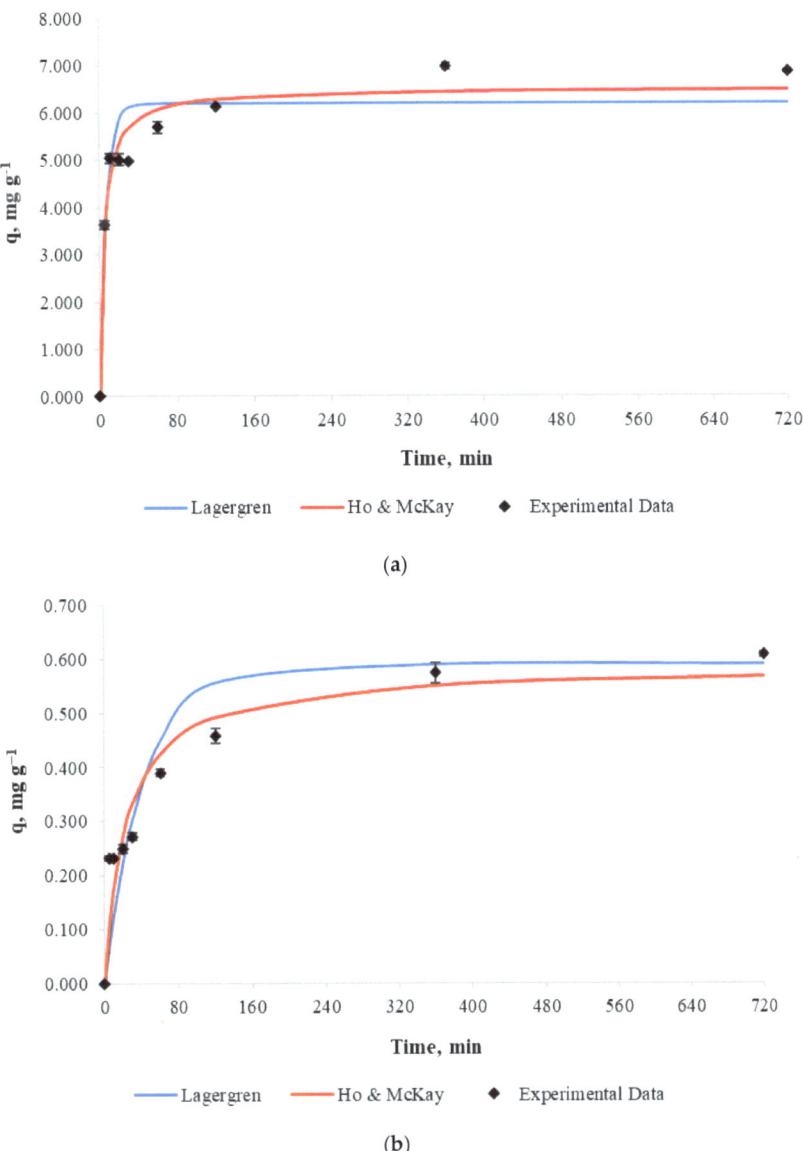

Figure 4. Biosorption kinetics (**a**): Initial copper concentration of 100 mg L^{-1}. (**b**): Initial copper concentration of 10 mg L^{-1}.

3.6. Adsorption Isotherm Determination

To determine the adsorption isotherms, experiments with 10, 25, 50, 75 and 100 mg L^{-1} copper concentration, a mass/volume ratio of 10 g L^{-1} and the same other conditions of the other experimental runs were carried out. Each mathematical model shown in Table 2 was used for adsorption isotherm determination that relates to the amount of copper adsorbed by the algae (retention capacity) and the equilibrium concentration in the solution.

To determine the parameters of the Freundlich, Sips and BET models, the Microsoft Excel SOLVER tool was used for data optimization, which uses the minimum squares error method. On the other hand, the Langmuir model parameters determination was made by using the linearization of the model, but in this case the parameter values were negative, so it can be concluded that the Langmuir model does not represent the process for the equilibrium concentrations used.

Experimental results for copper equilibrium retention capacity against copper equilibrium and the fitted Freundlich, Sips, BET and Langmuir adsorption isotherms are shown in Figure 5. Parameters and representative statistical values of the four models are summarized in Table 6.

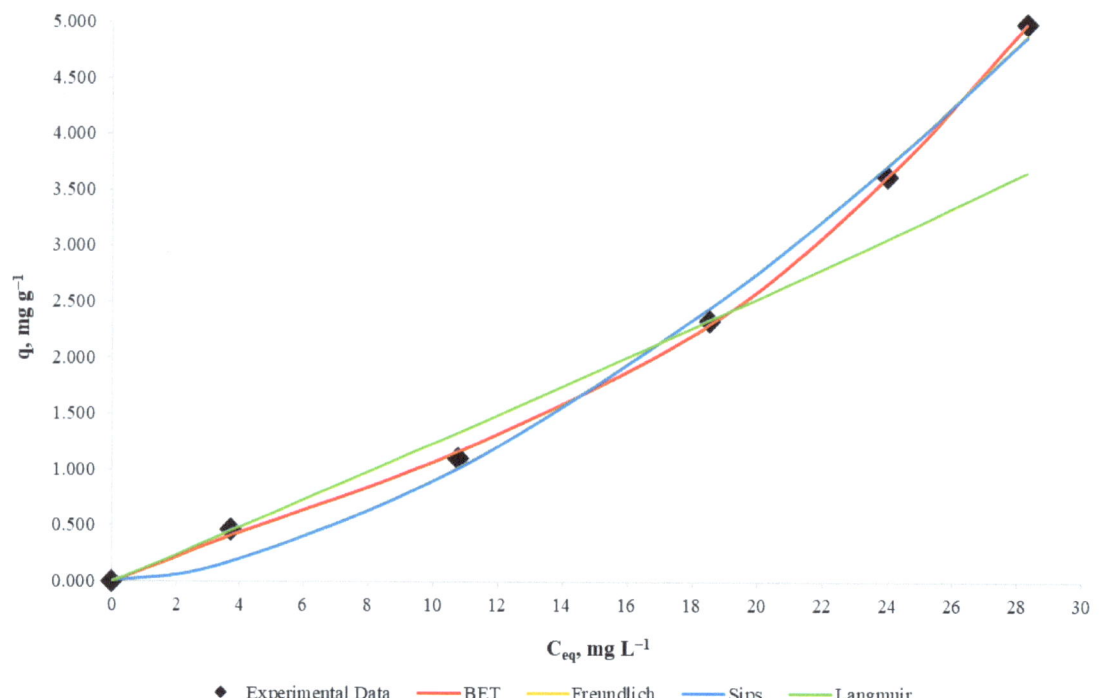

Figure 5. Adsorption isotherms for Cu biosorption with *D. Antarctica*. Cu equilibrium biomass uptake as a function of solution concentration at equilibrium.

Table 6. Isotherm model parameter values.

Model	Parameters (Units in Table 2)		Residuals Sum of Squares	Determination Coefficient R^2
Freundlich	k	0.021	1.247×10^{-1}	99.34%
	n	0.613		
Langmuir	q_m	−47.29	2.115	88.75%
	b	−0.00254		
Sips	q_m	623.4	1.247×10^{-1}	99.33%
	K_S	3.303×10^{-5}		
	n_S	1.638		
BET	q_m	3.955	7.010×10^{-3}	99.96%
	k_1	2.969×10^{-2}		
	k_2	2.821×10^{-2}		
	n	11.08		

From visual analysis of Figure 5, it can be noticed that the Langmuir model fits the data poorly, even when using its obtained parameters. On the other hand, the Freundlich model fits the data as well as the Sips model, and because their differences are not possible to detect, it is deduced that the Sips model is an overparameterization in this case. It can be noticed that the BET model has the best resemblance to the data, even better than the Freundlich model, which supports the idea that for the experimental conditions, the biosorbent is not saturated, so it can be used for more concentrated solutions under these conditions. In terms of model selection, the BET model presents a better determination coefficient than the Freundlich model (99.96% for BET and 99.34% for Freundlich), which could be assumed as negligible, but after performing a Fisher statistical test for model comparison [42], the p-value was below 3%. Therefore, there is statistical evidence that supports the BET model as the best fit for the analyzed experiment, so these algae can form multiple adsorption layers and have a monolayer adsorption capacity of 3.955 mg g^{-1}.

The maximum retention capacity should only be taken as an indicator of whether or not the biosorbent would be useful because the maximum retention capacity is never reached in an actual sorption wastewater treatment plant. Therefore, the retention capacities should be supplemented by the kinetic phenomena of the biosorption in order to estimate a decent residence time in a treatment process [10].

3.7. Regenerating Reagent Selection

This series of experiments was carried out for the selection of the regenerating reagent that removes the highest quantity of copper from the biosorbent. Among the regenerating reagent requirements, there should be (i) a high copper affinity, (ii) maintenance of biosorbent properties after contact, (iii) easy access, and (iv) low cost. Thus, the analyzed regenerating reagents were sulfuric acid and hydrochloric acid.

The results of this experiment aim to enhance the diffusion of copper from biosorbent to regenerating reagent because of its affinity to copper; therefore, the biosorbent holds a low copper concentration and is able to be reused for copper biosorption.

For assessing copper affinity with regenerating reagent, a copper sulfate solution was prepared, with 100 mg L^{-1} of copper as in the previous experiments, from which 500 mL was poured into every beaker and was in contact with 5 g of dried *D. antarctica* for 24 h. Then the biosorbent was rinsed twice with distilled water and 0.1 mol L^{-1} (pH 1) sulfuric acid or 0.1 mol L^{-1} (pH 1) hydrochloric was added for a determined time. After that time, the biosorbent was withdrawn from the solution and the amount of desorbed copper was determined by the difference of adsorbed copper mass from the first solution and desorbed mass from the second solution. Experimental results for copper desorption when applying H$_2$SO$_4$ and HCl with different contact times are shown in Figure 6.

It can be observed that for both regenerating reagents, copper is re-adsorbed in the biosorbent as contact time increases. In case of hydrochloric acid, desorption decreases drastically at 20 min, because copper is re-adsorbed rapidly, but in the case of sulfuric acid, the copper re-adsorption is lower and slower than hydrochloric acid desorption.

Because sulfuric acid (a) shows a more stable performance with copper as sorbate and *D. antarctica* as biosorbent than hydrochloric acid, (b) is widely available and (c) has a higher purity than hydrochloric acid, it is recommended as a regenerating reagent.

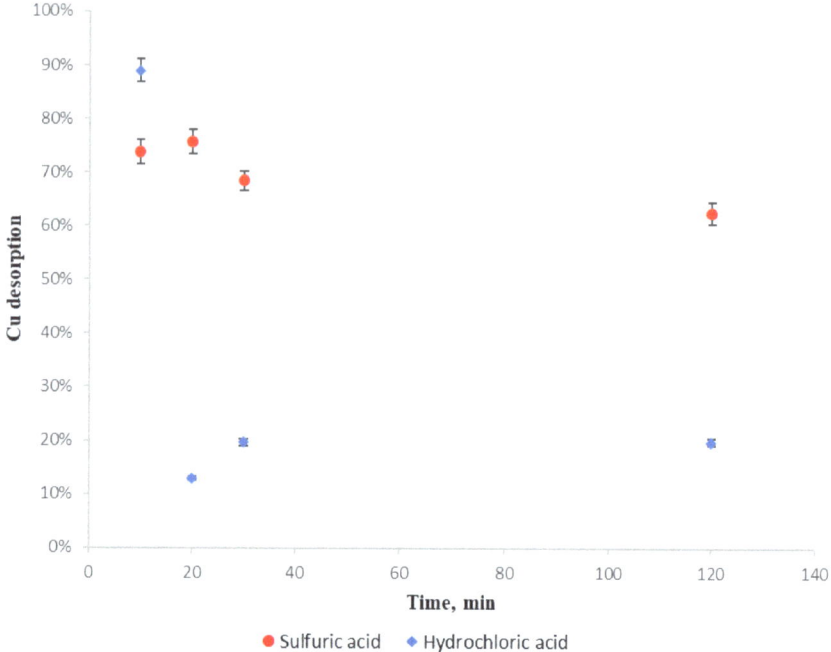

Figure 6. Regenerating reagents desorption capacity. (●): H_2SO_4. (♦): HCl.

4. Conclusions

From the results, it can be concluded that the brown seaweeds *Lessonia trabeculata* and *Durvillaea antarctica* have an important copper retention capacity in acidic solutions, where the capacity of *D. antarctica* is twice as high as that of *L. trabeculata*. Thus, *D. antarctica* is more recommended for continuous systems. Heavy metal removal efficiency varies with pH. From the analysis of two intervals, a pH between 4.5 and 5.0 gives better results in copper removal with the studied biosorbents.

For process kinetic parameters determination applying the Lagergren and Ho & McKay models, it is concluded that the Ho & McKay model fits the experimental data better. Concerning adsorption isotherms, the BET model shows the best fit, indicating that the biosorbent is not saturated. The chosen regenerating reagent is sulfuric acid as it presents higher copper removal values and shows no sign of metal re-adsorption before 30 min, whereas hydrochloric acid shows copper re-adsorption after 10 min.

Author Contributions: Conceptualization, H.K.H., C.G. (Claudia Gutiérrez), N.V. and C.G. (Claudia Gotschlich); methodology, N.V. and C.G. (Claudia Gotschlich); validation, H.K.H., A.L. and P.L.; formal analysis, H.K.H., C.G. (Claudia Gutiérrez) and R.O.-S.; investigation, N.V. and C.G. (Claudia Gotschlich); writing—original draft preparation, H.K.H., C.G. (Claudia Gutiérrez) and R.O.-S.; writing—review and editing, H.K.H., C.G. (Claudia Gutiérrez) and R.O.-S.; visualization, H.K.H., C.G. (Claudia Gutiérrez), A.L., P.L. and R.O.-S.; supervision, H.K.H. All authors have read and agreed to the published version of the manuscript.

Funding: This research was funded by the Chilean National Research Foundation (ANID) Fondecyt de Iniciación en Investigación, grant N° 11200189 and Fondecyt Regular, grant N° 1220712. The APC was funded by Metals.

Institutional Review Board Statement: Not applicable.

Informed Consent Statement: Not applicable.

Data Availability Statement: The data presented in this study are available on request from the corresponding author.

Conflicts of Interest: The authors declare no conflict of interest.

References

1. Newbold, J. Chile's environmental momentum: ISO 14001 and the large-scale mining industry—Case studies from the state and private sector. *J. Clean. Prod.* **2006**, *14*, 248–261. [CrossRef]
2. Hansen, H.K.; Lamas, V.; Gutierrez, C.; Nuñez, P.; Rojo, A.; Cameselle, C.; Ottosen, L.M. Electro-remediation of copper mine tailings. Comparing copper removal efficiencies for two tailings of different age. *Miner. Eng.* **2013**, *41*, 1–8. [CrossRef]
3. Gutiérrez, C.; Hansen, H.K.; Nuñez, P.; Jensen, P.E.; Ottosen, L.M. Electrochemical peroxidation as a tool to remove arsenic and copper from smelter wastewater. *J. Appl. Electrochem.* **2010**, *40*, 1031–1038. [CrossRef]
4. Hansen, H.K.; Arancibia, F.; Gutiérrez, C. Adsorption of copper onto agriculture waste materials. *J. Hazard. Mater.* **2010**, *180*, 442–448. [CrossRef] [PubMed]
5. Naja, G.M.; Murphy, V.; Volesky, B. Biosorption, metals. In *Encyclopedia of Industrial Biotechnology*; Wiley: Hoboken, NJ, USA, 2010; pp. 1–29.
6. Gutiérrez, C.; Hansen, H.K.; Hernández, P.; Pinilla, C. Biosorption of cadmium with brown macroalgae. *Chemosphere* **2015**, *138*, 164–169. [CrossRef]
7. Hansen, H.K.; Gutiérrez, C.; Madrid, A.; Jimenez, R.; Larach, H. Possible Use of the Algae *Lessonia nigrescens* as a Biosorbent: Differences in Copper Sorption Behavior Using Either Blades or Stipes. *Waste Biomass Valorization* **2017**, *8*, 1295–1302. [CrossRef]
8. Ramesh, B.; Saravanan, A.; Kumar, P.S.; Yaashikaa, P.R.; Thamarai, P.; Shaji, A.; Rangasamy, G. A review on algae biosorption for the removal of hazardous pollutants from wastewater: Limiting factors, prospects and recommendations. *Environ. Pollut.* **2023**, *327*, 121572. [CrossRef] [PubMed]
9. Thirunavukkarasu, A.; Nithya, R.; Sivashankar, R. Continuous fixed-bed biosorption process: A review. *Chem. Eng. J. Adv.* **2021**, *8*, 100188. [CrossRef]
10. Elgarahy, E.; Elwakeel, K.Z.; Mohammad, S.H.; Elshoubaky, G.A. A critical review of biosorption of dyes, heavy metals and metalloids from wastewater as an efficient and green process. *Clean. Eng. Technol.* **2021**, *4*, 100209. [CrossRef]
11. Franco, D.S.P.; Georgin, J.; Netto, M.S.; Fagundez, J.L.S.; Salau, N.P.G.; Allasia, D.; Dotto, G.L. Conversion of the forest species *Inga marginata* and *Tipuana tipu* wastes into biosorbents: Dye biosorption study from isotherm to mass transfer. *Environ. Technol. Innov.* **2021**, *22*, 101521. [CrossRef]
12. Znad, H.; Awual, M.R.; Martini, S. The Utilization of Algae and Seaweed Biomass for Bioremediation of Heavy Metal-Contaminated Wastewater. *Molecules* **2022**, *27*, 1275. [CrossRef] [PubMed]
13. Foday, E.H., Jr.; Bo, B.; Xu, X. Removal of toxic heavy metals from contaminated aqueous solutions using seaweeds: A review. *Sustainability* **2021**, *13*, 12311. [CrossRef]
14. El-Said, G.F.; El-Sikaily, A. Chemical composition of some seaweed from Mediterranean Sea coast, Egypt. *Environ. Monit. Assess.* **2013**, *185*, 6089–6099. [CrossRef]
15. Gao, X.; Guo, C.; Hao, J.; Zhao, Z.; Long, H.; Li, M. Adsorption of heavy metal ions by sodium alginate based adsorbent-a review and new perspectives. *Int. J. Biol. Macromol.* **2020**, *164*, 4423–4434. [CrossRef] [PubMed]
16. Caballero, E.; Flores, A.; Olivares, A. Sustainable exploitation of macroalgae species from Chilean coast: Characterization and food applications. *Algal Res.* **2021**, *57*, 102349. [CrossRef]
17. Kelly, B.; Brown, M. Variations in the alginate content and composition of *Durvillaea antarctica* and *D. Willana* from southern New Zealand. *J. Appl. Phycol.* **2000**, *12*, 317–324. [CrossRef]
18. Véliz, K.; Toledo, P.; Araya, M.; Gómez, M.F.; Villalobos, V.; Tala, F. Chemical composition and heavy metal content of Chilean seaweeds: Potential applications of seaweed meal as food and feed ingredients. *Food Chem.* **2023**, *398*, 133866. [CrossRef]
19. Dold, B. Evolution of Acid Mine Drainage Formation in Sulphidic Mine Tailings. *Minerals* **2014**, *4*, 621–641. [CrossRef]
20. Cid, H.; Ortiz, C.; Pizarro, J.; Barros, D.; Castillo, X.; Giraldo, L.; Moreno-Pirajan, J.C. Characterization of copper (II) biosorption by brown algae *Durvillaea antarctica* dead biomass. *Adsorption* **2015**, *21*, 645–658. [CrossRef]
21. Cid, H.A.; Flores, M.I.; Pizarro, J.F.; Castillo, X.A.; Barros, D.E.; Moreno-Pirajan, J.C.; Ortiz, C.A. Mechanisms of Cu^{2+} biosorption on Lessonia nigrescens dead biomass: Functional groups interactions and morphological characterization. *J. Environ. Chem. Eng.* **2018**, *6*, 2696–2704. [CrossRef]
22. Deniz, F.; Ersanli, E.T. An ecofriendly approach for bioremediation of contaminated water environment: Potential contribution of a coastal seaweed community to environmental improvement. *Int. J. Phytoremediation* **2018**, *20*, 256–263. [CrossRef] [PubMed]
23. Patel, G.G.; Dosh, H.V.; Thakur, M.C. Biosorption and Equilibrium Study of Copper by Marine Seaweeds from North West Coast of India. *J. Environ. Sci. Toxicol. Food Technol.* **2016**, *10*, 54–64.
24. Kleinübing, S.J.; da Silva, E.A.; da Silva, M.G.C.; Guibal, E. Equilibrium of Cu(II) and Ni(II) biosorption by marine alga Sargassum filipendula in a dynamic system: Competitiveness and selectivity. *Bioresour. Technol.* **2011**, *102*, 4610–4617. [CrossRef] [PubMed]
25. Patrón-Prado, M.; Acosta-Vargas, B.; Serviere-Zaragoza, E.; Méndez-Rodríguez, L.C. Copper and Cadmium Biosorption by Dried Seaweed *Sargassum sinicola* in Saline Wastewater. *Water Air Soil Pollut.* **2010**, *210*, 197–202. [CrossRef]

26. Ahmady-Asbchin, S.; Andrès, Y.; Gérente, C.; Le Cloirec, P. Biosorption of Cu(II) from aqueous solution by Fucus serratus: Surface characterization and sorption mechanisms. *Bioresour. Technol.* **2008**, *99*, 6150–6155. [CrossRef]
27. Mata, Y.N.; Blázquez, M.L.; Ballester, A.; González, F.; Muñoz, J.A. Characterization of the biosorption of cadmium, lead and copper with the brown alga *Fucus vesiculosus*. *J. Hazard. Mater.* **2008**, *158*, 316–323. [CrossRef]
28. Karthikeyan, S.; Balasubramanian, R.; Iyer, C.S.P. Evaluation of the marine algae *Ulva fasciata* and *Sargassum* sp. for the biosorption of Cu(II) from aqueous solutions. *Bioresour. Technol.* **2007**, *98*, 452–455. [CrossRef]
29. Perumal, S.V.; Joshi, U.M.; Karthikeyan, S.; Balasubramanian, R. Biosorption of lead (II) and copper (II) from stormwater by brown seaweed *Sargassum* sp.: Batch and column studies. *Water Sci. Technol.* **2007**, *56*, 277–285. [CrossRef]
30. Romera, E.; González, F.; Ballester, A.; Blázquez, M.L.; Muñoz, J.A. Comparative study of biosorption of heavy metals using different types of algae. *Bioresour. Technol.* **2007**, *98*, 3344–3353. [CrossRef]
31. Sheng, P.X.; Ting, Y.P.; Chen, J.P.; Hong, L. Sorption of lead, copper, cadmium, zinc, and nickel by marine algal biomass: Characterization of biosorptive capacity and investigation of mechanisms. *J. Colloid Interface Sci.* **2004**, *275*, 131–141. [CrossRef]
32. Davis, T.A.; Volesky, B.; Vieira, R.H.S.F. Sargassum seaweed as biosorbent for heavy metals. *Water Res.* **2000**, *34*, 4270–4278. [CrossRef]
33. Chen, J.P.; Hong, L.; Wu, S.; Wang, L. Elucidation of interactions between metal ions and Ca alginate-based ion-exchange resin by spectroscopic analysis and modeling simulation. *Langmuir* **2002**, *18*, 9413–9421. [CrossRef]
34. Figueira, M.M.; Volesky, B.; Mathieu, H.J. Instrumental analysis study of iron species biosorption by *Sargassum* biomass. *Environ. Sci. Technol.* **1999**, *33*, 1840–1846. [CrossRef]
35. Matheickal, J.T.; Yu, Q. Biosorption of lead(II) and copper(II) from aqueous solutions by pre-treated biomass of Australian marine algae. *Bioresour. Technol.* **1999**, *69*, 223–229. [CrossRef]
36. Murphy, V.; Hughes, H.; McLoughlin, P. Cu(II) binding by dried biomass of red, green and brown macroalgae. *Water Res.* **2007**, *41*, 731–740. [CrossRef] [PubMed]
37. Kratochvil, D.; Volesky, B. Advances in the biosorption of heavy metals. *Trends Biotechnol.* **1998**, *16*, 291–300. [CrossRef]
38. Venegas, M.; Matshuiro, B.; Edding, M.E. Alginate composition of *Lessonia trabeculata* (Phaeophyta:Laminariales) growing in exposed and sheltered hábitats. *Bot. Mar.* **1993**, *36*, 47–51. [CrossRef]
39. Miller, I. Alginate composition of some New Zeland brown seaweeds. *Phytochemistry* **1993**, *41*, 1315–1317. [CrossRef]
40. Davis, T.; Llanes, F.; Volesky, B.; Mucci, A. Metal selectivity of Sargassum spp. and their alginates in relation to their α-L-guluronic acid content and conformation. *Environ. Sci. Technol.* **2003**, *37*, 261–267. [CrossRef]
41. Kleinübing, S.J.; Vieira, R.S.; Beppu, M.M.; Guibal, E.; Da Silva, M.G.C. Characterization and evaluation of copper and nickel biosorption on acidic algae *Sargassum filipendula*. *Mater. Res.* **2010**, *13*, 541–550. [CrossRef]
42. Motulsky, H.J.; Ransnas, L.A. Fitting curves to data using nonlinear regression: A practical and nonmathematical review. *FASEB J.* **1987**, *1*, 365–374. [CrossRef] [PubMed]

Disclaimer/Publisher's Note: The statements, opinions and data contained in all publications are solely those of the individual author(s) and contributor(s) and not of MDPI and/or the editor(s). MDPI and/or the editor(s) disclaim responsibility for any injury to people or property resulting from any ideas, methods, instructions or products referred to in the content.

Article

Raw Eggshell as an Adsorbent for Copper Ions Biosorption—Equilibrium, Kinetic, Thermodynamic and Process Optimization Studies

Miljan Marković *, Milan Gorgievski, Nada Štrbac, Vesna Grekulović, Kristina Božinović, Milica Zdravković and Milovan Vuković

Technical Faculty in Bor, University of Belgrade, Vojske Jugoslavije 12, 19210 Bor, Serbia
* Correspondence: mmarkovic@tfbor.bg.ac.rs; Tel.: +381-62-826-4336

Abstract: The study on the biosorption of copper ions using raw eggshells as an adsorbent is presented in this paper. The influence of different process parameters, such as: initial pH value of the solution, initial Cu^{2+} ions concentration, initial mass of the adsorbent, and stirring rate, on the biosorption capacity was evaluated. The SEM-EDS analysis was performed before and after the biosorption process. SEM micrographs indicate a change in the morphology of the sample after the biosorption process. The obtained EDS spectra indicated that K, Ca, and Mg were possibly exchanged with Cu^{2+} ions during the biosorption process. The equilibrium analysis showed that the Langmuir isotherm model best describes the experimental data. Four kinetic models were used to analyze the experimental data, and the results revealed that the pseudo-first order kinetic model is the best fit for the analyzed data. Calculated thermodynamic data indicated that the biosorption process is spontaneous, and that copper ions are possibly bound to the surface of the eggshells by chemisorption. The biosorption process was optimized using Response Surface Methodology (RSM) based on the Box-Behnken Design (BBD), with the selected factors: adsorbent mass, initial metal ion concentration, and contact time.

Keywords: biosorption; response surface methodology; Box-Behnken design; eggshell; equilibrium; kinetics; thermodynamics; copper ions

Citation: Marković, M.; Gorgievski, M.; Štrbac, N.; Grekulović, V.; Božinović, K.; Zdravković, M.; Vuković, M. Raw Eggshell as an Adsorbent for Copper Ions Biosorption—Equilibrium, Kinetic, Thermodynamic and Process Optimization Studies. *Metals* 2023, 13, 206. https://doi.org/10.3390/met13020206

Academic Editors: Jean François Blais and Petros E. Tsakiridis

Received: 22 December 2022
Revised: 12 January 2023
Accepted: 18 January 2023
Published: 20 January 2023

Copyright: © 2023 by the authors. Licensee MDPI, Basel, Switzerland. This article is an open access article distributed under the terms and conditions of the Creative Commons Attribution (CC BY) license (https:// creativecommons.org/licenses/by/ 4.0/).

1. Introduction

Wastewater containing heavy metals, that originate from tanneries, batteries, mining and metallurgical operations, chemical manufactories, pesticides, and other sources, has been a major pollutant in the environment for many years. The non-biodegradability and persistent nature of these metals means they tend to enter the food chain and accumulate in the living organisms, causing numerous disorders and diseases [1,2].

Wastewater treatment methods can be classified into five groups, i.e., adsorption-, chemical-, membrane-, electric-, and photocatalytic- based treatments [3].

The adsorption-based separation methods are defined by the properties of the adsorbent, and the working conditions of the process, like temperature, pH value of the solution, adsorption time, etc. The adsorbents can be classified as carbon-based adsorbents, chitosan-based adsorbents, mineral adsorbents, magnetic adsorbents, and biosorbents [3].

Membrane-based filtration and separation is a wastewater treatment method that usually includes ultrafiltration, nanofiltration, microfiltration, reverse osmosis, forward osmosis, and electrodialysis [3].

Chemical-based separation methods for wastewater treatment polluted with heavy metals include precipitation, coagulation and flocculation, and flotation. These methods change the form of the dissolved metal into solid particles, to facilitate their sedimentation [3].

Electric-based separation methods for wastewater treatment include electrochemical reduction, electroflotation, electrooxidation and ion-exchange treatment [3].

Heavy metals are being removed from wastewater on the industrial scale by well-known conventional technologies. However, these conventional technologies have many disadvantages that include high operating costs, incomplete metal removal, continuous input of chemicals, and others. These disadvantages raise the question of finding a new method of wastewater treatment that could become an alternative to the existing conventional technologies, and improve the overall process [1,4].

Adsorption methods are a more suitable processes for wastewater treatment, due to the high metal recovery rate, no sludge production, low economic investments, the ability to regenerate the adsorbent, and many others [1].

In recent years, the scientific community has recognized biosorption as a potential, efficient and economically feasible alternative to conventional technologies for the removal of heavy metal ions from aqueous solutions. The scientific research is focused on examining the possibility of using many industrial and agricultural waste materials as biosorbents [5].

Since inactive biomass is usually used in biosorption processes, the mechanism of metal ions removal is based on adsorption, chelation, ion exchange, complexation, coordination, microprecipitation, electrostatic interaction, or the combination of the beforementioned mechanisms [6].

Many industries that produce and use eggs generate considerable amounts of waste in the form of eggshells. These by-products constitute approximately 6 g/egg, an amount which represents significant waste. Waste eggshells are considered useless and are disposed in landfills without any pre-treatment [7].

The aim of this work is to study the possibility of using waste raw eggshells as an adsorbent for copper ions removal from aqueous solutions, as well as to analyze the specifics of the process and the influence of certain parameters on its efficiency. The use of eggshells as an adsorbent for wastewater treatment could potentially solve two problems. First, it would reduce the amount of waste in landfills, thus directly help industries based on the use of eggs by reducing the costs of their disposal. And, secondly, it would contribute to solving the problem of watercourses contamination with heavy metals (in this case, copper).

The performed analysis in this work include:
- the influence of different process parameters (initial Cu^{2+} ions concentration, pH value of the solution, adsorbent mass and stirring rate) on the biosorption capacity;
- SEM-EDS analysis of the eggshells sample before and after the biosorption process;
- kinetic analysis of the biosorption process;
- equilibrium analysis of the biosorption process;
- thermodynamic analysis of the biosorption process;
- process optimization study by the mean of Response Surface Methodology based on Box-Behnken Design

2. Materials and Methods

Raw chicken eggshells (Figure 1), collected from local households (located in the city of Bor, in eastern Serbia), were washed with distilled water several times, ground, sieved, and the fraction ($-1 + 0.4$) mm was used for the biosorption experiments.

The eggshells samples were rinsed with 200 mL distilled water, prior to the biosorption experiments, in order to remove the physical impurities.

Biosorption experiments were conducted in batch conditions, using synthetic Cu^{2+} solutions, prepared with $CuSO_4 \cdot 5H_2O$ (p.a.). The concentrations of the solutions varied, based on the specifics of the performed experiment.

pH value of the solutions was adjusted using 0.1 M HNO_3 and 0.1 M KOH.

Figure 1. Chicken eggshells sample.

Process parameters, including contact time, initial copper ions concentration, temperature, stirring rate, initial mass of the adsorbent, and initial pH value were adjusted depending on the performed experiment.

All experiments were performed in batch conditions. A spectrophotometer (Spectroquant Pharo 300—Merck, Rahway, NJ, USA) was used to analyze the solutions for the remaining copper ions content. The SEM-EDS analysis was performed on a SEM scanning electron microscope (VEGA 3 LMU, Tescan, Brno, Czech Republic) with an integrated energy-dispersive X-ray detector (X act SDD 10 mm^2, Oxford Instruments, Abingdon, UK).

The biosorption capacity and the adsorption degree were calculated using the following equations:

$$q_t = \frac{c_i - c_t}{m} \cdot V \quad (1)$$

$$AD\% = \left(1 - \frac{c_t}{c_i}\right) \cdot 100 \quad (2)$$

where: q_t is the adsorbent capacity defined as mass of the adsorbed metal per unit mass of the adsorbent (mg g^{-1}) at time t; c_i is the initial metal ion concentration in the solution; c_t is the metal ion concentration in the solution at time t; m is the adsorbent mass; V is the volume of the solution; $AD\%$ is the adsorption degree.

3. Results and Discussions

3.1. The Influence of Different Process Parameters on the Adsorption Efficiency (Biosorption Capacity)

3.1.1. The Effect of ph Value on the Biosorption Capacity

In order to analyze the effect of the pH value on the biosorption capacity, a series of experiments was performed, using Cu^{2+} ion solutions, of different pH values, ranging from 2 to 5. The pH value of the solutions was adjusted by adding 0.1 M HNO$_3$ and 0.1 M KOH. The experiments were performed at room temperature, using solutions of initial Cu^{2+} concentration of 500 mg dm^{-3}. The suspension was stirred for 60 min. The obtained results are shown on Figure 2a.

As can be seen from Figure 2a, low pH values of the solution led to a low biosorption capacity. A rise in the pH value of the solution led to a rise in the biosorption capacity. At pH = 2 the biosorption capacity was determined to be around 10.82 mg g^{-1}, while at pH = 5 the biosorption capacity was almost twice as higher (q_t = 21.62 mg g^{-1}). The rise in the biosorption capacity of eggshells at higher pH values of the solution occurs due to the fact that chicken eggshell constitutes of about 95% of calcium carbonate and 5% of organic matter. The calcium carbonate favors precipitation of metal ions, as it dissociates to carbonate and calcium ions. The solubility of calcium carbonate in the eggshells varies, based on the pH level of the solution. Carbonate species appear in solutions as H_2CO_3, HCO_3^- and CO_3^{2-}. The latter two are presumably responsible for the formation of metal carbonates [5]. Considering the divalent nature of the Cu^{2+} ions in the solution at pH = 5 (Figure 3), it is assumed that the carbonate ions from the eggshell interact with the copper ions to form copper carbonates.

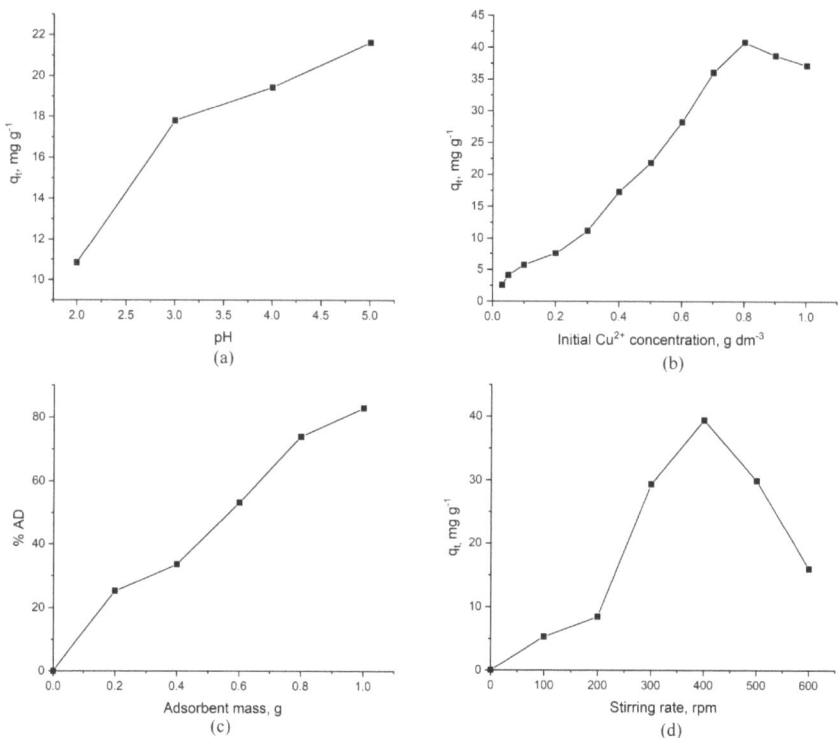

Figure 2. (a) pH value effect on the biosorption capacity, (b) initial Cu^{2+} concentration influence on the biosorption capacity, (c) adsorbent mass effect on the biosorption capacity, and (d) stirring rate influence on the biosorption capacity for copper ions biosorption onto chicken eggshells.

3.1.2. The Influence of the Initial Cu^{2+} Concentration on the Biosorption Capacity

The influence of the initial Cu^{2+} concentration on the biosorption capacity was examined by bringing into contact 1 g of chicken eggshell with 0.5 dm^{-3} copper ion solutions of different initial concentrations, ranging from 30 mg dm^{-3} to 1000 mg dm^{-3}, on a magnetic stirrer, for 60 min. The experiments were performed at room temperature. The obtained results are shown on Figure 2b. Figure 2b shows an increase in the biosorption capacity, with the rise in the initial Cu^{2+} ions concentration, up to 800 mg dm^{-3}, where it reaches the maximum value (q_t = 40.79 mg g^{-1}). With a further increase in the initial copper ions concentration, a decrease in the biosorption capacity is noted. It is assumed that this decrease occurs due to the saturation of the adsorbent with Cu^{2+}.

The fact that the adsorption process includes different simultaneous processes, among which are the diffusion in the liquid phase and adsorption in the solid phase, it is assumed that the increase in the initial metal ions concentration leads to an increase in the probability of their contact with the active sites in the structure of the adsorbent. The saturation of the active sites in the adsorbent structure leads to the decrease in the biosorption capacity, with the further increase in the initial metal ions concentration [7].

3.1.3. The Effect of the Adsorbent Mass on the Biosorption Capacity

The effect of the adsorbent mass on the biosorption capacity was determined by bringing into contact 0.5 dm^{-3} copper ion solutions (initial concentration 500 mg dm^{-3})

with different amounts of eggshells, ranging from 0.2 to 1 g. The suspension was stirred on a magnetic stirrer, under room temperature, for 60 min. The results of the performed analysis are shown on Figure 2c. As seen on Figure 2c, the adsorption degree increased from 25% to 82% with the increase in the adsorbent mass from 0.2 to 1 g, due to the higher number of available active sites on the adsorbent structure as a result of a larger amount of adsorbent available [8].

3.1.4. The Influence of the Stirring Rate on the Biosorption Capacity

The influence of the stirring rate on the biosorption capacity was analyzed by performing the following experiment 0.5 g of eggshells was brought into contact with 0.5 dm^{-3} copper ions solutions, and stirred at room temperature using different stirring rates from 100 to 600 rpm, for 60 min. The obtained results are shown on Figure 2d. The results show that the biosorption capacity increased with the increase in the stirring rate, up to 400 rpm, where it reached its maximum value. Further increase in the stirring rate resulted in a decrease of the biosorption capacity.

It is assumed that the increase in the stirring rate accelerates the diffusion of the metal ions through the liquid phase to the surface of the adsorbent, resulting in the rise of the biosorption capacity [9].

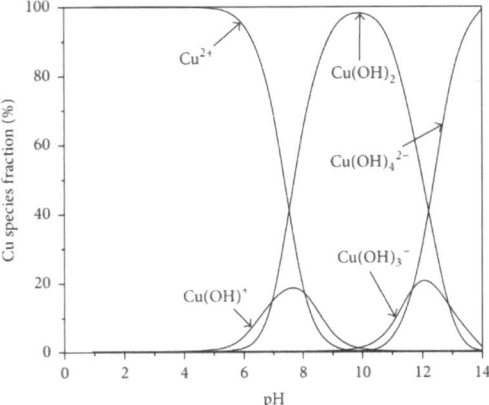

Figure 3. Cu species distribution at different pH values of the solution [10].

3.2. SEM-EDS Analysis

The SEM-EDS analysis of the eggshells was performed before and after the biosorption of copper ions in order to study the surface morphology and texture of the samples. The obtained results are shown on Figure 4.

Figure 4a shows a porous and dense surface structure of the untreated raw eggshell. Figure 4c shows a slight change to the surface morphology, with the surface becoming uneven, rough and heterogeneous, as a result of the incorporation of copper ions inside the structure of the eggshells sample. The interaction of eggshells with Cu^{2+} ions lead to the formation of flake-like deposits on the surface of the adsorbent [11,12].

The EDS analysis was performed by scanning multiple points on the surface of the untreated eggshell as well as the eggshell sample after the adsorption process. The EDS spectrum of the untreated eggshell (Figure 4b) showed peaks for O, Mg, K and Ca, with high O and Ca contents. The obtained spectrum after the adsorption process (Figure 4d) indicates the absence of the Mg peak, while the K and Ca peaks remained but were reduced. A new peak, corresponding to the adsorbed Cu ions appeared. Obtained EDS results indicate that Mg, K and Ca could potentially be exchanged with Cu during the adsorption process.

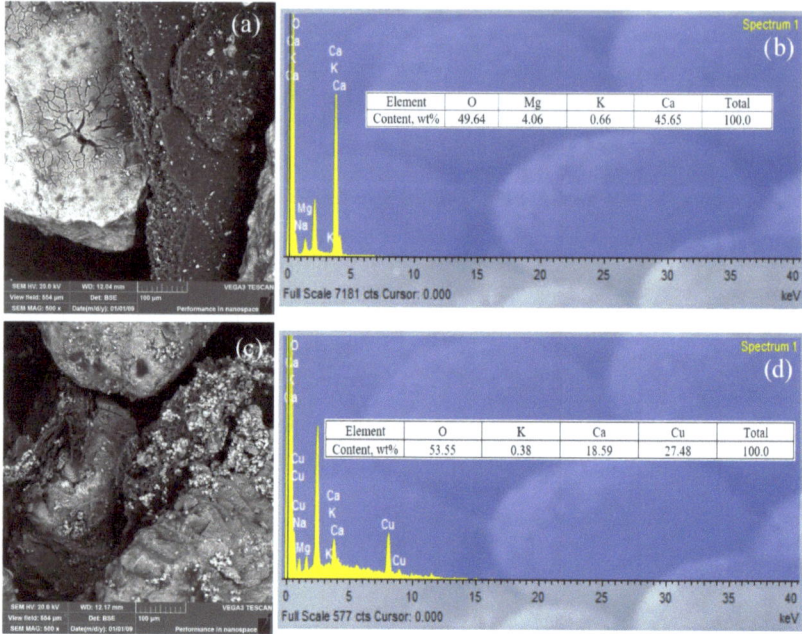

Figure 4. SEM micrographs of eggshells samples before Cu^{2+} biosorption (**a**) and after the biosorption process (**c**), with the corresponding EDS spectra before (**b**) and after the biosorption process (**d**).

3.3. Kinetic Study

Adsorption kinetic data provides insight into the mechanism of the adsorption process, it's rate, as well as information about the step that determines the overall rate of the process. In this work, the experimental data were modeled using the non-linear forms of the pseudo-first order kinetic model, pseudo-second order kinetic model, intraparticle diffusion kinetic model (Weber-Morris model), and the Elovich kinetic model.

In order to obtain the biosorption kinetic data, 50 mL of copper ion solutions (initial Cu^{2+} concentration 500 mg dm^{-3}) were brought into contact with 1 g of eggshells samples, for different process time (ranging from 1 to 90 min). The kinetic analysis is presented in Figure 4 along with the obtained kinetic data which are presented in Table 1.

3.3.1. Pseudo-First Order Kinetic Model

The pseudo-first order kinetic model is often used to describe the kinetics of a sorption process. According to this model, a type of sorbent reacts with one active center in the adsorbent structure, forming a sorption complex [13].

The non-linear form of the pseudo-first order kinetic model can be expressed as:

$$q_t = q_e \left(1 - e^{-k_1 t}\right) \qquad (3)$$

where: q_t is the adsorbent capacity defined as mass of the adsorbed metal per unit mass of the adsorbent (mg g^{-1}) at time t; q_e is the adsorbent capacity defined as mass of the adsorbed metal per unit mass of the adsorbent (mg g^{-1}) at equilibrium; k_1 is the adsorption rate constant for the pseudo-first order kinetic model (min^{-1}).

The experimental data were fitted using this model (Figure 5), and the kinetic parameters were determined and presented in Table 1.

Table 1. Kinetic model parameters for copper ions biosorption onto eggshells.

Model	Parameters	Values
Pseudo-first order kinetic model	k_1 (min^{-1}) $q_{e,exp}$ (mg g^{-1}) $q_{e,cal}$ (mg g^{-1}) R^2	0.018 22.84 28.34 0.999
Pseudo-second order kinetic model	k_2 (g mg^{-1} min^{-1}) $q_{e,exp}$ (mg g^{-1}) $q_{e,cal}$ (mg g^{-1}) R^2	$2.90 \cdot 10^{-4}$ 22.84 43.26 0.982
Intraparticle diffusion kinetic model (Weber-Morris model)	K_i (mg g^{-1} min$^{-0.5}$) C_i R^2	2.311 $9.99 \cdot 10^{-24}$ 0.955
Elovich kinetic model	α (mg g^{-1} min^{-1}) β (g mg^{-1}) R^2	0.603 0.067 0.983

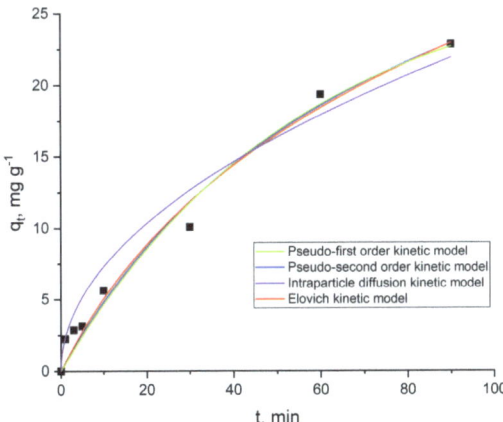

Figure 5. Non–linear kinetic models for copper ions biosorption onto chicken eggshells.

3.3.2. Pseudo-Second Order Kinetic Model

This model assumes that the kinetics of a sorption process simultaneously depends on the number of free active centers on the surface of the sorbent and the concentration of the sorbate in the solution [14].

The non-linear form of this model is given as:

$$q_t = \frac{q_e^2 k_2 t}{1 + k_2 t q_e} \quad (4)$$

where: q_t is the adsorbent capacity defined as mass of the adsorbed metal per unit mass of the adsorbent (mg g^{-1}) at time t; q_e is the adsorbent capacity defined as mass of the adsorbed metal per unit mass of the adsorbent (mg g^{-1}) at equilibrium; k_2 is the adsorption rate constant for the pseudo-second order kinetic model (g mg^{-1} min^{-1}).

The biosorption data were modeled using this model, and the results are shown on Figure 5 and in Table 1.

3.3.3. Intraparticle Diffusion Kinetic Model (Weber-Morris Model)

The Weber-Morris model assumes that the adsorption process does not take place only on the surface of the adsorbent, but that diffusion and adsorption inside the adsorbent structure also occur [15].

The non-linear form of the intraparticle diffusion kinetic model is given as:

$$q_t = K_i t^{0.5} + C_i \tag{5}$$

where: q_t is the adsorbent capacity defined as mass of the adsorbed metal per unit mass of the adsorbent (mg g^{-1}) at time t; K_i is the internal particle diffusion rate constant (mg g^{-1} min$^{-0.5}$); C_i is the boundary layer thickness constant.

The experimental data were fitted using the Weber-Morris model (Figure 5), and the kinetic parameters were determined and presented in Table 1.

3.3.4. Elovich Kinetic Model

The Elovich model is one of the most useful kinetic models for describing chemisorption [16].

The non-linear form of this model is given as:

$$q_t = \frac{1}{\beta} \ln(\alpha \beta t + 1) \tag{6}$$

where: q_t is the adsorbent capacity defined as mass of the adsorbed metal per unit mass of the adsorbent (mg g^{-1}) at time t; α is the starting adsorption rate (mg g^{-1} min^{-1}); β is the parameter that expresses the degree of surface coverage and activation energy for chemisorption (g mg^{-1}).

The obtained corresponding plot and kinetic data for this model are shown on Figure 5 and in Table 1.

The experimental data were fitted using four non-linear kinetic models, i.e., the pseudo-first order kinetic mode, pseudo-second order kinetic model, intraparticle diffusion kinetic model, and the Elovich kinetic model. Based on the obtained kinetic parameters (Table 1), it can be concluded that all the analyzed models show good agreement with the experimental data. However, the pseudo-first order kinetic model has proven to be the best fit for the analyzed data (R^2 = 0.999). Such results suggest that, in theory, copper ions react with active sites inside the structure of the eggshell, forming sorption complexes.

3.4. Equilibrium Study

Adsorption isotherm models are used to analyze experimental data in order to gain information about the mechanism of the adsorption process, it's equilibrium, and the maximum biosorption capacity. In this work, the non-linear Langmuir, Freundlich and Temkin isotherm models were used to analyze the equilibrium of the copper ions biosorption process onto chicken eggshells.

Biosorption isotherm data was obtained by performing the following experiment: 0.5 g of eggshells samples was brought into contact with 50 mL of copper ions solutions, of different initial Cu^{2+} concentrations (in the range from 30 to 400 mg dm^{-3}). The suspension was stirred on a magnetic stirrer for 90 min, assuming that is enough time to reach the equilibrium between phases [17]. The obtained experimental data was fitted using the mentioned non-linear isotherm models, and the results are presented in Figure 6 along with the isotherm parameters in Table 2.

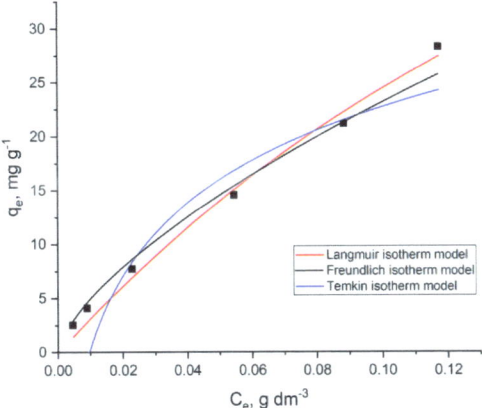

Figure 6. Biosorption isotherm data fitted using non–linear Langmuir, Freundlich and Temkin models.

Table 2. Adsorption isotherm model parameters for copper ions biosorption onto eggshells.

Model	Parameters	Values
Langmuir adsorption isotherm model	K_L (dm^3 mg^{-1}) $q_{e,exp}$ (mg g^{-1}) q_m (mg g^{-1}) R^2	3.49 28.3 94.59 0.989
Freundlich adsorption isotherm model	K_F $1/n$ R^2	108.5 0.671 0.931
Temkin adsorption isotherm model	B (J mol^{-1}) K_T (dm^3 g^{-1}) R^2	9.698 104.49 0.927

3.4.1. Langmuir Isotherm Model

The Langmuir isotherm model assumes that the adsorption process takes place in a monolayer, and that there are a finite number adsorption sites (each site can hold one adsorbate molecule). There is no interaction between the adsorbed molecules, and all adsorption sites are equivalent [18,19].

This model can be expressed as:

$$q_e = \frac{q_m K_L C_e}{1 + K_L C_e} \qquad (7)$$

where: q_e is the equilibrium biosorption capacity (mg g^{-1}); q_m is the maximum biosorption capacity (mg g^{-1}); Ce is the equilibrium concentration of metal ions in the solution (mg dm^{-3}); and K_L is the Langmuir equilibrium constant (dm^3 g^{-1}).

3.4.2. Freundlich Isotherm Model

The Freundlich model is a good representation of sorption processes at low and intermediate concentrations. This model can be applied to non-ideal and multilayer sorption on heterogeneous surfaces [20].

The Freundlich model can be represented as:

$$q_e = K_f C_e^{1/n} \qquad (8)$$

where: q_e is the equilibrium biosorption capacity (mg g^{-1}); Ce is the equilibrium concentration of metal ions in the solution (mg dm^{-3}); K_f is the Freundlich equilibrium constant ((mg g^{-1}) (dm^3 mg^{-1})$^{1/n}$).

The Freundlich constant n provides insight into the favorability of the adsorption process. When the value of n lays between 1 and 10 (i.e., $1/n$ is lower than 1), the adsorption process is favorable [21].

3.4.3. Temkin Isotherm Model

This model assumes that the adsorption heat of all the molecules in the layer shows a linear decrease with the coverage of molecules, and that adsorption is characterized by a uniform distribution of binding energies, up to a maximum binding energy [18].

The Temkin model is given as:

$$q_e = B \ln(K_T C_e) \tag{9}$$

where: q_e is the equilibrium biosorption capacity (mg g^{-1}); Ce is the equilibrium concentration of metal ions in the solution (mg dm^{-3}); $B = RT/b$ is the Temkin constant, which refers to the adsorption heat (J mol^{-1}); b is the variation of adsorption energy (J mol^{-1}); R is the universal gas constant (J mol^{-1} K^{-1}); T is the temperature (K); K_T is the Temkin equilibrium constant (dm^3 g^{-1}).

Based on the correlation coefficients (Table 2), it can be concluded that the biosorption process follows the Langmuir isotherm model, as it showed a very good agreement with the experimental data. This result indicates that there is a homogeneous distribution of active sites on the eggshell surface, while the adsorption process takes place in a monolayer, and that there are a finite number adsorption sites [18,19].

In addition, the Freundlich constant n suggests that the biosorption of copper ions onto chicken eggshells is a favorable process (n is between 1 and 10, i.e., $1/n$ is lower than 1) [21].

The performance of the adsorbent is usually defined by the maximum biosorption capacity. Based on the results in copper removal with various biosorbents reported by other workers (Table 3), it can be concluded that eggshells may play an important role as a cost-effective biosorbent for copper ions removal from aqueous environments.

Table 3. Cu^{2+} ions biosorption on eggshell in comparison with other adsorbents.

Biosorbent	Maximum Biosorbent Capacity (q_m, mg g^{-1})	Work
Eggshell	94.59	This work
Saccharomyces cerevisiae (brewer's yeast)	26.95	[22]
Carbonized sunflower stem	38.05	[23]
Sericin cross-linked with polyethylene glycol-diglycidyl ether	36.17	[24]
Sawdust of deciduous trees	9.9	[25]
Wheat straw	4.3	[17]
Chlorella pyrenoidosa (freshwater green algae)	11.88	[26]
Codium vermilara (codium seaweed)	14.4	[27]
Olive stone	1.96	[28]
Pine bark	11.35	[28]
Chitosan	103	[29]

3.5. Thermodynamic Study

The influence of temperature on the biosorption process was analyzed by bringing into contact 0.5 g of eggshells samples with 50 mL of synthetic Cu^{2+} solutions (initial concentration 500 mg dm^{-3}), at different temperatures (25 °C, 35 °C, and 45 °C). The suspension was stirred for 90 min. The obtained results (Figure 6) are analyzed in order to determine the thermodynamic parameters. The thermodynamic parameters are calculated using the Equations (10)–(13).

$$K_d = \frac{C_A}{C_S} \quad (10)$$

$$\Delta G^0 = -RT \ln K_d \quad (11)$$

$$\ln K_d = \left(\frac{\Delta S^0}{R}\right) - \left(\frac{\Delta H^0}{RT}\right) \quad (12)$$

$$\ln K_d = \left(\frac{-Ea}{RT}\right) + \ln A \quad (13)$$

where: K_d is the thermodynamic equilibrium constant; C_A is the concentration of the adsorbed adsorbate (mg dm^{-3}); C_S is the equilibrium concentration of the adsorbate in the solution (mg dm^{-3}); ΔG^0 is the Gibbs free energy (kJ mol^{-1}); R is the universal gas constant (J mol^{-1} K^{-1}); T is the temperature (K); ΔS^0 is the entropy change (J mol^{-1} K^{-1}); ΔH^0 is the enthalpy change (kJ mol^{-1}); Ea is the activation energy (kJ mol^{-1}).

The change of the Gibbs free energy was calculated using the Equation (11).

According to the Equation (12), the values of the enthalpy and entropy were calculated from the slope and intercept of the plot ln K_D vs. 1/T, which is shown on Figure 7 [20].

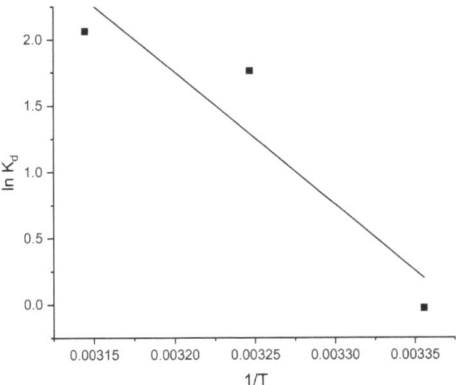

Figure 7. Thermodynamic plot ln KD vs. 1/T for copper ions biosorption onto eggshells.

The activation energy was calculated from the slope of the plot ln K_D vs. 1/T, using the Equation (13).

The thermodynamic parameters are given in Table 4.

Table 4. Thermodynamic parameters for copper ions biosorption onto eggshells.

T (K)	ΔG^0 (kJ mol^{-1})	ΔH^0 (kJ mol^{-1})	ΔS^0 (J mol^{-1} K^{-1})	Ea (kJ mol^{-1})
298	0.07	−9.98	33.68	82.97
308	−4.51			
318	−5.45			

The Gibbs free energy (Table 3) for the copper ions biosorption onto eggshells indicates that the process is spontaneous and favored at temperatures above 25 °C. Negative enthalpy change indicates that the process is exothermic. Positive entropy value indicates that there is an increased randomness at the solid/liquid interface during the adsorption process [30,31].

The activation energy (Ea) value of 82.97 kJ mol^{-1} indicates that the binding of copper ions onto eggshells took place mainly by chemisorption [32].

3.6. Process Optimization Study

The biosorption of copper ions using eggshells was optimized using an experimental design, in order to determine the effects of three selected (independent) variables on the percentage of Cu^{2+} removal (dependent variable). Response Surface Methodology (RSM) represents a set of techniques, which is useful for evaluating the relationships between a number of experimental factors and measured responses [33–35].

The RSM-BBD was applied to optimize the biosorption process using the Design Expert software (version: 22.0.0) [36].

The Box-Behnken factorial design, consisting of 17 experiments, coupled with Response Surface Methodology, was applied with the goal to optimize the biosorption process, comparing three factors: Adsorbent mass (A), initial copper ions concentration (B) and contact time (C). The experimental ranges and their levels in the design are given in Table 5. The experimental design matrix, as well as the response R (adsorption degree) are given in Table 6.

Table 5. Experimental ranges and levels in the experimental design.

Factors	Range Level		
	−1	0	1
A—Adsorbent mass (g)	0.5	1	1.5
B—Initial metal ion concentration (g/L)	0.5	1	1.5
C—Contact time (min)	10	60	90

Table 6. Box-Behken Design matrix for three factors along with observed response for Cu^{2+} biosorption onto eggshells.

Run	A: Adsorbent Mass (g)	B: Initial Cu^{2+} Ions Concentration (g/L)	C: Contact Time (min)	R: Adsorption Degree (%)
1	1.5	0.5	60	64.54
2	0.5	1	10	9.87
3	1	1.5	10	43.09
4	1	0.5	10	7.02
5	1.5	1	10	11.03
6	0.5	1.5	90	12.6
7	0.5	1	60	25.86
8	1.5	1	90	80.47
9	0.5	1.5	60	7.47
10	1	1	60	54.27
11	0.5	0.5	60	96.16

Table 6. Cont.

Run	A: Adsorbent Mass (g)	B: Initial Cu^{2+} Ions Concentration (g/L)	C: Contact Time (min)	R: Adsorption Degree (%)
12	1	1	60	51.14
13	1	1	60	52.16
14	1	1	60	49.67
15	1.5	1.5	60	38
16	1	0.5	90	97.06
17	0.5	1	90	89.74

The correlation between the following independent variables: linear (β_1, β_2, β_3), quadratic (β_{11}, β_{22}, β_{33}), interaction terms (β_{12}, β_{13}, β_{23}) and the response (R), was described by fitting the following polynomial equation [33]:

$$R = \beta_0 + \beta_1 A + \beta_2 B + \beta_3 C + \beta_{11} AA + \beta_{22} BB + \beta_{33} CC + \beta_{12} AB + \beta_{13} AC + \beta_{23} BC \quad (14)$$

The obtained results are displayed in Table 6. The biosorption of copper ions onto eggshells can be expressed using the following equation:

$$Y = 46.62 - 1.15A - 20.45B + 26.11C + 15.54A \cdot B - 2.61A \cdot C - 30.13B \cdot C + 6.38A \cdot A - 1.46B \cdot B - 5.22C \cdot C \quad (15)$$

The statistical significance of the applied model was evaluated by the ANOVA analysis and shown in Table 7. The significance of each coefficient is determined by the magnitude of the F-values and p-values (Table 7). The larger the F-value, and the smaller p-value, the corresponding coefficient is more significant. p-values less than 0.0500 indicate high significant regression at 95% confidence level [35].

Table 7. ANOVA analysis for response surface model in relation to Cu^{2+} biosorption onto eggshells.

Source	Sum of Squares	df	Mean Square	F-Value	p-Value	
Model	13,714.86	9	1523.87	5.03	0.0224	Significant
A-A	10.58	1	10.58	0.0349	0.8571	
B-B	3346.44	1	3346.44	11.04	0.0127	
C-C	5452.81	1	5452.81	17.99	0.0038	
AB	965.66	1	965.66	3.19	0.1174	
AC	27.20	1	27.20	0.0897	0.7732	
BC	3631.87	1	3631.87	11.98	0.0105	
A^2	171.32	1	171.32	0.5652	0.4767	
B^2	8.93	1	8.93	0.0295	0.8686	
C^2	114.79	1	114.79	0.3787	0.5578	
Residual	2121.64	7	303.09			
Lack of Fit	1571.72	3	523.91	3.81	0.1145	Not significant
Pure Error	549.92	4	137.48			
Cor Total	15,836.50	16				

The model F-value and p-value of 5.03 and 0.0224, respectively, indicate that the model is significant. p-values lower than 0.05 indicate that model terms are significant. In this study, B (initial Cu^{2+} ions concentration), C (contact time (min)), and BC (initial Cu^{2+} ions concentration combined with contact time (min)) are significant model terms. The suitability of the model was confirmed by the regression coefficients of the predicted and experimental values ($R^2 = 0.866$ and adj-$R^2 = 0.694$).

Figure 8 shows the relationship between the experimental responses and the responses predicted by the model.

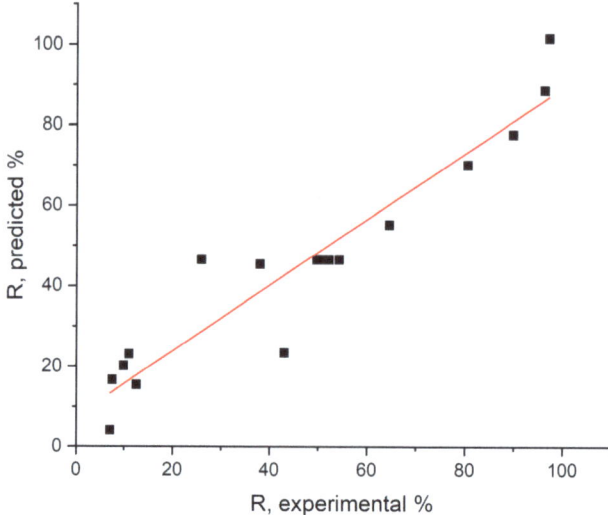

Figure 8. Plot of experimental and predicted responses.

Based on the data shown on Figure 8, and the correlation coefficient ($R^2 = 0.897$), it can be concluded that there is a good relationship between the experimental and predicted responses.

Response surface plots showing the influence of the analyzed parameters on the adsorption degree (R) are presented on Figures 9–11. Figure 9 indicates that lower initial metal ion concentration combined with lower adsorbent mass leads to a higher percentage of adsorbed metal (ANOVA analysis indicates that the combination of these two factors (A and B) is not significant). Figure 10 indicates that higher adsorbent mass and higher contact time leads to a higher response (metal removal%), while the ANOVA analysis does not classify the combination of these factors as significant. Figure 11 shows the interaction between factors B and C, i.e., the initial metal ion concentration and contact time, respectively. The ANOVA analysis indicates that the combination of these two factors is a significant model term. The corresponding Response surface plot indicates that high contact time and low initial metal ions concentration leads to the highest obtained response (adsorption degree).

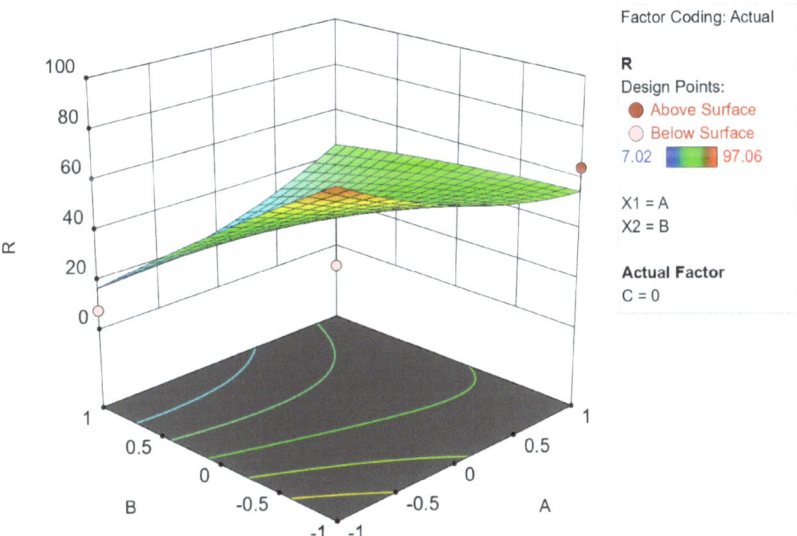

Figure 9. Response surface plot showing the interaction and influence of the adsorbent mass (A) and initial copper ions concentration (B) on the adsorption rate (R).

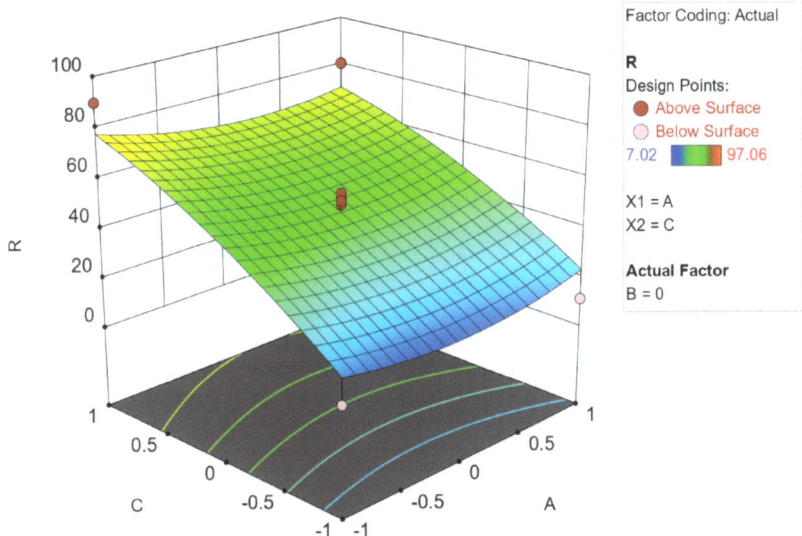

Figure 10. Response surface plot showing the interaction and influence on of the adsorbent mass (A) and contact time (C) on the adsorption rate (R).

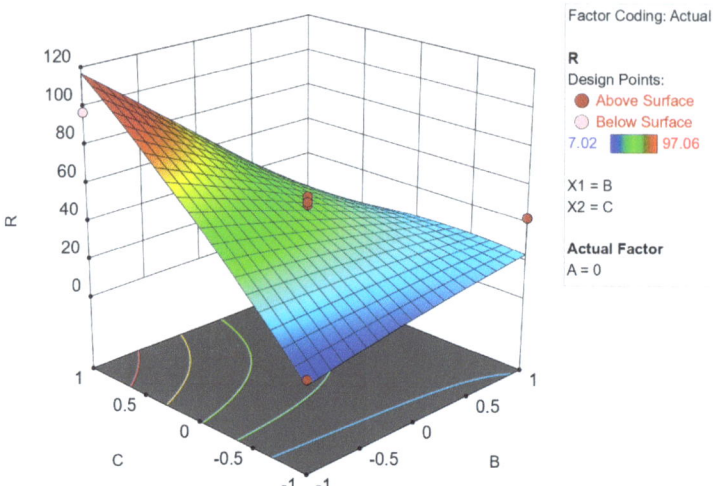

Figure 11. Response surface plot showing the interaction and influence of the initial copper ions concentration (B) and contact time (C) on the adsorption rate (R).

4. Conclusions

Biosorption of copper ions using chicken eggshells as an adsorbent was investigated and presented in this paper.

The influence of different process parameters on the biosorption process was evaluated. The biosorption capacity was found to increase with the increase in the pH value of the solution, reaching its maximum value at pH = 5. The influence of the pH value on the biosorption capacity could possibly be explained by the behavior of the carbonate species that originate from the eggshells (the source of the calcium carbonate) in the solution at different pH values, and their interaction with the divalent copper ions.

The analysis of the influence of initial copper ions concentration showed an increase in the biosorption capacity, with the rise in the initial Cu^{2+} ions concentration, up to 800 mg dm^{-3}, where it reaches the maximum value (q_t = 40.79 mg g^{-1}).

The initial mass of the adsorbent showed a significant influence on the biosorption efficiency. The adsorption degree increased from 25% to 82% with the increase in the adsorbent mass up to 1 g, due to the higher number of available active sites on the adsorbent structure as a result of a larger amount of adsorbent available.

The results of the analysis also showed that the biosorption capacity increased with the increase in the stirring rate, up to 400 rpm, where it reached its maximum value. Further increase in the stirring rate resulted in a decrease of the biosorption capacity.

The SEM-EDS analysis was performed on eggshells samples before and after the biosorption process. The SEM analysis showed a slight change to the surface morphology of the eggshells sample after the biosorption process, to an uneven, rough and heterogeneous nature. This change could be contributed to the incorporation of copper ions inside the structure of the eggshells sample. The interaction of eggshells with Cu^{2+} ions lead to the formation of flake-like deposits on the surface of the adsorbent. The EDS analysis of the eggshells samples before and after the biosorption process indicated that Mg, K and Ca could potentially be exchanged with Cu ions during the adsorption process.

Biosorption kinetics were analyzed using four empirical kinetic non-linear models, namely, the pseudo-first order kinetic model, pseudo-second order kinetic model, intraparticle diffusion kinetic model, and the Elovich kinetic model. The obtained kinetic parameters led to a conclusion that the pseudo-first order best fits the analyzed process, suggesting that copper ions possibly react with active sites inside the eggshell structure, forming sorption complexes in the process.

Three empirical adsorption isotherm models, namely, the Langmuir, Freundlich and Temkin model, in their non-linear form, were used to evaluate the equilibrium of the biosorption process. The performed analysis indicated that the Langmuir model showed the best fit with the experimental data. The Freundlich constant n also suggested that the biosorption of copper ions onto chicken eggshells is a favorable process.

The thermodynamic parameters of the biosorption process were calculated. The Gibbs free energy change indicated that the biosorption of copper ions onto chicken eggshell is a spontaneous process, and favored at temperatures above room temperature. The obtained enthalpy and entropy values indicated that the process is exothermic and that there is increased randomness at the solid/liquid interface during the biosorption.

Copper ions biosorption onto eggshells was optimized using Response Surface Methodology, based on Box-Behnken Design. The influence of three parameters (adsorbent mass, initial metal ions concentration and contact time) was investigated. The obtained data indicates that the used model is statistically significant. The data shows that initial Cu^{2+} ions concentration, contact time, and initial Cu^{2+} ions concentration combined with contact time are significant model terms. This model indicated that the optimal biosorption conditions are: adsorbent mass = 1 g; initial Cu^{2+} ions concentration = 0.5 g dm^{-3}; and contact time = 90 min.

Author Contributions: M.G., N.Š., V.G. and M.V. directed the project; M.G., M.M., K.B. and V.G. provided the samples for the biosorption experiments; M.M., K.B. and M.Z. performed the experiments and the analysis of the obtained data; M.M., M.G. and K.B. prepared the original draft. All authors have read and agreed to the published version of the manuscript.

Funding: This paper within the funding of the scientific research work at the University of Belgrade, Technical Faculty in Bor, according to the contract with registration number 451-03-68/2022-14/200131.

Data Availability Statement: The data presented in this study are available on request from the corresponding author.

Acknowledgments: The research presented in this paper was done with the financial support of the Ministry of Education, Science and Technological Development of the Republic of Serbia.

Conflicts of Interest: The authors declare no conflict of interest.

References

1. Xin, Y.; Li, C.; Liu, J.; Liu, J.; Liu, Y.; He, W.; Gao, Y. Adsorption of heavy metal with modified eggshell membrane and the in situ synthesis of Cu-Ag/modified eggshell membrane composites. *R. Soc. Open Sci.* **2018**, *5*, 180532. [CrossRef] [PubMed]
2. Zhang, W.; Dong, X.; Mu, X.; Wang, Y.; Chen, J. Constructing adjacent phosphine oxide ligands confined in mesoporous Zr-MOFs for uranium capture from acidic medium. *J. Mater. Chem. A* **2021**, *9*, 16685. [CrossRef]
3. Qasem, N.A.A.; Mohammed, R.H.; Lawal, D.U. Removal of heavy metal ions from wastewater: A comprehensive and critical review. *Npj Clean Water* **2021**, *4*, 36. [CrossRef]
4. Long, M.; Jiang, H.; Li, X. Biosorption of Cu^{2+}, Pb^{2+}, Cd^{2+} and their mixture from aqueous solutions by *Michelia figo* sawdust. *Sci. Rep.* **2021**, *11*, 11527. [CrossRef] [PubMed]
5. Vijayaraghavan, K.; Joshi, U.M. Chicken eggshells remove Pb(II) ions from synthetic wastewater. *Environ. Eng. Sci.* **2013**, *30*, 67–73. [CrossRef]
6. Cordeiro, C.M.M.; Hincke, M.T. Recent patents on eggshell: Shell and membrane applications. *Recent Pat. Food Nutr. Agric.* **2011**, *3*, 1–8. [CrossRef]
7. Su, W.; Yang, Y.; Dai, H.; Jiang, L. Biosorption of heavy metal ions from aqueous solution o Chinese fir bark modified by sodium hypochlorite. *Bioresources* **2015**, *10*, 6993–7008. [CrossRef]
8. Imessaoudene, A.; Cheikh, S.; Bollinger, J.C.; Belkhiri, L.; Tiri, A.; Bouzaza, A.; El Jery, A.; Assadi, A.; Amrane, A.; Mouni, L. Zeolite waste characterization and use as low-cost, ecofriendly, and sustainable material for malachite green and methylene blue dyes removal: Box-behnken design, kinetics and thermodynamics. *Appl. Sci.* **2022**, *12*, 7587. [CrossRef]
9. Murithi, G.; Onindo, C.O.; Muthakia, G.K. Kinetic and equilibrium study for the sorption of Pb(II) ions from aqueous phase by water hyacinth (*Eichhornia crassipes*). *Bull. Chem. Soc. Ethiop.* **2012**, *26*, 181–193. [CrossRef]
10. Yang, B.; Tong, X.; Deng, Z.; Lv, X. The adsorption of Cu species onto pyrite surface and its effect on pyrite flotation. *J. Chem.* **2016**, *2016*, 4627929. [CrossRef]

11. Putra, W.P.; Kamari, A.; Yusoff, S.N.M.; Ishak, C.F.; Mohamed, A.; Norhayati, H.; Isa, I.M. Biosorption of Cu (II), Pb (II) and Zn (II) ions from aqueous solutions using selected waste materials: Adsorption and characterization studies. *JEAS* **2014**, *4*, 43523. [CrossRef]
12. Anantha, R.K.; Kota, S. An evaluation of the major factors influencing the removal of copper ions using the egg shell (Dromaius novaehollandiae): Chitosan (Agaricus bisporus) composite. *Biotech* **2016**, *6*, 83. [CrossRef] [PubMed]
13. Lagergren, S. About the theory of so-called adsorption of soluble substances. *Sven. Vetenskapsakad. Handingarl.* **1898**, *241*, 1–39.
14. Ho, Y.S.; McKay, G. Pseudo-second order model for sorption processes. *Process Biochem.* **1999**, *34*, 451–465. [CrossRef]
15. Cheung, W.H.; Szeto, Y.S.; McKay, G. Intraparticle diffusion processes during acid dye adsorption onto chitosan. *Bioresour. Technol.* **2007**, *98*, 2897–2904. [CrossRef] [PubMed]
16. Elkady, M.F.; Ibrahim, A.M.; El Latif, M.M.A. Assessment of the adsorption kinetics, equilibrium, and thermodynamic for the potential removal of reactive red dye using eggshel biocomposite beads. *Desalination* **2011**, *278*, 412–423. [CrossRef]
17. Gorgievski, M.; Božić, D.; Stanković, V.; Štrbac, N.; Šerbula, S. Kinetics, equilibrium and mechanism of Cu^{2+}, Ni^{2+} and Zn^{2+} ions biosorption using wheat straw. *Ecol. Eng.* **2013**, *58*, 113–122. [CrossRef]
18. Tonk, S.; Majdik, C.; Szep, R.; Suciu, M.; Rapo, E.; Nagy, B.; Niculae, A.G. Biosorption of Cd(II) ions from aqueous solution onto eggshell waste, kinetic and equilibrium isotherm studies. *Rev. Chim.* **2017**, *68*, 1951–1958. [CrossRef]
19. Bouchelkia, N.; Mouni, L.; Belkhiri, L.; Bouzaza, A.; Bollinger, J.C.; Madani, K.; Dahmoun, F. Removal of lead(II) from water using activated carbon developed from jujube stones, a low cost adsorbent. *Sep. Sci. Technol.* **2016**, *51*, 1645–1653. [CrossRef]
20. Metwally, S.S.; Rizk, H.E.; Gasser, M.S. Biosorption of strontium ions from aqueous solution using modified eggshell materials. *Radiochim. Acta* **2017**, *105*, 1021–1031. [CrossRef]
21. Saha, P.D.; Chowdhury, S.; Mondal, M.; Sinha, K. Biosorption of direct red 28 (congo red) from aqueous solutions by eggshells: Batch and column studies. *Sep. Sci. Technol.* **2014**, *47*, 112–123. [CrossRef]
22. Savastru, E.; Bulgariu, D.; Zamfir, C.I.; Bulgariu, L. Application of *Saccharomyces cerevisiae* in the biosorption of Co (II), Zn (II), and Cu (II) ions from aqueous media. *Water* **2022**, *14*, 976. [CrossRef]
23. Sireesha, C.; Subha, R.; Sumithra, S. Biosorption of copper ions in aqueous solution by carbonized sunflower stem. *Rasayan J. Chem.* **2022**, *15*, 2267–2273. [CrossRef]
24. Marques, G.S.; Dusi, G.G.; Drago, F.; Gimenes, M.L.; da Silva, V.R. Biosorption of Cu(II) ions using sericin cross-linked with polyethylene glycol-diglycidyl ether. *Desalination Water Treat.* **2022**, *270*, 153–162. [CrossRef]
25. Božić, D.; Stanković, V.; Gorgievski, M.; Bogdanović, G.; Kovačević, R. Adsorption of heavy metal ions by sawdust of deciduous trees. *J. Hazard. Mater.* **2009**, *171*, 684–692. [CrossRef]
26. Moreira, V.R.; Lebron, Y.A.R.; Freire, S.J.; Santos, L.V.S.; Palladino, F.; Jacob, R.S. Biosorption of copper ions from aqueous solution using *Chlorella pyrenoidosa*: Optimization, equilibrium and kinetic studies. *Microchem. J.* **2019**, *145*, 119–129. [CrossRef]
27. Fawzy, M.A. Biosorption of copper ions from aqueous solution by *Codium vermilara*: Optimization, kinetic, isotherm and thermodynamic studies. *Adv. Powder. Technol.* **2020**, *31*, 3724–3735. [CrossRef]
28. Blazquez, G.; Martin-Lara, M.A.; Dionsio-Ruiz, E.; Tenorio, G.; Calero, M. Evaluation and comparison of the biosorption process of copper ions onto olive stone and pine bark. *J. Ind. Eng. Chem.* **2011**, *17*, 824–833. [CrossRef]
29. Modrzejewska, Z.; Rogacki, G.; Sujka, W.; Zarzycki, R. Sorption of copper by chitosan hydrogel: Kinetics and equilibrium. *Chem. Eng. Process.* **2016**, *109*, 104–113. [CrossRef]
30. Awwad, K.M.; Farhan, A.M. Equilibrium, kinetic and thermodynamics of biosorption of lead (II) copper (II) and cadmium (II) ions from aqueous solutions onto olive leaves powder. *Am. J. Chem.* **2012**, *2*, 238–244. [CrossRef]
31. Thilagan, J.; Gopalakrishnan, S.; Kannadasan, T. Thermodynamic study on adsorption of copper (II) ions in aqueous solution by chitosan blended with cellulose & cross linked by formaldehyde, chitosan immobilised on red soil, chitosan reinforced by banana stem fibre. *IJSRET* **2013**, *2*, 28–36.
32. Ozel, H.U.; Gemici, B.T.; Ozel, H.B.; Berberler, E. Evaluating forest waste on adsorption of Cd(II) from aqueous solution: Equilibrium and thermodynamic studies. *Pol. J. Environ. Stud.* **2019**, *28*, 3829–3836. [CrossRef] [PubMed]
33. Choinska-Pulit, A.; Sobolczyk-Bednarek, J.; Laba, W. Optimization of copper, lead and cadmium biosorption onto newly isolated bacterium using a Box-Behnken design. *Ecotoxicol. Environ. Saf.* **2018**, *149*, 275–283. [CrossRef] [PubMed]
34. Mammar, A.C.; Mouni, L.; Bollinger, J.C.; Belkhiri, L.; Bouzaza, L.; Bouzaza, A.; Assadi, A.A.; Belkacemi, H. Modeling and optimization of process parameters in elucidating the adsorption mechanism of gallic acid on activated carbon prepared from date stones. *Sep. Sci. Technol.* **2020**, *55*, 3113–3125. [CrossRef]
35. Turkyilmaz, H.; Kartal, T.; Yigitarslan Yildiz, S. Optimization of lead adsorption of mordenite by response surface methodology: Characterization and modification. *J. Environ. Health Sci. Eng.* **2014**, *12*, 5. [CrossRef]
36. *Design-Expert® Software*, version 22.0.0; Stat-Ease, Inc.: Minneapolis, MN, USA, 2022. Available online: https://www.statease.com (accessed on 22 December 2022).

Disclaimer/Publisher's Note: The statements, opinions and data contained in all publications are solely those of the individual author(s) and contributor(s) and not of MDPI and/or the editor(s). MDPI and/or the editor(s) disclaim responsibility for any injury to people or property resulting from any ideas, methods, instructions or products referred to in the content.

Article

Use of Ion-Exchange Resins to Adsorb Scandium from Titanium Industry's Chloride Acidic Solution at Ambient Temperature

Eleni Mikeli [1,*], Danai Marinos [1], Aikaterini Toli [1], Anastasia Pilichou [1], Efthymios Balomenos [2] and Dimitrios Panias [1]

[1] Laboratory of Metallurgy, School of Mining and Metallurgical Engineering, National Technical University of Athens (NTUA), Heroon Polytechniou 9, 15780 Zografou, Greece; dmarinou@metal.ntua.gr (D.M.); katerinatoli@metal.ntua.gr (A.T.); anastasiapilichou@mail.ntua.gr (A.P.); panias@metal.ntua.gr (D.P.)

[2] Mytilineos S.A, Metallurgy Business, Agios Nikolaos, 32003 Voiotia, Greece; efthymios.balomenos-external@alhellas.gr

* Correspondence: elenamikeli@mail.ntua.gr; Tel.: +30-210-7722-177

Abstract: Scandium metal has generated a lot of interest during the past years. This is due to the various crucial applications it has found ground in and the lack of production in countries outside China and Russia. Apart from rare earth ores, scandium is present in a variety of wastes and by-products originating from metallurgical processes and is not currently being sufficiently valorised. One of these processes is the production of titanium dioxide, which leaves an acidic iron chloride solution with a considerably high concentration of scandium (10–140 ppm) and is currently sold as a by-product. This research aims to recover scandium without affecting the solution greatly so that it can still be resold as a by-product after the treatment. To achieve this, two commercial ion-exchange resins, VP OC 1026 and TP 260, are used in the column setup. Their breakthrough curves are plotted with mathematical modelling and compared. Results indicate that VP OC 1026 resin is the most promising for Sc extraction with a column capacity of 1.46 mg/mL, but Zr, Ti, and V coextract have high capacities, while Fe does not interfere with the adsorption.

Keywords: scandium; titanium dioxide industry; ion-exchange resins; VPOC1026; TP260; nonlinear regression

1. Introduction

Scandium (Sc) is the most valuable element among the rare earth elements (REEs) because of its application in many cutting-edge technologies such as metal halide lamps, high strength Al-Sc alloys (aerospace industry, baseball bats, military and medical purposes, and bicycle frames), scandia-stabilised zirconia as a high-efficiency electrolyte in solid oxide fuel cells, and as a tracer in cruel-oil refinery [1–3]. In the most recent report by the EU, Scandium is considered a Critical Raw Material (CRM) because of its high supply risk and economic importance [4]. The production of Sc has been modest to date (~16.3 t/y for 2018), while the demand for this metal is expected to gain a significant increase (up to 300 t/y) by 2028 [5]. At the same time, the Sc supply in Europe depends solely on imports, mainly from China (66%), Russia (28%), and Ukraine (7%) [4]. The abundance of scandium in the earth's crust is comparatively high (20–30 ppm), but it is sparsely dispersed and found in trace amounts in a variety of ores such as ilmenite, rutile, bauxite, nickel-laterite, tungsten, tin, and uranium [1,6–9]. During the industrial processing of these ores, Sc is usually accumulated in the waste streams (solid, liquid, or gas). Many researchers have studied the potential use of these wastes to recover Sc [3,10].

A large amount of the Sc_2O_3 produced in China originates from the titanium oxide (TiO_2) pigments production industries [6,11]. Other countries, such as European ones, which are currently solely dependent on imports of Sc, can benefit from the TiO_2 pigment industries to produce Sc as a by-product. Both in the sulfuric and chloride process of

TiO$_2$ production, most of the Sc ends up in the iron-rich residual solution (10–140 ppm of Sc) [1,6,8,12,13] together with other CRMs such as vanadium (V) and niobium (Nb) [13]. The ferrous chloride solution is usually sold to chemical industries and used as a coagulation/flocculation agent and for odour control, or its neutralised filter cake is landfilled without valorising the present CRMs.

Purification of scandium-containing streams is a challenging task. Unwanted elements, defined as impurities, are typically present in more than 10 times higher concentrations. Precipitation, which is the traditional approach for metal liquor purification, is prohibited as a stand-alone method for scandium extraction since metal impurities often coprecipitate, resulting in poor Sc purity [14,15]. In general, solvent extraction (SX) is the most employed separation/purification technique for recovering Sc, as it shows high extraction capacities and is easily scalable. The Sc that is recovered from the titanium acidic side streams is also separated with SX. The most common extraction systems that have been examined by many researchers and are currently applied in China's scandium extraction plants are a mixture of D2EHPA and TBP [1,6,8,9,12,16] Despite the benefits, the SX also possesses some drawbacks, the most important one being the losses of the organic extractants which usually contaminate the solution as well [17,18].

Among the technologies for the separation and purification of aqueous solutions, ion exchange (IX) has some benefits compared with SX techniques: less contamination of the pregnant solution, lower operation costs [18], etc., and thus, it has attracted a lot of attention for scandium extraction [7,15,17–27]. The stable oxidation state of scandium in an aqueous solution is Sc(III) [28]. Scandium-hydrated ions are Pearson hard acids because of their high oxidation state, and they choose to form complexes with hard ligands, including hydroxide, fluoride, sulfate, and phosphate [28]. According to Wang et al. [14], organophosphorus compounds such as phosphoric, phosphonic, and phosphinic acid, as well as neutral phosphine oxides, have been effective in the quantitative extraction of scandium from aqueous solution through solvating and the cation-exchange mechanisms.

Along with the wide variety of commercial and custom-made resins that have been used, Lewatit VP OC 1026 and Lewatit TP 260 seem promising [18,21–23,29] for the selective extraction of Sc. Mostajeran et al. [23] studied the use of VP OC 1026 in a sulfuric acid pregnant leaching solution (PLS), which resulted from coal fly ash leaching and reported fast adsorption kinetics for Sc and high selectivity for Sc towards Fe and Al. Reynier et al. [29] conducted bioleaching of uranium tailings and utilised a variety of IX resins. Among them, VP OC 1026 and TP 260 showed high extraction of Sc but with high coextraction of U and Th and low coextraction of Fe. In another study [18], the TP 260 IX resin was used to extract REEs from phosphoric acid solutions; the results showed very high recovery rates for Sc (80%) and reported the ions of Al and Fe to be the most interfering in terms of coextraction. Bao et al. [22] studied the recovery of Sc from sulfuric acid solutions using IX resins and reported a slow adsorption rate for TP 260. In addition, they showed that the selectivity order for TP 260 resin is Sc(III) > Fe(III) > Al(III) > Fe(II). Rychkov et al. [21] conducted kinetic experiments with TP 260 in sulfuric acid solutions and showed that Th, Fe, and Al are sorbed in the intradiffusion mode while Sc, Ti, and Zr are sorbed in the mixed diffusion mode.

This paper examines the potential utilisation of a by-product of the TiO$_2$ industry for Sc extraction through the ion-exchange purification technique. More particularly, a real solution derived from a European TiO$_2$ industry was loaded directly into the column setup, imitating an actual ion-exchange process. Two different commercial resins are examined and compared using a fixed-bed column setup: Lewatit VP OC 1026, a bis(2-ethylhexyl)phosphoric acid (D2EHPA) in a cross-linked polystyrene matrix and Lewatit TP 260, a macroporous cross-linked polystyrene with an aminomethyl-phosphonic acid functional group.

The evaluation of such experiments can be challenging, from data analysis to experimental and analytical point of view, as the feed solution is a multicomponent industrial-grade solution, but also the column setup is a dynamic procedure. For this reason, triplicate experiments were conducted, and mathematical modelling was applied for the data analysis. Assessing the ion-exchange procedure was focused not only on the loading behaviour of Sc but also on its selectivity towards Zr, Ti, V, and Fe because the literature showed that these metals might impede Sc extraction. Even though the combined effect of these metals has not been studied using these two resins, studies in similar extraction systems indicate that Sc extraction is influenced when these metal impurities are present since they exhibit similar chemical behaviour and extraction rates [21,26,30].

2. Materials and Methods
2.1. Materials and Experimental Procedure

Two different commercial resins were used in this study provided by Lenntech (Delfgauw, The Netherlands). In Table 1, the main characteristics of the resins are listed, and in Figure 1, the functional groups of each resin are illustrated. Before the column experiments, an adequate quantity of Lewatit TP 260 resin was conditioned in three cycles with 1.5 M HCl solution, in a ratio of 1/10 for 2 h to be converted into the hydrogen form, according to the available datasheets. Lewatit VP OC 1026 resin was received in hydrogen form; hence, no conditioning step was required.

Table 1. Resins characteristics.

Resin Name	Lewatit VP OC 1026	Lewatit TP 260
Functional group	bis(2-ethylhexyl)phosphoric acid (D2EHPA)	aminomethyl-phosphonic acid (AMPA)
Matrix	Cross-linked polystyrene	Cross-linked polystyrene
As received form	H^+	Na^+
Bead size range	0.3–1.6 mm	0.4–1.25 mm

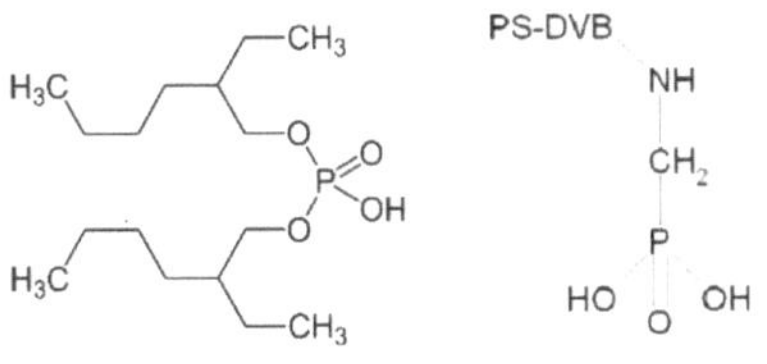

Figure 1. Functional groups of the two commercial resins (a) Lewatit VP OC 1026 and (b) Lewatit TP 260.

The feed solution used in this study is an acidic $FeCl_2$ solution that originated from the titanium dioxide pigment industry of Europe. The chemical analysis of the main elements of the solution is illustrated in Table 2.

Table 2. Chemical analysis of $FeCl_2$ feed solution.

Metals	Fe (II)	Fe (III)	Mn	Al	Mg	Na	V	Zr	Ca	Cr	Ti	Sc
Concentration (g/l)	110.5	1.5	19.09	9.11	8.34	6.41	4.20	3.68	2.36	1.69	0.528	0.130

For the fixed-bed column experiments, laboratory borosilicate glass columns with 20 mL bed volume, 1.5 cm × 10 cm dimensions were used. The top and bottom of the column had fixed caps with a luer lock inlet and outlet fittings for tubing attachment. A known weight of the resin was packed into the column. The $FeCl_2$ solution continuously passed through the column in an upward constant flow of 0.93 BVh^{-1}, with the use of a peristaltic pump LabDos, from Hitec Zang. Samples of the effluent solution were fractionally collected from the top of the column to perform a chemical analysis of the metals of interest (Fe, Ti, Sc, V, and Zr).

Wet chemical analysis of the samples was carried out using the inductively coupled plasma–optical emission spectrometry (ICP-OES), Optima 8000 by Perlin Elmer (Waltham, MA, USA). Calibration standard solutions were prepared from commercially available ICP, Ti, Sc, V, Zr, and Fe standards (1000 ppm) obtained from Merck (Darmstadt, Germany). The standard solutions were prepared in a suitable concentration and diluted further with 1% v/v analytical grade nitric acid (65% wt.) as required for working standards. High purity deionized water (18.2 MΩ/cm) and argon of special purity (99.999%) were used. Re was used as the Internal Standard (IS) for samples and standards in the case of V and Zr analysis in order to correct the signal drift and reduce the interferences from Fe.

2.2. Mathematical Modelling

2.2.1. Mathematical Description of the Fixed-Bed Columns

The total capacity of the column can be calculated for every metal of interest, q_{Me} (mg M_e per mL of resin), by integrating the plot of the adsorbed metal concentration C_{ad} (mg/L) against the effluent volume, V_{Ef} (mL) [31]. The area, A, under this integrated plot is substituted in Equation (1):

$$q_{Me} = \frac{A}{1000 V_{resin}} = \frac{\int_{V_i=0}^{V_i=total} C_{ad} dV_i}{1000 V_{resin}} \quad (1)$$

where C_{ad} is the difference between the initial concentration and the instant concentration of the effluent solution ($C_{ad} = C_0 - C_i$), and V_{resin} is the volume of resin packed into the column (mL).

The total removal efficiency (E%) of each metal can be calculated from Equation (2):

$$E\% = 100 \frac{m_{total,ads}}{m_{total,in}} \quad (2)$$

where $m_{total,ads}$ is the cumulative amount of metal adsorbed in the resin, and $m_{total,in}$ is the cumulative amount of metal entering the resin.

2.2.2. Modelling of Experimental Results through Regression Method

The performance of fixed-bed column adsorption is typically evaluated through the plot of the dimensionless concentration of the effluent solution versus the eluted volume or time [32–35]. This plot is referred to as the breakthrough curve. In the literature, many models have been developed to describe the breakthrough curve of a fixed-bed column, either based on empirical relationships or based on different assumptions regarding the kinetics and mechanism of extraction. In our study, the Thomas model was chosen as the most extensively reported mathematical model for predicting breakthrough curves [36]. This model was developed from the equation of mass conservation in a flow system assuming plug flow, equilibrium described by Langmuir isotherm, and second-order reversible kinetics. Particularly, it neglects internal and external diffusion effects [32]. The equation for the Thomas model is demonstrated in Equation (3):

$$\frac{C_i}{C_0} = \frac{1}{1 + \exp\left[\left(\frac{K_{Th}}{Q}\right)\left(q_{Th} V_{resin} - \frac{C_0 V_i}{1000}\right)\right]} \quad (3)$$

where Q is the volumetric flow rate (mL/min), V_{resin} is the volume of resin packed into the column (mL), and C_0 and C_i represent the initial concentration of the feed solution and the effluent concentration (mg/L), respectively, V_i is the effluent volume (mL), K_{Th} (mL/min g) is the Thomas rate constant, and q_{Th} (mg/mL) represents the adsorption capacity of the fixed beds.

Linear regression is the most typical method for fitting experimental results [31,37,38], where the calculation of the model parameters becomes associated with the slope and intercept of a straight line. The linearised form of the Thomas model is illustrated in Equation (4):

$$\ln\left(\frac{C_0}{C_i} - 1\right) = \frac{K_{Th} q_{Th} V_{resin}}{Q} - \frac{K_{Th}}{Q1000} C_0 V_i \quad (4)$$

It is evident that the error distortion created by linearisation may lead to misleading results [39]. The linear transformation strongly modifies error distribution and alters the weight associated with each point, either for the worse or for the better [40]. An alternative to linear regression is nonlinear regression. In this method, model parameters are first estimated and, through a trial and error procedure, evolve towards the values that minimise a given error function, i.e., the sum of squares error (SSE) Equation (5), based on a selected algorithm.

With the advancement of computing technology, nonlinear least-squares regression is now easier to implement with common mathematical tools such as Microsoft Excel's Solver. Moreover, nonlinear regression has been shown to be more stable and less susceptible to experimental error [32,34]:

$$SSE = \sum_{i=1}^{n}(y_c - y_e)_i^2 \quad (5)$$

where n is the number of experimental data points, y_c is the predicted (calculated) data with the Thomas model, and y_e is the experimental data.

3. Results and Discussion

3.1. Breakthrough Curve Data Fitting

Normally, fixed-bed column adsorption is a dynamic procedure where both the kinetics of the reaction as well as the mass balance are constantly changing [41]. The analysis of such experiments can be challenging as both these factors should be considered. A valid method for evaluating such studies that have been utilised in the literature is through mathematical modelling [31,34,41–43]. In this work, Thomas's model was chosen for data fitting through nonlinear regression analysis.

In Table 3, the results from data fitting of the fixed-bed column loading experiments using the FeCl$_2$ solution for the two different resins are illustrated. The high R^2 values obtained from all curved through nonlinear regression indicate a good fit of the Thomas model to experimental data. Only in the case of V with the VP OC 1026 resin the correlation coefficient is lower than 0.95, showing a slight deviation from the Thomas model's theoretical assumptions. The plotted breakthrough curves, as well as the extraction curves for Sc, Ti, Zr, and V, are depicted in Figures 2 and 3 for VP OC 1026 and TP 260 resins, accordingly.

Table 3. The results of nonlinear regression analysis with Thomas mathematical model for VP OC 1026 and TP 260 resin.

Resin	Metal	R^2	K_{Th} (mL/min g)	q_{Th} (mg/mL)
VP OC 1026	Sc	0.981	0.069	1.46
	Ti	0.965	0.036	2.2
	Zr	0.959	0.005	15.9
	V	0.887	0.009	4.06
TP 260	Sc	0.991	0.310	0.46
	Ti	0.978	0.043	1.98
	Zr	0.988	0.013	8.7
	V	0.973	0.039	4.86

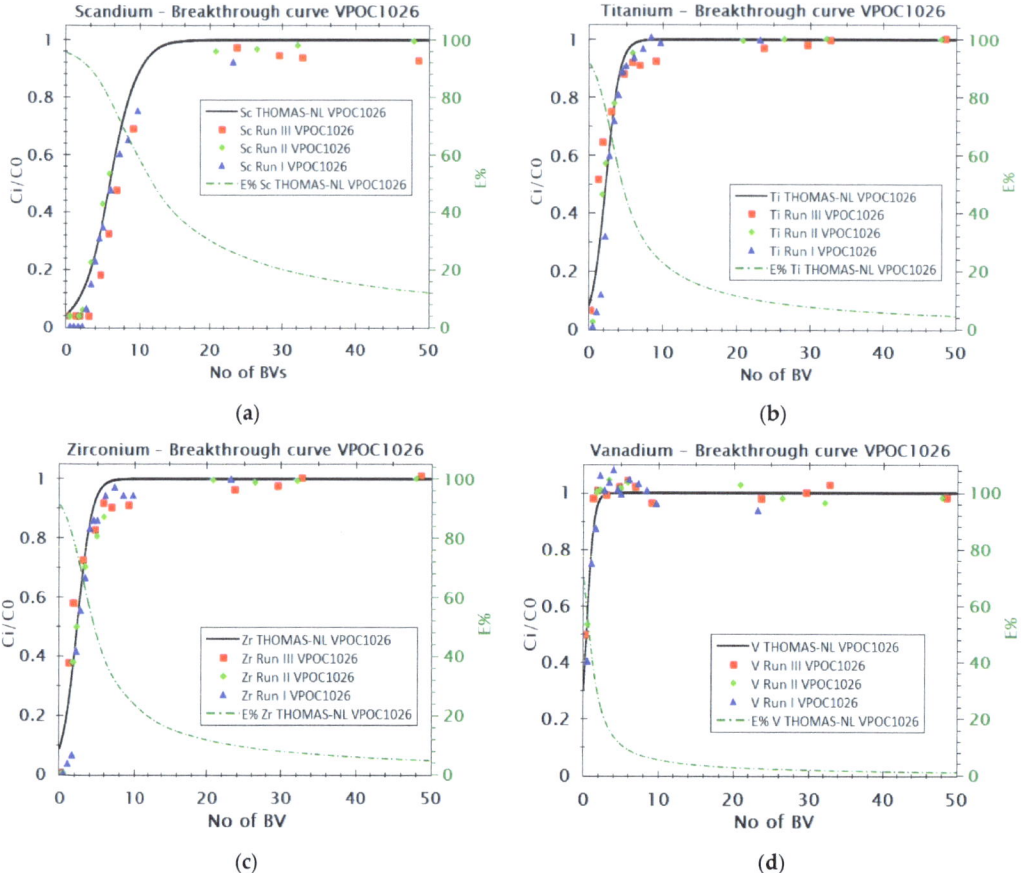

Figure 2. Extraction and breakthrough curves for VP OC 1026 resin for (**a**) Sc, (**b**) Ti, (**c**) Zr, and (**d**) V were measured in three different experimental runs and their Thomas nonlinear (NL) prediction mathematical model.

For the VP OC 1026 resin, the time to reach exhaustion ($C_i/C_0 = 0.9$) for Sc is estimated at 10BV for Zr and Ti 4.3BV, and 1BV for V according to Thomas nonlinear model (Figure 2). When the volume of effluent does not exceed 5.8 BV, the extraction of Sc from the $FeCl_2$ solution using VPOC1026 resin is greater than 80%, while the coextraction of other metals is at 9.5% for V, 39% for Ti and 40% for Zr.

According to the Thomas nonlinear model for TP 260 resin, the time to attain exhaustion ($Ci/C0 = 0.9$) for Sc is estimated at 3BV and 2BV for Zr, 4BV for Ti, and 1BV for V (Figure 3). Sc extraction from $FeCl_2$ solution using this resin is higher than 80% when the effluent volume is less than 2.6 BV, while the coextraction of other metals is 77% for Ti, 56.5% for Zr, and 27.5% for V.

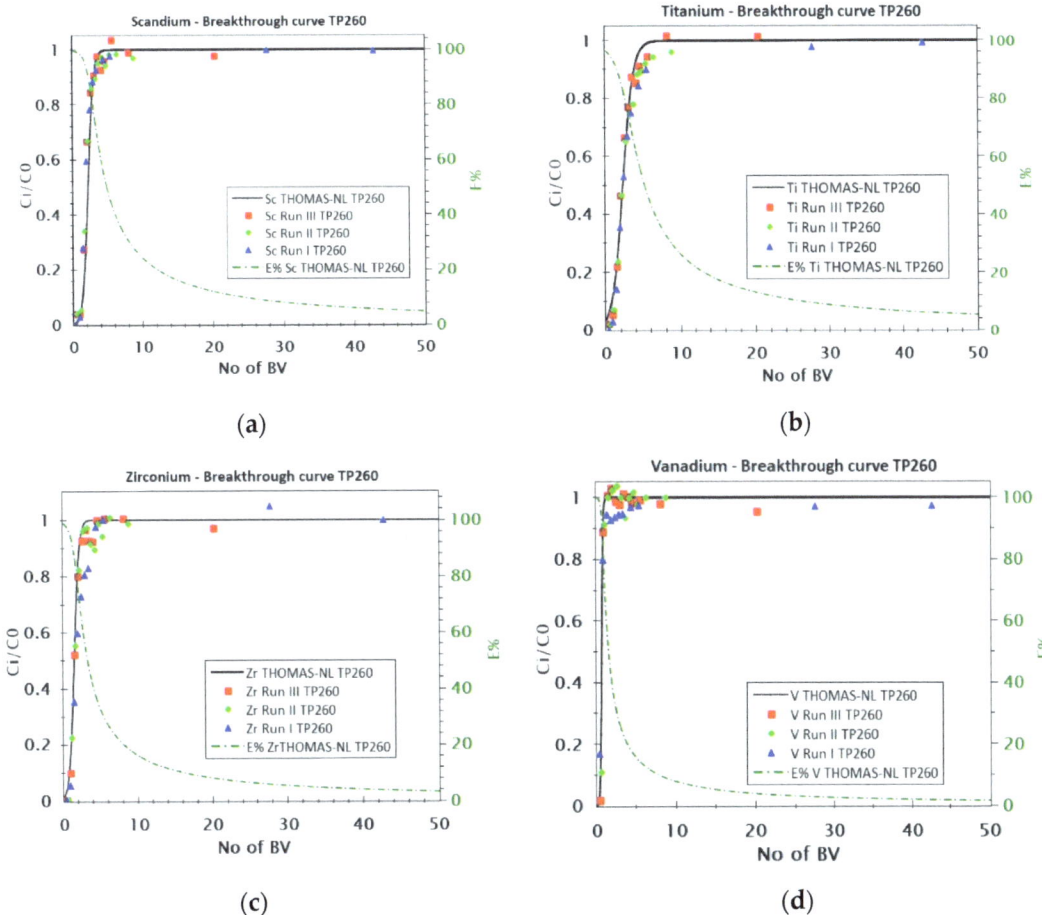

Figure 3. Extraction and breakthrough curves for TP 260 resin for (**a**) Sc, (**b**) Ti, (**c**) Zr, and (**d**) V were measured in three different experimental runs and their Thomas nonlinear (NL) prediction mathematical model.

3.2. Resins Adsorption Behavior

3.2.1. Adsorption Behavior of Sc, Ti, Zr, and V from $FeCl_2$ Solution with VP OC 1026 Resin

In Figure 4, the breakthrough curves for VP OC 1026 are compared for all the metals of interest. For V, the curve is steeper, and the breakpoint appears early, indicating a fast uptake making it the weakest adsorbed component. The curve plot is almost identical for Zr and Ti. Even though there are no references available on the adsorption of Zr and Ti on VP OC 1026 resin, it is reported that these metals interfere with Sc extraction in solvent extraction systems with D2EHPA, which is also the resin's functional group [26]. Sc breakpoint and saturation appear later than the other metals, showing that Sc is a strongly adsorbed component, confirming the high affinity of Sc with D2EHPA reported in other studies [44].

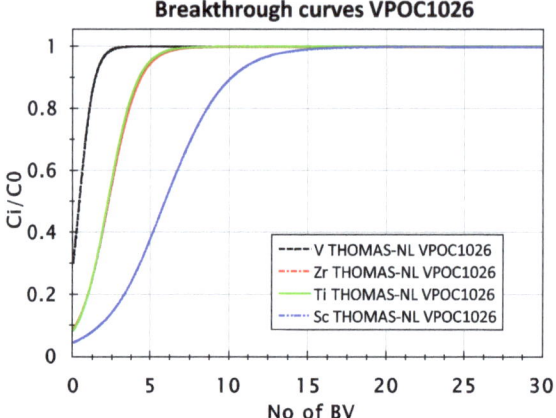

Figure 4. The comparative breakthrough curve for the metals of interest for VP OC 1026 resin according to Thomas nonlinear (NL) prediction mathematical model.

3.2.2. Adsorption Behavior of Sc, Ti, Zr, and V from $FeCl_2$ Solution with TP 260 Resin

TP 260 resin adsorbs the metals at a similar rate without showing any selectivity towards any metal (Figure 5). This observation is confirmed by the kinetic study of Rychkov et al. in sulfuric acid solutions using TP 260 resin, where it was observed that Zr, Ti, and Sc are adsorbed by the resin following the same kinetic model [21].

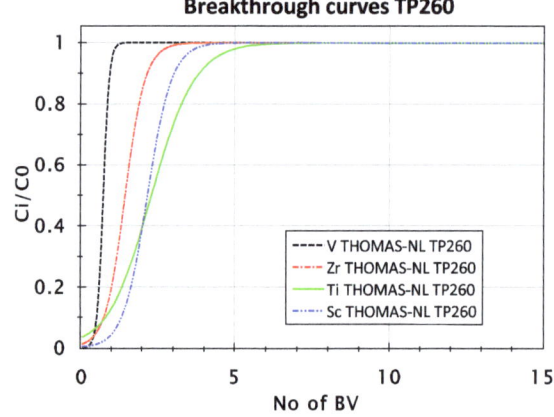

Figure 5. The comparative breakthrough curve for the metals of interest for TP 260 resin according to Thomas nonlinear (NL) prediction mathematical model.

3.2.3. Adsorption Behavior of Fe with VP OC 1026 and TP 260 Resins

In this section, the results regarding the adsorption behaviour of Fe are illustrated for both VP OC 1026 and TP 260 resin. The experimental breakthrough curves of Fe for both resins can be seen in Figure 6.

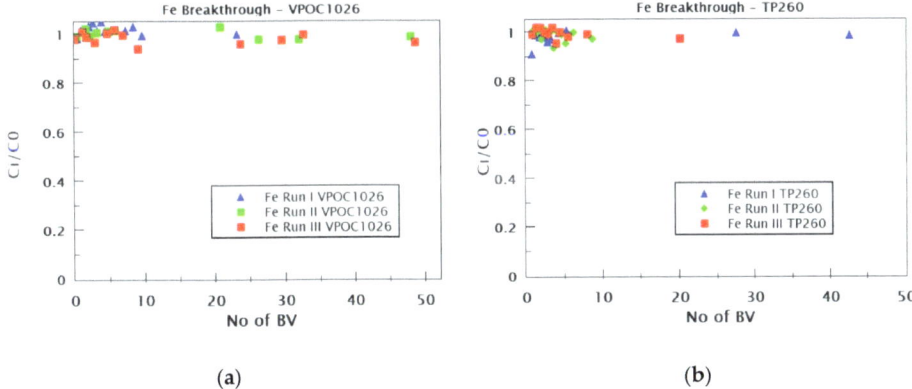

Figure 6. Chemical analysis results of Fe for VP OC 1026 (**a**) and TP 260 (**b**) resin throughout adsorption experiments.

Despite the high content of Fe in the feed solution, the results presented in Figure 6 show that no significant adsorption profile can be identified. This behaviour provides a huge advantage for the potential utilization of the examined resins, as the resins can operate by leaving the Fe content in the solution unchanged and separating it from Sc. The above observation is not in line with the research of Heres et al. [18], where iron was always coextracted with REEs from phosphate media, and it significantly reduced the loading of REEs, including Sc. This is probably due to either the different matrix of the solution examined (phosphoric media vs. chloride media) or the presence of trivalent Fe in these solutions. According to the study by Bao et al. [22], Fe(III) coextracts with Sc in cation solvent extraction and ion exchange, most likely because iron (III) competes more strongly with Sc than Fe(II) because of its similar charge density and size (II).

3.2.4. Comparison of VP OC 1026 and TP 260 Resins' Loading Capacity

The column capacity q (mg/mL), as a function of the effluent volume, can be integrated through the combination of Equations (1) and (3). The diagrams in Figure 6 compare the two resins for each metal to demonstrate the loading of the column as a function of the volume of effluent solution, expressed in bed volume (BV).

Mostajeran et al. [23] studied the adsorption behaviour of VP OC 1026 resin from synthetic sulfuric acid solution and defined 8 mg Sc/mL resin loading from a feed solution with 100–150 ppm of Sc. According to Mostajeran et al., this value is the maximum capacity of the resin in such a solution as a 1:1 molar ratio reaction between Sc and solvent molecules was confirmed. In our study, the highest Sc column capacity achieved with VP OC 1026 resin from a feed solution with 130 ppm Sc estimated 1.46 mg/mL resin (Figure 7.), less than what was expected from the cited reference. This reduced loading can be mainly justified by the high coextraction of the other metals that are present in the $FeCl_2$ solution.

In our study, TP 260 has a maximum Sc loading capacity of 0.31 mg/mL resin (Figure 7). The performance of TP 260 resin is much lower than what was expected from other studies. Bao et al. [22] determined that the loading of TP 260 resin is 30 mg Sc/mL resin at pH 2.5 and 16 mg Sc/mL at pH 1 from a pure synthetic sulfuric acid solution, confirming the adsorption dependence on pH [33]. Antisel et al. [7] determined the loading capacity of TP 260 6 mg Sc/mL resin from a polymetallic sulfuric pregnant leach solution containing Al, Co, Fe, Mn, and Ni at a pH of 3.5. These findings suggest that the adsorption capacity of the resin is decreased substantially when other metals coexist in the feed solution. The comparatively low performance of TP 260 resin for Sc adsorption, reported in our work, can be attributed to the high acidity of the feed solution but also to the presence of the other metals Ti, Zr, and V that are coloaded in the resin.

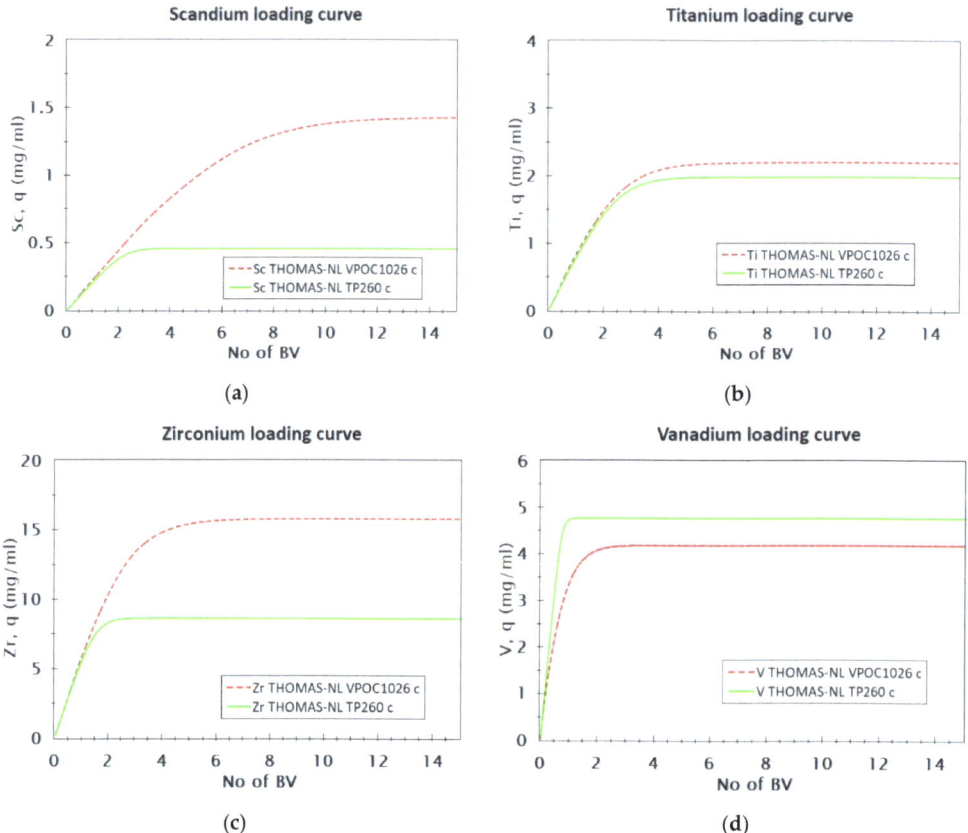

Figure 7. Breakthrough curve of (**a**) Sc, (**b**) Ti, (**c**) Zr, and (**d**) V for VP OC 1026 and TP 260 resin according to Thomas nonlinear (NL) prediction mathematical model.

Table 4 shows the ratio of metal concentration in VP OC 1026 resin to concentration in TP 260 resin.

Table 4. Ratios of metal's capacity in VP OC 1026 resin to TP 260 resin.

Metal, Me	$Me_{VPOC1026}/Me_{TP260}$
Sc	3.16
Zr	1.83
Ti	1.11
V	0.84

In VP OC 1026 resin, Sc loading (1.46 mg/mL resin) is more than three times higher, and Zr loading (15.9 mg/mL resin) is almost two times higher compared with TP 260 resin which is 0.46 mg/mL resin and 8.7 mg/mL resin, accordingly. The Ti loading is similar for both resins, 2 mg/mL resin, while V loading is slightly higher in TP 260 4.86 mg/mL resin compared with VP OC 1026 resin 4.06 mg/mL resin.

The mass ratios of Ti, Zr and V to Sc (Me/Sc) for the initial $FeCl_2$ solution and the two loaded resins are shown in Table 5. The ratio's lower values suggest a higher concentration of scandium.

Table 5. Ratios of metal impurities concentration to Sc concentration in the three examined phases: $FeCl_2$ solution, VP OC 1026 resin and TP 260 resin.

Ratio, Me/Sc	$FeCl_2$	VP OC 1026	TP 260
Zr/Sc	27.69	10.91	18.89
Ti/Sc	4.06	1.51	4.30
V/Sc	32.31	2.79	10.59

As shown in Table 5, the Me/Sc ratio is greater than 1 in every case, indicating metal impurities are always in excess. However, VP OC 1026 exhibits high selectivity for Sc extraction towards Zr, Ti, and V. The ratio Zr/Sc decreases from 27.69 in the initial feed solution to 10.91 in VP OC 1026 resin and for Ti is reduced from 4.06 to 1.51. Moreover, in VP OC 1026, the ratio of V/Sc is significantly reduced from 32.31 to 2.79. This suggests that the resin has a decreased affinity for V, despite its relatively high loading concentration in the resin (4.06 mg V/mL resin), which might be due to the high concentration of this metal in the feed solution.

The Ti/Sc ratio for TP 260 resin is raised to 4.3, compared to 4.06 in the original feed solution, showing that TP 260 resin has a larger affinity for Ti extraction towards Sc. All the metal ratios towards Sc are higher with TP 260 resin compared with the VP OC 1026, meaning that the latter is overall more selective for Sc in the given feed solution.

4. Conclusions

In this study, the Sc extraction through ion exchange technique was examined. The feed solution was an acidic $FeCl_2$ solution obtained from the TiO_2 industry. Two different commercial resins, VP OC 1026 and TP 260, were examined and compared under a standard loading experiment in a fixed-bed column setup. The capacity, the breakthrough, and the exhaustion of the column for Sc, but also for the basic metal impurities Zr, Ti and V were determined through mathematical modelling of the column. The Thomas model provided a very good fit to the experimental data, with R^2 higher than 0.98 for Sc, higher than 0.95 for Zr and Ti and higher than 0.88 for V.

For both resins, the Fe adsorption is insignificant, suggesting that the resins could be utilised for separating Sc from the solution without affecting the initial Fe content of the feed solution, which was an important objective of this study. The results indicate that VP OC 1026 is the most promising resin for scandium extraction, showing a higher capacity of Sc 1.46mg/mL in the given experimental conditions. The metals Zr, V, and T present in the initial solution are also coextracted in both resins' tests. It is important to note that the performance of the resins for Sc extraction should not be assessed merely based on the column's capacity. Even though Zr, V, and Ti have higher capacity values in the loaded resins, especially in VP OC 1026, the upgrade in Sc concentration is higher than the other metals. This indicates a high affinity of VP OC 1026 for Sc, while TP 260 resin shows a higher affinity for Ti.

Author Contributions: Conceptualization, E.M., D.M, A.T. and D.P.; methodology, E.M., A.T. and D.M.; validation, D.M, A.T. and A.P.; data curation E.M. and A.P.; writing—original draft preparation, E.M., D.M. and A.P.; writing—review and editing, E.B. and D.P.; supervision E.B. and D.P.; All authors have read and agreed to the published version of the manuscript.

Funding: This research was funded by EIT-KIC, ScaVanger Project (2021–2024), project number 20093.

Data Availability Statement: The data presented in this study are available on request from the corresponding author.

Acknowledgments: The resins used in this paper were provided by MEAB, Germany.

Conflicts of Interest: The authors declare no conflict of interest.

References

1. Shibata, J.; Murayama, N. Solvent Extraction of Scandium from the Waste Solution of TiO$_2$ Production Process. *Trans. Indian Inst. Met.* **2016**, *70*, 471–477. [CrossRef]
2. Hartley, C.J.; Hazen, W.W.; Baughman, D.R.; Bemelmans, C.M.A.; Belits, P.F.; Lanyk, T.J.; Porter, B.F.; Liao, L.; McAllister, J.; Yang, M.S.-Y. Methods of Recovering Scandium from Titanium Residue Streams. U.S. Patent No. 9,102,999, 11 August 2015.
3. Botelho Junior, A.B.; Espinosa, D.C.R.; Vaughan, J.; Tenório, J.A.S. Recovery of scandium from various sources: A critical review of the state of the art and future prospects. *Miner. Eng.* **2021**, *172*, 107148. [CrossRef]
4. European Commision. *Critical Raw Materials Resilience: Charting a Path towards greater Security and Sustainability*; European Commision: Brussels, Belgium, 2020.
5. Petrakova, O.; Kozyrev, A.; Suss, A.; Panov, A.; Gorbachev, S.; Perestoronina, M.; Vishnyakov, S. BR04-Industrial Trials Results of Scandium Oxide Recovery from Red Mud at UC RUSAL Alumina Refineries. *Ser. Geol. Tech. Sci.* **2020**, *4*, 156–165.
6. Zhang, L.; Zhang, T.-A.; Lv, G.; Zhang, W.; Li, T.; Cao, X. Separation and Extraction of Scandium from Titanium Dioxide Waste Acid. *JOM* **2021**, *73*, 1301–1309. [CrossRef]
7. Altinsel, Y.; Topkaya, Y.; Kaya, Ş.; Şentürk, B. *Extraction of Scandium from Lateritic Nickel-Cobalt Ore Leach Solution by Ion Exchange: A Special Study and Literature Review on Previous Works*; Springer: Cham, Switzerland, 2018; pp. 1545–1553. [CrossRef]
8. Li, Y.; Li, Q.; Zhang, G.; Zeng, L.; Cao, Z.; Guan, W.; Wang, L. Separation and recovery of scandium and titanium from spent sulfuric acid solution from the titanium dioxide production process. *Hydrometallurgy* **2018**, *178*, 1–6. [CrossRef]
9. Chen, Y.; Ma, S.; Ning, S.; Zhong, Y.; Wang, X.; Fujita, T.; Wei, Y. Highly efficient recovery and purification of scandium from the waste sulfuric acid solution from titanium dioxide production by solvent extraction. *J. Environ. Chem. Eng.* **2021**, *9*, 106226. [CrossRef]
10. Chernoburova, O.; Changes, A. The Future of Scandium Recovery from Wastes. In Proceedings of the International Conference on Raw Materials and Circular Economy, Athens, Greece, 5–9 September 2021.
11. Zhou, J.; Ning, S.; Meng, J.; Zhang, S.; Zhang, W.; Wang, S.; Chen, Y.; Wang, X.; Wei, Y. Purification of scandium from concentrate generated from titanium pigments production waste. *J. Rare Earths* **2020**, *39*, 194–200. [CrossRef]
12. Qiu, H.; Wang, M.; Xie, Y.; Jianfeng, S.; Huang, T.; Li, X.-M. From trace to pure: Recovery of scandium from the waste acid of titanium pigment production by solvent extraction. *Process Saf. Environ. Prot.* **2018**, *121*, 118–124. [CrossRef]
13. Yagmurlu, B.; Orberger, B.; Dittrich, C.; Croisé, G.; Scharfenberg, R.; Balomenos, E.; Panias, D.; Mikeli, E.; Maier, C.; Schneider, R.; et al. Sustainable Supply of Scandium for the EU Industries from Liquid Iron Chloride Based TiO$_2$ Plants. *Mater. Proc.* **2021**, *5*, 86.
14. Wang, W.; Pranolo, Y.; Cheng, C.Y. Metallurgical processes for scandium recovery from various resources: A review. *Hydrometallurgy* **2011**, *108*, 100–108. [CrossRef]
15. Zhou, G.; Li, Q.; Sun, P.; Guan, W.; Zhang, G.; Cao, Z.; Zeng, L. Removal of impurities from scandium chloride solution using 732-type resin. *J. Rare Earths* **2018**, *36*, 311–316. [CrossRef]
16. Shaoquan, X.; Suqing, L. Review of the extractive metallurgy of scandium in China (1978–1991). *Hydrometallurgy* **1996**, *42*, 337–343. [CrossRef]
17. Van Nguyen, N.; Iizuka, A.; Shibata, E.; Nakamura, T. Study of adsorption behavior of a new synthesized resin containing glycol amic acid group for separation of scandium from aqueous solutions. *Hydrometallurgy* **2016**, *165*, 51–56. [CrossRef]
18. Hérès, X.; Blet, V.; Di Natale, P.; Ouaattou, A.; Mazouz, H.; Dhiba, D.; Cuer, F. Selective Extraction of Rare Earth Elements from Phosphoric Acid by Ion Exchange Resins. *Metals* **2018**, *8*, 682. [CrossRef]
19. Smirnov, A.; Titova, S.; Rychkov, V.; Bunkov, G.; Semenishchev, V.; Kirillov, E.; Poponin, N.; Svirsky, I. Study of scandium and thorium sorption from uranium leach liquors. *J. Radioanal. Nucl. Chem.* **2017**, *312*, 277–283. [CrossRef]
20. Ivanov, V.; Abilmagzhanov, A.; Shokobayev, N.; Adelbayev, I.; Nurtazina, A. Scandium Extraction By Phosphorus-Containing Sorbents. *NAS RK. Ser. Geol. Tech. Sci. Sci. J.* **2020**, *4*, 156–165.
21. Rychkov, V.N.; Nalivayko, K.A.; Titova, S.M.; Abakumova, E.V.; Yakovleva, O.V.; Kirillov, E.V.; Skripchenko, S.Y. Kinetics of scandium sorption from sulfuric acid solutions by ampholyte Lewatit TP260. *AIP Conf. Proc.* **2020**, *2313*, 050028. [CrossRef]
22. Bao, S.; Hawker, W.; Vaughan, J. Scandium Loading on Chelating and Solvent Impregnated Resin from Sulfate Solution. *Solvent Extr. Ion Exch.* **2018**, *36*, 100–113. [CrossRef]
23. Mostajeran, M.; Bondy, J.-M.; Reynier, N.; Cameron, R. Mining value from waste: Scandium and rare earth elements selective recovery from coal fly ash leach solutions. *Miner. Eng.* **2021**, *173*, 107091. [CrossRef]
24. Meshkov, E.Y.; Akimova, I.; Bobyrenko, N.; Solov'ev, A.; Klochkova, N.; Savel'ev, A. Separation of Scandium and Thorium in the Processing of Scandium Rough Concentrate Obtained from Circulating Solutions of In-Situ Leaching of Uranium. *Radiochemistry* **2020**, *62*, 652–657. [CrossRef]
25. Zhu, L.; Liu, Y.; Chen, J.; Liu, W. Extraction of Scandium(III) Using Ionic Liquids Functionalized Solvent Impregnated Resins. *J. Appl. Polym. Sci.* **2011**, *120*, 3284–3290. [CrossRef]
26. Wang, W.; Cheng, C.Y. Separation and purification of scandium by solvent extraction and related technologies: A review. *J. Chem. Technol. Biotechnol.* **2011**, *86*, 1237–1246. [CrossRef]
27. Roosen, J.; Van Roosendael, S.; Borra, C.R.; Van Gerven, T.; Mullens, S.; Binnemans, K. Recovery of scandium from leachates of Greek bauxite residue by adsorption on functionalized chitosan–silica hybrid materials. *Green Chem.* **2016**, *18*, 2005–2013. [CrossRef]

28. Wood, S.A.; Samson, I.M. The aqueous geochemistry of gallium, germanium, indium and scandium. *Ore Geol. Rev.* **2006**, *28*, 57–102. [CrossRef]
29. Reynier, N.; Gagné-Turcotte, R.; Coudert, L.; Costis, S.; Cameron, R.; Blais, J.-F. Bioleaching of Uranium Tailings as Secondary Sources for Rare Earth Elements Production. *Minerals* **2021**, *11*, 302. [CrossRef]
30. Sokolova, Y.V. Sorption of Sc(III) on phosphorus-containing cation exchangers. *Russ. J. Appl. Chem.* **2006**, *79*, 573–578. [CrossRef]
31. Li, Y.; Zhu, Y.; Zhu, Z.; Zhang, X.; Wang, D.; Xie, L. Fixed-Bed Column Adsorption Of Arsenic(V) By Porous Composite Of Magnetite/Hematite/Carbon With Eucalyptus Wood Microstructure. *J. Environ. Eng. Landsc. Manag.* **2018**, *26*, 38–56. [CrossRef]
32. González-López, M.E.; Laureano-Anzaldo, C.M.; Pérez-Fonseca, A.A.; Arellano, M.; Robledo-Ortíz, J.R. A discussion on linear and non-linear forms of Thomas equation for fixed-bed adsorption column modeling. *Rev. Mex. De Ing. Química* **2021**, *20*, 875–884. [CrossRef]
33. Helfferich, F.G. *Ion Exchange*; Courier Corporation: Toronto, Canada, 1995.
34. Han, R.; Wang, Y.; Zou, W.; Wang, Y.; Shi, J. Comparison of linear and nonlinear analysis in estimating the Thomas model parameters for methylene blue adsorption onto natural zeolite in fixed-bed column. *J. Hazard. Mater.* **2007**, *145*, 331–335. [CrossRef]
35. Goel, J.; Kadirvelu, K.; Rajagopal, C.; Kumar Garg, V. Removal of lead(II) by adsorption using treated granular activated carbon: Batch and column studies. *J. Hazard. Mater.* **2005**, *125*, 211–220. [CrossRef]
36. Thomas, H.C. Heterogeneous Ion Exchange in a Flowing System. *J. Am. Chem. Soc.* **1944**, *66*, 1664–1666. [CrossRef]
37. Gallindo, A.D.A.S.; Silva Junior, R.A.D.; Rodrigues, M.G.F.; Ramos, W.B. Modelling and simulation of the ion exchange process for Zn^{2+}(aq) removal using zeolite NaY. *Res. Soc. Dev.* **2021**, *10*, e310101220362. [CrossRef]
38. Lin, L.-C.; Li, J.-K.; Juang, R.-S. Removal of Cu(II) and Ni(II) from aqueous solutions using batch and fixed-bed ion exchange processes. *Desalination* **2008**, *225*, 249–259. [CrossRef]
39. Xiao, Y.; Azaiez, J.; Hill, J.M. Erroneous Application of Pseudo-Second-Order Adsorption Kinetics Model: Ignored Assumptions and Spurious Correlations. *Ind. Eng. Chem. Res.* **2018**, *57*, 2705–2709. [CrossRef]
40. Tran, H.N.; You, S.-J.; Hosseini-Bandegharaei, A.; Chao, H.-P. Mistakes and inconsistencies regarding adsorption of contaminants from aqueous solutions: A critical review. *Water Res.* **2017**, *120*, 88–116. [CrossRef]
41. Weber, T.W.; Chakravorti, R.K. Pore and solid diffusion models for fixed-bed adsorbers. *AIChE J.* **1974**, *20*, 228–238. [CrossRef]
42. Lim, A.P.; Aris, A.Z. Continuous fixed-bed column study and adsorption modeling: Removal of cadmium (II) and lead (II) ions in aqueous solution by dead calcareous skeletons. *Biochem. Eng. J.* **2014**, *87*, 50–61. [CrossRef]
43. Fallah, N.; Taghizadeh, M. Continuous fixed-bed adsorption of Mo(VI) from aqueous solutions by Mo(VI)-IIP: Breakthrough curves analysis and mathematical modeling. *J. Environ. Chem. Eng.* **2020**, *8*, 104079. [CrossRef]
44. Wang, W.; Pranolo, Y.; Cheng, C.Y. Recovery of scandium from synthetic red mud leach solutions by solvent extraction with D2EHPA. *Sep. Purif. Technol.* **2013**, *108*, 96–102. [CrossRef]

Review

Methacrylate-Based Polymeric Sorbents for Recovery of Metals from Aqueous Solutions

Aleksandra Nastasović [1], Bojana Marković [1], Ljiljana Suručić [2] and Antonije Onjia [3,*]

[1] Institute of Chemistry, Technology and Metallurgy, University of Belgrade, Njegoševa 12, 11000 Belgrade, Serbia; aleksandra.nastasovic@ihtm.bg.ac.rs (A.N.); bojana.markovic@ihtm.bg.ac.rs (B.M.)
[2] Faculty of Medicine, University of Banja Luka, Save Mrkalja 14, 78000 Banja Luka, Bosnia and Herzegovina; ljiljana.surucic@med.unibl.org
[3] Faculty of Technology and Metallurgy, University of Belgrade, Karnegijeva 4, 11000 Belgrade, Serbia
* Correspondence: onjia@tmf.bg.ac.rs

Abstract: The industrialization and urbanization expansion have increased the demand for precious and rare earth elements (REEs). In addition, environmental concerns regarding the toxic effects of heavy metals on living organisms imposed an urgent need for efficient methods for their removal from wastewaters and aqueous solutions. The most efficient technique for metal ions removal from wastewaters is adsorption due to its reversibility and high efficiency. Numerous adsorbents were mentioned as possible metal ions adsorbents in the literature. Chelating polymer ligands (CPLs) with adaptable surface chemistry, high affinity towards targeted metal ions, high capacity, fast kinetics, chemically stable, and reusable are especially attractive. This review is focused on methacrylate-based magnetic and non-magnetic porous sorbents. Special attention was devoted to amino-modified glycidyl methacrylate (GMA) copolymers. Main adsorption parameters, kinetic models, adsorption isotherms, thermodynamics of the adsorption process, as well as regeneration of the polymeric sorbents were discussed.

Keywords: magnetic sorbents; EGDMA; heavy metal(loid)s; kinetics; isotherms; adsorption mechanisms; desorption; reusability

1. Introduction

Due to the rapid technological development, large amounts of heavy, precious, and rare metals cause contamination of water resources [1,2]. Major pollutants of marine and ground waters are toxic heavy metals such as nickel, chromium, lead, zinc, arsenic, cadmium, selenium, and uranium, originating from mining, metal processing, pesticides, pharmaceuticals, etc. [3]. Moreover, since they exist in a dissolved ion state in wastewaters, heavy metals accumulate in living organisms, causing serious pollution and health problems [3,4].

Precious metals (primarily gold, silver, platinum, palladium, and rhodium) find their applications in specialized fields (catalysis, electronic devices, and jewelry) [5]. Rare earth elements (REEs) signify a group of 17 chemically similar elements representing about 17% of the total quantity of all naturally occurring elements [6]. They have wide applications in electronics, optics, catalysis, green energy technologies, etc. The growing demand for these metals, key ingredients of modern technologies, is a source of numerous and severe ecological and economic issues. Thus, the efficient removal and recovery of these metals prior to their discharge into the environment is nowadays the main focus of scientific research [3,4,7–9].

Various conventional techniques for the removal and recovery of metal ions, as well as environmental detoxification, are adsorption, chemical precipitation, electrochemical extractions, membrane filtration, biosorption, ion exchange, evaporation, flotation, and oxidation [3,8,10,11].

Citation: Nastasović, A.; Marković, B.; Suručić, L.; Onjia, A. Methacrylate-Based Polymeric Sorbents for Recovery of Metals from Aqueous Solutions. *Metals* **2022**, *12*, 814. https://doi.org/10.3390/met12050814

Academic Editor: Felix A. Lopez

Received: 31 March 2022
Accepted: 3 May 2022
Published: 8 May 2022

Publisher's Note: MDPI stays neutral with regard to jurisdictional claims in published maps and institutional affiliations.

Copyright: © 2022 by the authors. Licensee MDPI, Basel, Switzerland. This article is an open access article distributed under the terms and conditions of the Creative Commons Attribution (CC BY) license (https://creativecommons.org/licenses/by/4.0/).

The most popular among them is adsorption, which is proven to be economical and effective. A great variety of articles published on the usage of different adsorbents can be found in the literature, such as: agro-industrial waste [12–14]; polymers and polymer membranes [15–18]; organic-inorganic hybrid polymers [19–21]; bioadsorbents [22]; chitosan; modified chitosan and chitosan-based hybrid composites [23–26]; biogenic iron compounds [27]; polymer hydrogels [28,29]; acrylamide-based microgels [30,31]; geopolymers [32,33]; silica [34]; biochar [35]; carbon nanotubes and graphene [7]; green activated magnetic graphitic carbon oxide [36]; magnetic nanoparticles [1]; nanomaterials [37], etc.

2. Methodology

In order to find the literature related to this topic, we used keywords such as "polymeric metal ions sorbents," "methacrylate-based sorbents," "post-functionalization," "magnetic polymers and composites," "non-linear and linear kinetic models," "adsorption isotherms," etc., in online searching tools including Web of Science, Scopus, Google scholar, MDPI, and ScienceDirect websites. We predominantly focused on studies published in the last 15 years, citing the older papers as well, if they were relevant. In addition, we equally consulted and cited scientific research papers as well as Reviews. Most of these articles were cited in the Literature section, as a guide for researchers in this field and as a starting point for their further research.

3. Natural Polymers, Inorganic Materials, and Polymer/Inorganic Composite as Metal Ions Sorbents

Natural materials, such as aluminosilicates and clays [38], have been used as metal ions adsorbents due to their favorable characteristics, such as high ion exchange capacity, chemical inertness, and low toxicity) [39]. Đolić et al. tested Cu^{2+} and Zn^{2+} sorption on different natural materials (clays and clay minerals), natural-modified (activated carbon and alumina), and synthetic (zeolite, titanium dioxide (TiO_2), and ion exchange resin) [38].

Smičiklas et al. investigated sorption performances of zeolite and hydroxyapatite (HAP) towards Cu(II) ions [40]. The effect of zeolite particle size and sorption parameters such as initial meal concentration, agitation speed, and adsorbent mass on Cu(II) sorption kinetics was examined. The volumetric mass transfer coefficient (k_{fa}) and effective diffusion coefficient (D_{eff}) using single resistance mass transfer models were calculated. Their results showed that the controlling step in the case of sorption with HAP was only pore diffusion, while sorption with zeolite sorption was governed by film diffusion within the first sorption stage (up to 10 min), followed by diffusion inside the pores. As a continuation of their research, Smičiklas et al. examined the possibility of enhancing of sorption performances of biogenic HAP (BHAP) for lead, copper, nickel, cadmium, and zinc (Pb, Cu, Ni, Cd, and Zn) by the corresponding functionalization, i.e., condensation reaction of surface hydroxyl groups in BHAP and hydroxyl groups from caffeic acid (CA) and 3,4-dihydroxybenzoic acid (3,4-DHBA) [41]. The sorption results showed selectivity towards Pb ions from mixed equimolar solutions of investigated metal ions (Pb, Cu, Ni, Cd, and Zn ions) by all investigated sorbents. It was observed that the sorption capacities of functionalized BHAP were higher in comparison with unmodified ones.

Meseldžija et al. used agroindustrial waste, i.e., unmodified lemon peel, to test Cu(II) removal efficiency from aqueous solutions and wastewater [13]. The effects of sorption parameters (pH, adsorption time, initial ion concentration, and adsorbent dose) on sorption efficiency were studied in batch experiments. The maximum Langmuir adsorption capacity was 13.2 mg/g at an optimum contact time of 15 min. A high value for Cu(II) ions removal efficiency (89%) from mining wastewater at natural pH (pH 3.0) was observed.

One of the most commonly known natural polymeric sorbents is chitosan, which originates from crustacean shells (crabs and prawns) [42]. The presence of $-NH_2$ and -OH groups on the polymeric chains provides chelating and reaction sites [43]. The main advantages of chitosan are the high density of functional groups, easy functionalization, non-toxicity, biocompatibility, biodegradability, etc. Chitosan can be used as raw material, as well as after physical (preparation in the shape of membranes, fibers, and spherical beads of different sizes and porosities) or chemical modifications (impregnation, crosslinking, graft polymerization, and composite preparation) [43,44].

Since chitosan is soluble in most diluted minerals and organic acids, it has to be chemically stabilized by crosslinking. Laus et al. modified chitosan by incorporating epichlorohydrin (via a covalent crosslinking reaction) and triphosphate (via an ionic crosslinking reaction), and is used for Cu(II), Cd(II), and Pb(II) ions adsorption and desorption [45]. The optimum adsorption pH values were 6.0 for Cu(II), 7.0 for Cd(II), and 5.0 for Pb(II). The adsorption process was best fitted with the pseudo-second-order and Langmuir isotherm models. Maximum Cu(II), Cd(II), and Pb(II) ions adsorption capacities were 130.72, 83.75, and 166.94 mg/g, respectively. The best desorption was observed with nitric and hydrochloric acid.

Chitosan-poly(maleic acid) nanomaterial, obtained by grafting poly(maleic acid) onto chitosan and crosslinked with glutaraldehyde, was used for mercury, lead, copper, cadmium, cobalt, and zinc (Hg(II), Pb(II), Cu(II), Cd(II), Co(II), and Zn(II)) ions adsorption [46]. The obtained material was selective for Hg(II) ions, with a maximum sorption capacity of 1044 mg/g at pH 6.0.

The increasing interest in mesoporous silica originates from its favorable porous characteristics, i.e., large surface area in the range 600–1000 m^2/g, narrow pore-size distributions, and large and controlled pore size (5–30 nm), which results in fast kinetics of metal ions adsorption [34]. The sorption performances could be improved by co-condensation and post-synthesis grafting functionalization.

Lee et al. tested mesoporous silica materials functionalized with amino and mercapto groups (fiber-like, rod-like, and platelets) as Cu(II) and Pb(II) adsorbents [47]. It was concluded that thiol-functionalized mesoporous silica adsorbents have a better affinity for Pb(II) compared to amino-mesoporous silica. On the other hand, amino-mesoporous silica has a stronger affinity for Cu(II) ions.

However, grafting could reduce the pore size of the modified mesoporous materials, particularly when grafting is performed with bulky functional groups. This might cause a decrease in diffusion to the adsorption sites and, consequently, reduce adsorption capacity. Mureseanu et al. [48] observed a sharp decrease in silica pore volume and surface area after grafting. For example, a surface area decrease of 53% was observed for aminopropyl functionalized mesoporous silica compared to an initial mesoporous silica support.

Shiraishi et al. studied the adsorption of Cu(II) on various inorganic adsorbents (silica gel, aluminum oxide, etc.), functionalized with ethylenediaminetetraacetic acid (EDTA) and diethylenetriaminepentaacetic acid (DTPA) [49]. The Cu(II) removal capacity of DTPA-modified silica was considerably lower compared to EDTA-modified silica, suggesting that the adsorption was restricted due to the decrease in pore sizes and pore blockage when the bulkier DTPA was attached to the material.

The composite of mesoporous silica functionalized with (3-chloropropyl) triethoxysilane with incorporated tetrakis(4-hydroxyphenyl) porphyrin was used as Pb(II) ions adsorbent [50]. The optimal pH values for Pb(II) and for Cu(II) ions sorption were 2–6 and 5, respectively. The maximum Pb(II) adsorption capacity of the composite was 134 mg/g. It was observed that the presence of porphyrin in silica causes a significant increase in heavy metal ion adsorption. The adsorption process is well fitted with Langmuir isotherm, while the kinetics obeys the pseudo-second-order kinetics.

Wang et al. used magnetic multiwall magnetic carbon nanotubes (6O-MWCNTs@Fe_3O_4) as Cd(II), Ni(II), Zn(II), Cu(II), and Pb(II) sorbent. The rapid sorption was observed, with 30 min needed to attain equilibrium [51]. It was shown that 6O-MWCNTs@Fe_3O_4 exhibits good selective Pb(II) adsorption performances, with a high Pb(II) maximum adsorption capacity of 215.05 mg/g, much higher than the existing adsorption capacity of this type of adsorbent. The adsorption capacities for Cu(II) and Cd(II) were 87.1 mg/g and 57.3 mg/g, respectively.

4. Polymeric Sorbents

A great variety of homopolymers, copolymers, and polymer nanocomposites synthesized by free radical polymerization, radiation polymerization, graft polymerization, oxidation polymerization, dispersion/suspension, etc., have been used as metal ions sorbents.

For example, membrane-supported crosslinked poly(acrylamide-2-methylpropane sulfonic acid) hydrogel [17], amidoxime chelating polymer [18], conjugated polymers [19], aminated-glycidyl methacrylate polypropylene adsorbent [52], magnetic glycidyl methacrylate-based polymer grafted with diethylenetriamine [53] were used as heavy metal ions sorbents.

Numerous research articles, reviews, and patents have been published regarding the synthesis and usage of polymeric sorbents for the sorption of precious metals [54,55]. According to hard-soft acid-base (HSAB) theory, functional groups bearing S and N donor atoms can strongly interact with precious metals [56]. Therefore, polymeric sorbents selective for precious metals often possess groups such as thiourea [57,58], thiazole [59], dithiocarbonate [60], amino [61], imino [62], etc.

Numerous new functional polymeric sorbents selective for rare earth elements (REEs) have been designed in the last decade. For example, terpolymer of styrene-divinylbenzene and glycidyl methacrylate (GMA) with diglycolamic acid ligands was used for adsorption of neodymium (Nd(III)) and dysprosium (Dy(III)) ions [63]. Galhoum et al. used methylene phosphonic groups grafted on poly(glycidyl methacrylate) with incorporated diethylenetriamine groups for adsorption of lanthanum (La(III) and yttrium (Y(IIII)) ions [64].

The properties of polymeric sorbents, such as functional groups and surface chemistry, particle size, porosity, hydrophobicity, polymer chain size, and molecular weight distribution, can be controlled by the synthesis and/or functionalization conditions and parameters [65]. In addition, polymeric sorbents can be regenerated and reused in a number of sorption/desorption cycles, which justifies the production costs.

For example, hydrogels are very interesting polymeric materials, insoluble due to the presence of chemical or physical crosslinks but swellable. They could incorporate various functional groups and, thus, be used for heavy metal ions adsorption. Stajčić et al. prepared polyethersulfone membranes with an integrated negatively-charged poly(acrylamido-2-methylpropane sulfonic acid) hydrogel [17]. An intramembrane diffusion model was used to describe Cu(II) and Cd(II) sorption kinetics. The calculated apparent diffusion coefficients were $6.26 \cdot 10^{-10}$ m^2/s for Cd(II) and $7.15 \cdot 10^{-10}$ m^2/s for Cu(II), i.e., 2–3 times larger than in commercial ion-exchange resins.

Poly(2-hydroxyethyl acrylate-*co*-itaconic acid), P(HEA/IA), hydrogels synthesized using free radical crosslinking/copolymerization and used as Pb(II) sorbents from aqueous solutions [66]. It was observed that the parameters such as metal ions' initial concentration, pH, adsorbent dose, ionic strength, and temperature, strongly influenced the metal sorption. The best fit was obtained with the Redlich–Peterson isotherm and pseudo-second-order kinetic model. The maximum sorption capacities for Pb(II) ions were 392.2 and 409.8 mg/g for hydrogel samples with IA mole fractions of 2.0 and 10.0 P(HEA/2IA) and P(HEA/10IA), respectively. In multi-component system, selectivity decreased in the following order: Pb(II) > Cu(II) > Zn(II) > Cd(II) > Ni(II) > Co(II). Sorption/desorption experiments showed that the P(HEA/IA) hydrogels could be reused without significant loss after three adsorption-desorption cycles. Maximum desorption of 95.2 was observed for Pb(II) at 0.1 M HNO$_3$.

Chelating polymer ligands (CPLs) recently became attractive as promising efficient sorbents for metal ions. Up to this moment, various CPLs such as linear, branched, crosslinked, grafted polymers, dendrimers, star-shaped, and hyperbranched polymers with almost endless possibilities of CPLs design, rich coordination chemistry, their high affinity for various metals depending on the type of the ligand, combined with chemical stability, regenerability, and reusability have been used for metal sorption [15].

The synthesis of CPLs polymers includes three basic approaches: (co)polymerization of monomers with chelating fragments; monomers that already have ligand in their structure, polycondensation, and post-polymerization (PPM), in order to introduce ligand groups in previously synthesized (co)polymer.

(Co)polymerization of monomers with chelating fragments is considered to be the simplest method that can be performed by free-radical (co)polymerization, living/controlled radical polymerization, metathesis polymerization, grafted polymerization, etc. The major polymerization techniques for the preparation of CPLs include bulk, precipitation, suspension, emulsion, and dispersion polymerization. More details could be found in the literature [15,67]. PPM, or polymer-analogous modification, is an approach that enables the polymerization of monomers with functional groups inert towards the polymerization conditions, which can be converted through an additional reaction step into a variety of different functional groups [68]. Thus, the obtained functional polymers have identical average chain lengths and chain-length distributions and diverse functional groups.

5. Adsorption of Metal Ions on Polymeric Sorbents—General Remarks

Metal sorption from aqueous solutions proceeds through the following mechanisms: coordination, precipitation, ion exchange, Van der Waals, and electrostatic interactions, depending on solution pH, type of the active sites on the sorbent surface, point of zero charge of polymeric sorbent, etc. (Figure 1) [5,69].

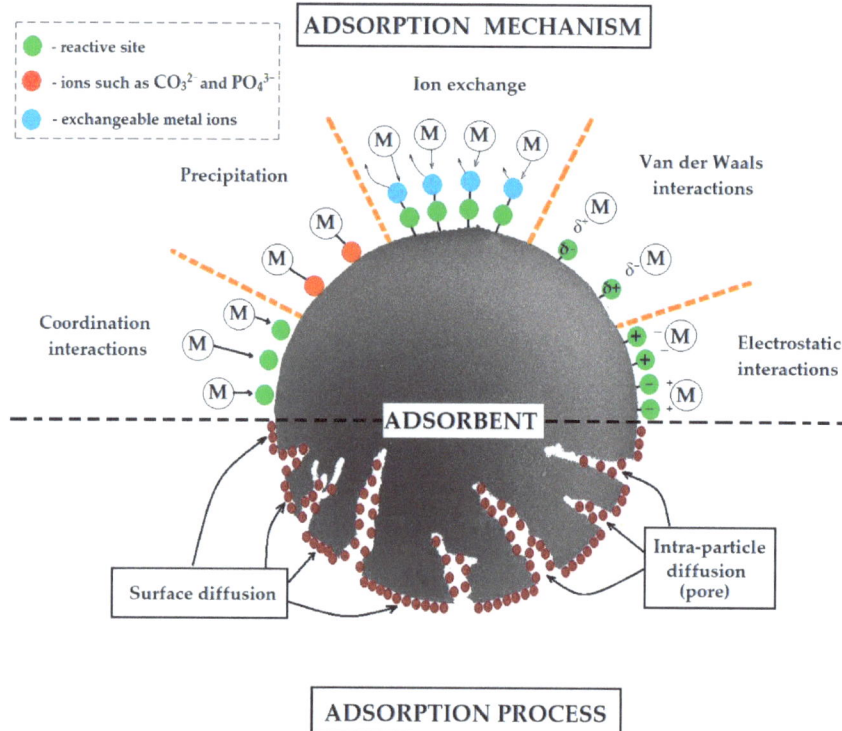

Figure 1. Schematic representation of possible metal adsorption mechanisms by polymer sorbents.

Generally, adsorption experiments include determining maximum adsorption capacity under static and dynamic conditions. In order to optimize adsorption conditions, experiments should be performed under different conditions, such as adsorption time, pH, adsorbent mass, initial metal ions concentrations, and temperature. Adsorption capacity (q_t) is calculated according to the Equation (1):

$$q_t = \frac{(C_i - C_t) * V}{m} \quad (1)$$

where C_i is the initial concentration and C_t is the concentration at time t for metal ions in aqueous solution, V is the volume of the aqueous solution, and m is the mass of the adsorbent.

The research on metal ions removal also requires the determination of adsorption equilibrium and kinetics, i.e., analysis of experimental data with adsorption isotherms and kinetic models. Calculation of thermodynamic parameters such as the standard free energy, enthalpy, and entropy change enables the prediction of the nature of the adsorption process, endothermal or exothermal. Last but not least, experiments in several adsorption/desorption cycles provide insight into the possibility of regeneration and repeated use of the adsorbent.

The most important criteria for the evaluation of sorbent applicability and efficiency are the nature and location of functional groups in a sorbent, sorbent capacity, selectivity, and the rate of complexation of metal ions [42]. In order to facilitate sorption kinetics, it is preferable for functional groups to be located at the surface or nearby. For example, amino-functionalized macroporous polymers have accessible functional groups on the particle surface, which promote the metal sorption process. Therefore, several parameters must be considered when sorbent usability and capabilities are assessed, such as adsorption isotherms and kinetics, selectivity, adsorption thermodynamics, and capability of the metals' desorption and recovery.

5.1. Adsorption Isotherms Models

By definition, adsorption isotherm is the relationship between the adsorbate in the liquid phase and the adsorbate adsorbed on the adsorbent surface at equilibrium at a constant temperature. Modeling experimental data from adsorption processes is a valuable method for determining potential interactions between adsorbents and adsorbates and predicting the mechanisms in diverse adsorption systems [70,71]. Adsorption isotherms are mathematical models that illustrate the distribution of metals between adsorbate and adsorbent. Equilibrium data obtained from initial concentration could be analyzed using numerous mathematical isotherm models, reported in the literature in detail [71]. For a single-component system, the most commonly used are two parameters, Langmuir, Freundlich, and Temkin isotherm models, or three-parameters Redlich–Peterson, Sips, and Toth isotherms [70]. The most frequently used adsorption isotherm models are listed in Table 1.

Table 1. Two and three parameters adsorption isotherm models (Nonlinear and linear form) [70,72–75].

Models	Non-Linear Form	Linear Form	Parameters
Langmuir	$q_t = \frac{q_{max} K_L C_e}{1 + K_L C_e}$	$\frac{C_e}{q_e} = \frac{1}{K_L q_{max}} + \frac{C_e}{q_{max}}$	q_{max}—maximum monolayer coverage capacity, K_L—Langmuir isotherm constant (binding energy of adsorption)
Freundlich	$q_e = K_F C_e^{1/n}$	$\log q_e = \log K_F + \frac{1}{n} \log C_e$	K_F—Freundlich constant indicator of adsorption capacity, n—Freundlich constant related to adsorption intensity
Temkin	$q_e = \frac{RT}{b_T} \ln(A_T C_e)$	$q_e = \frac{RT}{b_T} \ln A_T + \frac{RT}{b_T} \ln C_e$	b_T—Temkin isotherm constant related to the heat of adsorption, A_T—Temkin isotherm equilibrium binding constant
Elovich	$\frac{q_e}{q_m} = K_E C_e e^{\frac{q_e}{q_m}}$	$\ln \frac{q_e}{C_e} = \ln K_E q_m - \frac{q_e}{q_m}$	K_E—Elovich equilibrium constant, q_m—Elovich maximum adsorption capacity
Dubinin–Radushevich	$q_e = q_{DR} e^{-A_{DR} \varepsilon^2}$	$\ln q_e = \ln q_{DR} - A_{DR} \varepsilon^2$	q_{DR}—Dubinin–Radushevich maximum adsorption capacity, A_{DR}—constant representing mean adsorption energy

Table 1. Cont.

Models	Non-Linear Form	Linear Form	Parameters
Halsey	$q_e = e^{\frac{\ln K_H - \ln C_e}{n_H}}$	$\log q_e = \left(\frac{1}{n_H}\right) \ln K_H - \left(\frac{1}{n_H}\right) \ln C_e$	K_H—Halsey isotherm constant, n_H—Halsey isotherm exponent
Harkin-Jura	$q_e = \left(\frac{A_{HJ}}{B_{HJ} - \log C_e}\right)^{1/2}$	$\frac{1}{q_e^2} = \frac{B_{HJ}}{A_{HJ}} - \frac{1}{A_{HJ}} \log C_e$	A_{HJ}—Harkins–Jura isotherm parameter, B_{HJ}—Harkins–Jura isotherm constant
Jovanovic	$q_e = q_{J,\max}(1 - e^{K_J C_e})$	$\ln q_e = \ln q_{J,\max} - K_J C_e$	K_J—Jovanovic isotherm constant, $q_{J,\max}$—Jovanovic maximum adsorption capacity
Redlich–Peterson	$q_e = \frac{K_P C_e}{1 + a_P C_e^g}$	$\ln\left(K_P \frac{C_e}{q_e} - 1\right) = g \ln C_e + \ln a_P$	K_P—Redlich–Peterson model constant, a_P—Redlich–Peterson model isotherm constant, g—Redlich–Peterson model exponent
Sips	$q_e = \frac{q_{m,S} K_S C_e^m}{1 - K_S C_e^m}$	$\ln\left(\frac{q_e}{q_{m,S} - q_e}\right) = m \ln C_e - \ln K_S$	K_S—Sips isotherm model constant, m—Sips model exponent, $q_{m,S}$—Sips maximum adsorption capacity
Toth	$q_e = \frac{q_m K_T C_e}{\left(1 + (K_T C_e)^t\right)^{1/t}}$	$\ln\left(\frac{q_e}{q_m - q_e}\right) = t \ln K_T + t \ln C_e$	K_T—Toth isotherm model constant, q_m—Toth maximum adsorption capacity, t—Toth model exponent
Hill	$q_e = \frac{q_H C_e^{n_H}}{K_D + C_e^{n_H}}$	$\log\left(\frac{q_e}{q_H - q_e}\right) = n_H \log C_e - \log K_D$	K_D—Hill isotherm model constant, q_H—Hill maximum adsorption capacity, n_H—Hill model exponent

Real wastewaters are multi-component systems that contain various pollutants, making the adsorption system more complicated. Thus, interaction and competition between the adsorbate molecules must be taken into account. In order to understand the adsorption mechanism of such complex systems, single-component models were modified. Consequently, non-modified, modified, and extended Langmuir and Freundlich model, Redlich–Peterson model, Sheindorf–Rebuhn–Sheintuch equation, and extended Sips isotherm model were developed [76].

5.2. Adsorption Kinetics

Adsorption kinetics controls the rate of adsorption and determines the time required to attain equilibrium for the adsorption process, giving valuable information on probable adsorption mechanisms [77]. The adsorption on the solid–liquid interface proceeds through the following stages: bulk diffusion (adsorbate transport from the solution bulk to the liquid film around the sorbent particle); external diffusion (adsorbate diffuses through the liquid film on the particle surface); intraparticle diffusion of the adsorbate from the liquid film to the particle surface, by pore diffusion and surface diffusion and interaction with the surface sites (by physisorption or chemisorption). The overall adsorption rate is determined by the slowest of the above stages. The first and the last steps are faster than the second and third ones. Determination of the adsorption mechanism and the rate-controlling stages are the critical factors for selecting the optimum operating conditions of the adsorption system.

Various surface-reaction and particle diffusion-based kinetic models are widely applied for the determination of the adsorption process dynamics. The most frequently used kinetic models are listed in Table 2. The models and calculations of characteristic parameters are comprehensively explained in the literature.

Table 2. Kinetics and mechanism of adsorption (nonlinear and linear form with the description of parameters) [78–82].

Models	Non-Linear Form	Linear Form	Parameters
Pseudo-first-order	$q_t = q_e\left(1 - e^{-k_1 t}\right)$	$\log(q_e - q_t) = \log(q_e) - \frac{k_1}{2.303}t$	q_e—adsorption capacity at equilibrium, k_1—pseudo-first-order rate constant
Pseudo-second-order	$q_t = \frac{q_e^2 k_2 t}{1 + q_e k_2 t}$	$\frac{t}{q_t} = \frac{1}{k_2 q_e^2} + \frac{t}{q_e}$	q_e—adsorption capacity at equilibrium, k_2—pseudo-second-order rate constant
Elovich	$q_t = \frac{1}{\beta}\ln(\alpha\beta t + 1)$	$q_t = \frac{1}{\beta}\ln(\alpha\beta) + \frac{1}{\beta}\ln(t)$	α—initial adsorption rate, β—desorption constant
Avrami	$q_t = q_e(1 - \exp[-k_{AV} t^n])$	$\ln\left(\ln\left(\frac{q_e}{q_e - q_t}\right)\right) = n \ln k_{AV} + n \ln t$	k_{AV}—Avrami fractional-order constant rat, n—Avrami fractional kinetic order
Fractional power	$q_t = K t^v$	$\log q_t = \log K + v \log t$	K—fractional power constant, v—fractional power rate constant
Intraparticle diffusion	$q_t = k_{id} t^{0.5} + C_{id}$		k_{id}—intraparticle diffusion rate constant, C_{id}—thickness of the adsorbent
Liquid film diffusion	$q_t = q_e(1 - e^{-k_{LFD} t})$	$\ln(1 - F) = -k_{LFD} t + C, F = \frac{q_t}{q_e}$	F—fraction of solute adsorbed at any time t, k_{LFD}—equilibrium fractional attainment
Bangham	$q_t = k_B t^\alpha$	$\log \log\left(\frac{C_0}{C_0 - q_t m}\right) = \log \frac{k_B m}{2.303 V} + \alpha \log t$	k_B—constant rate of Bangham model, α—constant (indicates the adsorption intensity)
Boyd	$F < 0.85$ $q_t = q_e \frac{3}{\pi}\left[1 - \left(1 - \frac{(Bt)^2}{\sqrt{\pi}}\right)^2\right]$ $F > 0.85$ $q_t = q_e\left(1 - \frac{6}{\pi^2} e^{(-Bt)}\right)$	$F < 0.85$ $Bt = \left(\sqrt{\pi} - \sqrt{1 - \frac{\pi F}{3}}\right)^2$ $F > 0.85$ $Bt = -0.4977 - \ln(1 - F)$	Bt—mathematical function of F

The linear regression correlation coefficient (R^2) values are frequently compared to evaluate the best fit model. However, to assess the best kinetic fitting model, besides regression coefficient (R^2), statistical error validity models such as average relative error, normalized standard deviation, hybrid fractional error function, a derivative of Marquardt's percent standard deviation, and standard deviation of relative error should also be used [71]. Another criterion that should be taken into account is the closeness, i.e., the agreement between the experimental (Q_e^{exp}) and calculated (Q_e^{calc}) value of adsorption capacity.

5.3. Adsorption Thermodynamics

One of the important parameters of the adsorption process is temperature, i.e., the endothermal and exothermic character of the process, determined by the increase or decrease of the temperature all through the adsorption process.

In order to evaluate the feasibility of the adsorption process, thermodynamic parameters such as the standard free energy (ΔG^0), enthalpy change (ΔH^0), and entropy change (ΔS^0) were estimated. The Gibb's free energy change of adsorption was calculated from the following equation:

$$\Delta G^0 = -RT \ln K_c \qquad (2)$$

where R is the ideal gas constant (8.314 J/mol K), T (K) is the absolute temperature and K_c is the thermodynamic equilibrium constant that is expressed as:

$$K_c = \frac{C_a}{C_e} \qquad (3)$$

where C_a (mg/L) is the amount of metal ion adsorbed at equilibrium, and C_e (mg/L) is the concentration of metal ions in solution at equilibrium. Gibb's free energy is also related to the enthalpy change and entropy change at constant temperature by the Van't Hoff equation as follows [83,84]:

$$\ln K_c = -\frac{\Delta G^0}{RT} = \frac{\Delta S^0}{R} - \frac{\Delta H^0}{RT} \qquad (4)$$

The ΔH^0 and ΔS^0 values can be calculated from the slope and intercept of the plot of $\ln K_c$ versus $1/T$. The negative values of ΔG^0 indicate that the adsorption process is spontaneous. In the case of positive entropy change, the randomness at the moment of adsorption rises, while in the case of negative free energy change, adsorption is spontaneous or favorable. In the case of positive enthalpy change, the reaction is endothermic, meaning adsorption efficiency rises with the temperature increase. When the ΔH^0 value is in the range 2.1–20.9 kJ/mol, the adsorption process is physical. On the other hand, the ΔH^0 value in the range of 80–200 kJ/mol suggests chemisorption [85].

5.4. Desorption of Metals and Reusability of Polymeric Sorbents

An effective adsorbent should have a high adsorption/desorption capacity. The ability of the adsorbent to regenerate makes the adsorbent desirable and the adsorption process economical. In order to ensure successful adsorbent regeneration and reusability, a suitable stripping agent should be carefully chosen. It has to be cost-effective, highly efficient, and non-damaging to the adsorbent.

Various stripping agents such as acids hydrochloric, sulphuric, nitric, formic, and acetic acid; bases: sodium and potassium hydroxide, sodium carbonate and bicarbonate, potassium carbonate; salts: sodium and potassium chloride, ammonium sulfate and nitrate, calcium chloride, potassium nitrate, deionized water, chelating agents (ethylene diamine tetraacetic acid, EDTA); and buffer solutions (bicarbonate, phosphate and tris) were used in literature.

Desorption of heavy metal ions seems to be rapid and higher in acidic than in basic and neutral media [86]. In order to reduce the consumption of acids and bases, the use of other chemicals was investigated. A comprehensive review that summarizes the removal efficiency of various adsorbents, desorption efficiency of various stripping agents, and recovery of heavy metals can be found in the literature [87].

Nastasović et al. published a desorption study on Cu(II), Ni(II), and Pb(II) loaded poly(glycidyl methacrylate) and ethylene glycol dimethacrylate (PGME-en) [88]. Regeneration experiments with 2 M H_2SO_4 as desorption eluent showed that PGME-en could be reused in several sorption/desorption cycles. For example, a capacity loss of 8% after four cycles of Cu(II) sorption was observed.

Marković et al. performed a desorption study on chromium(VI)-loaded copolymer of glycidyl methacrylate and ethylene glycol dimethacrylate functionalized with hexamethylene diamine (PGME-HD) [89]. It was concluded that Cr(VI) sorption was reversible. PGME-HD can be easily regenerated with 0.1 M NaOH up to 90% recovery in the fourth sorption/desorption cycle, while in the fifth cycle, a considerable sorption loss of 37% was noted.

Galhoum et al. used nitric acid solutions for testing the desorption efficiency of La(III) and Y(III) loaded polyaminophosphonic acid-functionalized polyglycidyl methacrylate (PGMA) [64]. It was concluded that metal-loaded sorbent could be regenerated over six successive sorption/desorption cycles with 0.5 M HNO$_3$ solutions. The sorption and desorption efficiencies decreased by less than 7% after the sixth cycle.

Piĺsniak-Rabiega et al. tested the desorption ability of vinylbenzyl chloride/divinylbenzene copolymer (VBC/DVB) with attached 2-mercapto-1-methylimidazole and guanylthiourea ligands loaded with silver (Ag(I)) ions [58]. As desorption eluents, solutions of sodium thiosulphate, thiourea, potassium cyanide, potassium cyanide in hydrogen peroxide, sodium hydroxide and ammonium buffer were used. It was observed that Ag(I)-loaded sorbents can be effectively regenerated with 1% potassium cyanide solution in 0.5% hydrogen peroxide solution at 50 °C. In addition, polymer sorbents with mercapto-1-methylimidazole and guanylthiourea ligands retained their Ag(I) capacity in five consecutive sorption/desorption cycles.

6. Methacrylate-Based Sorbents

The results regarding the usage of various polymers as metal ions sorbents were published in numerous studies. However, since it is impossible to cover all of them, we summarized methacrylate-based sorbents in Table 3.

Table 3. Adsorption capacity of metal ions using different methacrylate sorbents.

Adsorbents	Metals *	pH	Concentration	Adsorption Capacity	Reference
Poly(methyl methacrylate)/poly(ethyleneimine) core-shell nanoparticles	Cu(II)	5	5 mg/L	14 mg/g	[90]
Amidoximated acrylonitrile/methyl acrylate copolymer	Hg(II)	2	0.25–5 mmol/L	4.97 mmol/g	[91]
Poly(N-2-methyl-4-nitrophenyl maleimide-maleic anhydride-methyl methacrylate) terpolymers	Cd(II)	7	0.2–100 mg/L	77.56 mg/g	[92]
Boehmite/poly(methyl methacrylate) nanocomposites	Cu(II)	4	10 mg/L	9.43 mg/g	[93]
Poly(methyl methacrylate-2-hydroxyethyl methacrylate-ethylene glycol dimethacrylate)	Co(II) Pb(II) Cu(II)	6	10–50 mg/L	28.8 mg/g 31.4 mg/g 31.2 mg/g	[94]
Nano-magnetic poly(methyl methacrylate-glycidyl methacrylate-divinylbenzen)-ethylenediamine	Cu(II) Cr(VI)	6 2	100–300 mg/L 50–1000 mg/L	87.72 mg/g 136.98 mg/g	[95]
Nano-magnetic poly(methyl methacrylate-glycidyl methacrylate-divinylbenzen)-diethylenetriamine	Cu(II) Cr(VI)	6 2	100–300 mg/L 50–1000 mg/L	94.30 mg/g 149.25 mg/g	[95]
Nano-magnetic poly(methyl methacrylate-glycidyl methacrylate-divinylbenzen)-triethylenetetramine	Cu(II) Cr(VI)	6 2	100–300 mg/L 50–1000 mg/L	92.60 mg/g 204.08 mg/g	[95]
Nano-magnetic poly(methyl methacrylate-glycidyl methacrylate-divinylbenzen)-tetraethylenepentamine	Cu(II) Cr(VI)	6 2	100–300 mg/L 50–1000 mg/L	116.80 mg/g 370.37 mg/g	[95]
Poly(methyl methacrylate-co-ethyl acrylate) membrane	Cu(II) Fe(III)	6	0.1–1.0 mmol/L	8.20 mmol/L 2.51 mmol/L	[96]
Poly(methyl methacrylate-co-buthyl methacrylate) membrane	Cu(II) Fe(III)	6	0.1–1.0 mmol/L	1.35 mmol/g 2.41 mmol/g	[96]
Poly(methyl methacrylate-glycidyl methacrylate-ethylene glycol dimethacrylate)-1,9-nonanedithiol	Ag(I)	-	10 mg/L	21.7 mg/g	[97]
Poly(glycidyl methacrylate-co-ethylene glycol dimethacrylate)-ethylenediamine	Cu(II) Cd(II) Zn(II) Fe(II) Mn(II) Pb(II) Cr(III) Pt(IV)	5.5 5.5 2.1 1.25 1.25 1.25 1.25 5.5	0.05 mol/L	1.10 mmol/g 0.67 mmol/g 0.25 mmol/g 0.14 mmol/g 0.12 mmol/g 1.06 mmol/g 0.47 mmol/g 1.30 mmol/g	[88]

Table 3. Cont.

Adsorbents	Metals *	pH	Concentration	Adsorption Capacity	Reference
Magnetic poly(glycidyl methacrylate/divinylbenzene)-tetraethylenepentamine resin	Mo(VI)	2	8×10^{-3} mol/L	5.6 mmol/g	[98]
Magnetic poly(glycidyl methacrylate/N,N'-methylenebisacrylamide)-tetraethylenepentamine resin	Mo(VI)	2	8×10^{-3} mol/L	7.6 mmol/g	[98]
Poly(glycidyl methacrylate)-pyromellitic acid	Pb(II)	2	100 mg/L	206.71 mg/g	[99]
2-Aminothiazole-functionalized poly(glycidyl methacrylate) microspheres	Au(III)	4	200–700 mg/g	440.84 mg/g	[59]
Poly(glycidyl methacrylate-co-trimethylolpropane trimethacrylate)-diethylenetriamine	Cu(II) Co(II) Ni(II) Zn(II) Cd(II)	5	0.4–4.0 mmol/L	1.25 mmol/g 0.54 mmol/g 0.71 mmol/g 0.69 mmol/g 0.67 mmol/g	[100]
Poly(glycidyl methacrylate) modified with trithiocyanuric acid microsphere	Ag(I)	5	200–600 mg/L	225.23 mg/g	[101]
Poly(glycidyl methacrylate–styrene–N,N'-methylenebisacrylamide) functionalized with –tetraethylenepentamine	Pb(II) Hg(II)	4.6 5.7	4.83 mmol/L 4.97 mmol/L	4.76 mmol/g 4.80 mmol/g	[102]
Magnetic-poly(glycidyl methacrylate-N,N'-methylene bisacrylamide)-ethylenediamine	Th(IV)	3.5	100 mg/L	60 mg/g	[103]
Magnetic-poly(glycidyl methacrylate-N,N'-methylene bisacrylamide)-diethylenetriamine	Th(IV)	3.5	100 mg/L	84 mg/g	[103]
Magnetic-poly(glycidyl methacrylate-co-ethylene glycol dimethacrylate)-diethylenetriamine	V(V)	5	1.36–146 mg/L	11.23 mg/g	[104]
Macroporous poly(glycidyl methacrylate-co-ethylene glycol dimethacrylate)-ethylenediamine	Pt(IV)	5.5	0.05 mol/L	1.30 mmol/g	[61]
Cu(II) ion-imprinted poly(methacrylic acid/ethylene glycol dimethacrylate)	Cu(II)	6	15 mg/L	1058 μg/g	[105]
Poly(glycidyl methacrylate-ethyleneglycol dimethacrylate)-polyethylene imine	U(VI)	6	25–500 mg/L	71.4 mg/g	[106]
Poly(glycidyl methacrylate-ethyleneglycol dimethacrylate)-tris(2-aminoethyl) amine	U(VI)	6	25–500 mg/L	88.9 mg/g	[106]
As ion-imprinted poly(styrene/ethylene glycol dimethacrylate)	As	6	-	482 μg/g	[107]
Se ion-imprinted poly(styrene/ethylene glycol dimethacrylate)	Se	7	-	447 μg/g	[107]
Polyaminophosphonic acid-functionalized poly(glycidyl methacrylate)	Y(III) La(III)	5	25–400 mg/L	0.73 mmol/g 0.79 mmol/g	[64]
Poly(methacrylic acid/ethylene glycol dimethacrylate)	Gd(III)	6	5–200 mg/L	19.4 mg/g	[108]
Poly(glycidyl methacrylate) functionalized with 2,6-diaminopyridine	Au(III)	4	-	459.28 mg/g	[109]
Hypercrosslinked poly(styrene-glycidyl methacrylate-iminodiacetic acid)	Tb(III)	5.8	-	145.6 mg/g	[110]
Poly(glycidyl methacrylate-co-ethylene glycol dimethacrylate)-diethylenetriamine	Mo(VI)	2	0.1 mol/L	3.58 mmol/g	[111]
Praseodymium ion imprinted Chloromethyl-8-hydroxyquinoline functionalized poly(Hydroxyethyl methylacrylate)/SiO$_2$	Pr(III)	4.5	0.001–0.01 mol/L	0.15 mmol/g	[112]
Ethanediamine-modified magnetic poly(glycidyl methacrylate) microspheres	Cd(II)	6.5	0.178 mmol/L	189.89 mg/g	[113]
Fe$_3$O$_4$@polyglyceryl methacrylate-graft-triethylenetetramine-beta-cyclodextrin microspheres	Pb(II) Cd(II)	-	50–200 mg/L	229.41 mg/g 210.65 mg/g	[114]
Polypropylene/poly(glycidyl methacrylate)	Sc(III)	2	1–20 mg/L	3.13 mg/g	[115]
Magnetic(glycidyl methacrylate-N,N'-methylenebisacrylamide)- diethylenetriamine	Hg(II)	4	5 mmol/L	2.81 mmol/g	[53]

Table 3. Cont.

Adsorbents	Metals *	pH	Concentration	Adsorption Capacity	Reference
Magnetic-poly(glycidyl methacrylate-co-divinylbenzene)-tetraethylenepentamine	U(IV)	4.5	0.525 mM	1.68 mmol/g	[116]
Polyethylenimine/poly(glycidyl methacrylate) magnetic microspheres	Cr(VI)	2	25–500 mg/L	492.61 mg/g	[117]
Poly(glycidyl Methacrylate-co-ethylene glycol dimethacrylate)-diethylenetriamine	Pb(II) Cu(II) Cd(II)	2 4 4	0.05 mol/L	164 mg/g 120 mg/g 152 mg/g	[118]
Magnetic-poly(glycidyl methacrylate)	Hg(II)	-	5–2500 mg/L	543 mg/g	[119]
Magnetic-poly(methyl methacrylate-co-maleic anhydride)	Co(II) Cr(III) Zn(II) Cd(II)	6	20–100 mg/L	90.09 mg/g 90.91 mg/g 109.89 mg/g 111.11 mg/g	[120]

* Cu-Copper, Hg-Mercury, Cd-Cadmium, Co-Cobalt, Pb-Lead, Cr-Chromium, Fe-Iron, Ag-Silver, Zn-Zinc, Mn-Manganese, Pt-Platinum, Mo-Molybdenum, Au-Gold, Ni-Nickel, Th-Thorium, V-Vanadium, U-Uranium, As-Arsenic, Se-Selenium, Y-Yttrium, La-Lanthanum, Gd-Gadolinium, Tb-Terbium, Pr-Praseodymium, Sc-Scandium.

In further text, the main focus will be set on macroporous non-magnetic and magnetic, amino-functionalized glycidyl methacrylate (GMA). Macroporous polymeric sorbents consisting of crosslinked copolymers (solid support) and functional groups (ligands) are potentially very attractive as selective sorbents for precious and heavy metal ions with some advantages over other sorbents, being highly efficient, cost-effective, and reusable [110,117].

7. Adsorption on Amino-Functionalized Glycidyl Methacrylate-Based Polymers

GMA based macroporous copolymers crosslinked with ethylene glycol dimethacrylate (EGDMA), PGME, and trimethyl trimethylolpropane trimethacrylate (TMPTMA), PGMT, in the shape of regular beads are very interesting due to the ability of the epoxy group to react with nucleophiles, such as amines, thiols, azide, carboxylic acids, etc. (Scheme 1) [68,111,121,122]. This type of copolymers can be synthesized by suspension copolymerization in the presence of a pore-forming agent (inert component, porogen), having a permanent well-developed porous structure even in the dry state, and macropores with diameters larger than 50 nm [123].

Two different approaches can be used in order to attach ligands. The first one is a copolymerization of a suitable monomer already carrying the required functional group, and the second one or by post-polymerization functionalization, i.e., to perform an additional reaction in order to introduce selective chelating groups. The latter method is more practical and efficient since it could be assumed that all the groups are accessible and reactive.

The separation of metal ions on macroporous amino-functionalized GMA copolymers is determined by the sorption parameters (pH, presence of other ions which compete for the active sites), structural properties of the chelating copolymers (particle size, porosity, specific surface area), ligand structure as well as kinetic and thermodynamic stability of the formed metal complexes with the chemically bonded amine ligands [61].

Amino-functionalized GMA based copolymers were proven to be adaptable sorbents for removal of precious [61] and heavy metals [88,118,124,125], technetium-99 [126,127], as well as chromium [89,128,129], molybdenum and rhenium [111,130], and vanadium [104].

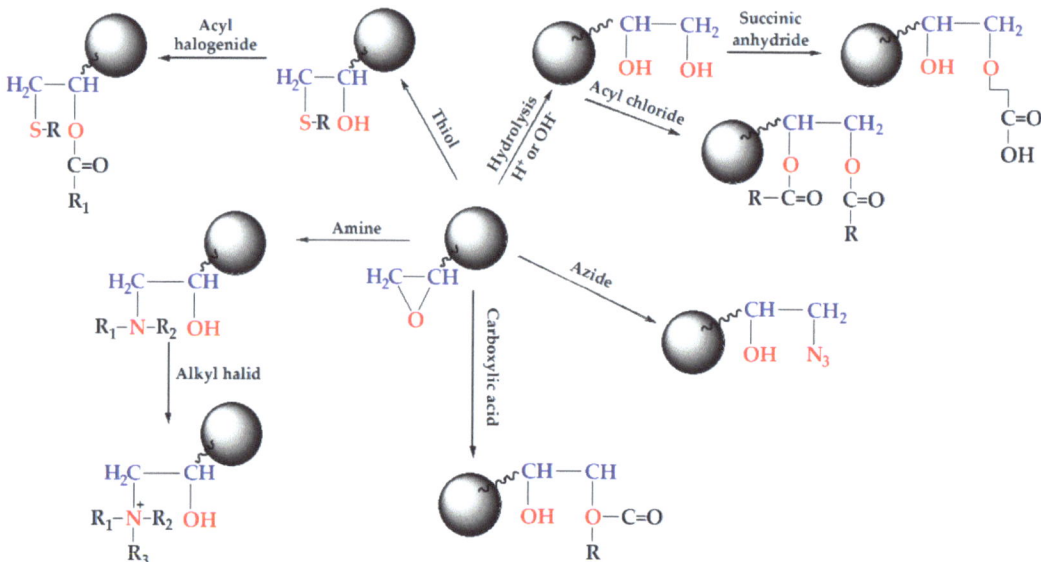

Scheme 1. Functionalization reactions of GMA-based polymers.

Nastasović et al. investigated sorption performances of amino-functionalized PGME towards heavy metals and precious metals [61,118,124,128,131–133]. Macroporous crosslinked copolymer of glycidyl methacrylate and ethylene glycol dimethacrylate with different porosity parameters functionalized with ethylenediamine, PGME-en, towards copper, iron, manganese, cadmium, zinc, lead, chromium, and platinum (Cu(II), Fe(II), Mn(II), Cd(II), Zn(II), Pb(II), Cr(III), and Pt(IV)) ions were studied [88]. It was observed that the sorption rate of PGME-en for Cu(II) ions determined under non-competitive conditions was relatively rapid, i.e., the maximum capacity was reached within 30 min. In addition, strong pH dependence of PGME-en selectivity was observed. Namely, a considerably higher sorption capacity for Pt(IV) in comparison to Cu(II), Co(II), Ni(II), and Pb(II) ions at pH 2.1 was established. On the other hand, at pH 5.5, the metal sorption capacities of PGME-en decreased in the order: Cu(II) > Co(II) > Pt(IV) ≈ Ni(II) > Pb(II). Reusability was also proven since the regeneration with 2 M H_2SO_4 of the Cu(II), Ni(II), and Pb(II) loaded PGME-en showed that copolymer can be reused in several sorption/desorption cycles.

The sorption ability of macroporous PGME-en towards rhodium, gold, and platinum (Rh(III), Au(III), and Pt(IV)) [61] ions were also examined. A faster Rh(III) uptake than those of Au(III) and Pt(IV) was observed. After 5 min of sorption, approx. 90% Rh(III), 57% of Pt(IV), and 46% of Au(III) had been sorbed. These different sorption rates of Rh(III), Au(III), and Pt(IV) enabled PGME-en application for the selective separation of platinum metals. Due to strong coordination with the modified copolymer, Rh(III) desorption was very difficult and incomplete, even with strong acids (such as HCl and H_2SO_4). It was shown by the same research group that PGME-en can sorb 5–8 times more Pt(IV) than Cu(II) and Ni(II) ions from single-component solutions and 5 times more Pt(IV) than Cu(II) from their mixed chloride solutions [134].

The sorption performances of macroporous PGME with attached diethylene triamine, PGME-deta, for Cu(II), Cd(II), and Pb(II) sorption were determined in batch static experiments at room temperature [118]. The sorption half time was approx. 5 min for all metal ions. High capacities were observed, i.e., PGME-deta after 30 min reached approx. 90% and after 180 min, 95% of maximum capacity. It was shown that PSO kinetic model best fitted the Cu(II), Cd(II), and Pb(II) sorption, suggesting that the sorption rate is controlled by both sorbent capacity and concentration, with the influence of intraparticle diffusion.

Şenkal et al. used glycidyl methacrylate-based copolymer with acetamide groups as Hg(II) sorbents [135]. The Hg(II) ions sorption capacity was approx. 2.2 mmol/g in nonbuffered solutions. On the other hand, the adsorption capacities of sorbent for Cd(II), Pb(II), Zn(II), and Fe(III) were relatively low (0.2–0.8 mmol/g). It was shown that Hg(II) ions could be regenerated by repeated treatment with hot acetic acid without hydrolysis of the amide groups, up to 2.0 mmol/g, i.e., 86% of the capacity of fresh polymer.

Atia et al. reported Cu(II) and Pb(II) sorption performances of poly(glycidyl methacrylate-co-divinylbenzene) functionalized by ethylene diamine [136]. The results showed that metal-sorbent interaction proceeds via surface and diffusion mechanisms. The most suitable pH value for metal sorption was 5.8, while maximum adsorption capacities for Cu(II) and Pb(II) were 1.24 and 0.32 mmol/g, respectively. Regeneration efficiency of 97% was achieved in 10 sorption/desorption cycles with 0.5 HNO_3. Haratake et al. tested the triethylene tetra amine-functionalized poly (glycidyl methacrylate-co-ethylene glycol dimethacrylate) as sorbents for Cu(II), Zn(II), Co(II), and Ni(II) ions sorption from seawater [137].

Malović et al. investigated the influence of the porosity parameters, particle size, and type of the ligand on the uptake of heavy metals on macroporous amino-functionalized PGME [133]. Sorption capacities and rates of PGME-en, PGME-deta, and PGME-teta for Cu(II) ions were determined. In addition, the selectivity of PGME-deta and PGME-teta towards individual metal ions under competitive conditions was investigated as a function of pH and particle size. The Cu(II) sorption was rapid, i.e., the sorption half time for PGME-deta and PGME-teta, was approximately 3 min. In addition, the high selectivity of PGME-deta for Cu(II) over Cd(II) of 3:1 and for Cu(II) over Ni(II) and Co(II) of 6:1 was observed. The decrease in particle size of PGME-teta resulted in the increase of sorption capacities for all metal ions.

Suručić et al. published studies on the theoretical modeling of metal ions sorption [124,125,138]. The sorption of Cu(II) ion on PGME-en, PGME-deta, and PGME-teta were successfully modeled by quantum chemical calculations [125]. Higher maximum sorption capacities (Q_{max}) were obtained for deta- and teta-copolymers, due to their abilities to form binuclear complexes. The study offers an explanation of the experimentally obtained trend for Cu(II) by applying theoretical techniques to predict the selectivity of ligands. A comparison of the Gibbs free energy (ΔG_{aq}) for mononuclear and binuclear tetaOH complex suggests that the formation of mononuclear complexes is a slightly more favorable (spontaneous) process. The results indicate that the amines with three nitrogen ligator atoms were preferable (due to the possibility of binuclear complex formation). On the other hand, the more ligator atoms it contains, the amine diffusion inside the polymer is more difficult. In addition, a higher number of ligator atoms increases the strain of chelate rings and reduces the stability of amino-functionalized complex with the sorbed ion.

In a very complex and detailed study, Suručić et al. used quantum chemical calculation for modeling Cu(II), Cd(II), Co(II), and Ni(II) sorption by PGME-teta [124]. Cambridge Structural Database (CSD) was a source of geometries of aqua complexes of the studied metal ions and coordination modes of the teta ligand in crystal structures. Cd(II), Co(II), and Ni(II) ions form complexes with octahedral geometry, while Cu(II) ion forms complexes with the coordination number 5. The agreement of theoretical and experimental results was achieved when mononuclear tetaOH complexes of Cu(II) and Cd(II) were compared with mononuclear complexes of Ni(II) and Co(II) with monoprotonated teta OH ligand (tetaOH). It was observed that the inclusion of the solvation effect was needed since the sorption takes place in an aqueous solution. In addition, it was shown that solvation energy contributions significantly improve the stability of ions in an aqueous solution.

Macroporous PGME copolymers amino-functionalized with ethylene diamine, diethylene triamine and hexamethylene diamine (PGME-en, PGME-deta, PGME-HD) were tested as potential Cr(VI) oxyanion sorbents from aqueous solutions [89,128,129].

Maksin et al. studied kinetics and temperature dependence of Cr(VI) sorption by PGME-deta in the temperature range 25–70 °C [129]. Pseudo-first order, pseudo-second-order, Elovich, intraparticle diffusion, and Bangham kinetic models were used for sorption behavior analysis. Equilibrium data were tested with Langmuir, Freundlich, and Tempkin adsorption isotherm models. Langmuir model was the most suitable, while thermodynamic data suggested spontaneous and endothermic Cr(VI) adsorption onto PGME-deta. The best kinetic results fit was observed with the pseudo-second-order model, with a definite influence of pore diffusion.

Additionally, the authors proposed electrostatic interactions Cr(VI) sorption mechanism by PGME-deta at acidic media. Namely, at acidic pH, the amino groups were in the protonated cationic form (NH_3^+), which results in stronger attraction with negatively charged ions in the solution ($HCrO_4^-$). Consequently, electrostatic interaction between the adsorbent and anions and high chromium removal is observed.

The sorption kinetics of Cr(VI), Cu(II), Co(II), Cd(II), and Ni(II) by PGME-en and PGME-deta, was studied under single-component and mixed metal salt solutions [128]. The competitive sorption was analyzed for the following mixed solutions: Cu(II) and Cr(VI); Cu(II), Co(II), Cd(II), and Ni(II); Cr(VI), Cu(II), Co(II) and Cd(II) solutions. Very rapid uptake of Cr(VI) ions under non-competitive conditions (the sorption half time ≤ 1 min) was observed. The Cr(VI) and Cu(II) sorption were much slower from their binary solutions (the sorption half time for Cr(VI) and Cu(II) were 11 and 45 min) than from single-component solutions (observed sorption half time for Cr(VI) and Cu(II) were 0.5 and 3 min) presumably due to the competition of metal ions for the active sites on the copolymer surface. As a result of kinetics analysis, it was concluded that the best fit for investigated heavy metals sorption by PGME-en and PGME-deta provides pseudo-second-order kinetics.

Marković et al. tested macroporous PGME functionalized with hexamethylene diamine (PGME-HD) as Cr(VI) oxyanion sorbent from aqueous solutions [89]. Kinetic data were analyzed using chemical reaction particle and diffusion models kinetic models (pseudo-first order, pseudo-second-order, Elovich), intraparticle diffusion, Bangham, Boyd, and McKay). The monolayer sorption Langmuir model was the most suitable to fit the experimental equilibrium data. Calculated thermodynamic parameters suggested that Cr(VI) adsorption onto PGME-HD was spontaneous and endothermic. In addition, PGME–HD was found to be easily regenerated with 0.1 M NaOH and reusable in four sorption/desorption cycles, with up to 90% recovery.

PGME–deta was studied as a potential recovery agent for molybdenum, Mo(VI) oxyanions, by varying pH, time, initial concentration, and temperature [111,130]. Calculated thermodynamic parameters revealed that both chemical adsorption and intraparticle diffusion were rate-controlling, with chemisorption as predominant. This could be the consequence of the transition metal nature of molybdenum. Namely, during the sorption process, d orbitals become filled with free electron pairs from amino or hydroxy groups of PGME-deta. Among seven chemical-reaction and particle-diffusion kinetic models (pseudo-first-order, pseudo-second-order, Elovich, intraparticle diffusion, Bangham, Boyd and Mckay), the best fit was observed with pseudo-second-order, with the considerable effect of intraparticle diffusion. The maximum Mo(VI) sorption capacity for PGME–deta was 3.58 mmol/g at 343 K.

8. Adsorption on Magnetic Polymeric Sorbents

There is a growing interest in the application of magnetic nanoparticles for the removal of heavy metals from wastewater since they possess a whole range of advantages in comparison with traditional sorbents, such as small particle size, large surface area, and easy separation after treatment by applying an external magnetic field [1,36,70–73]. However, magnetic nanoparticles tend to aggregate, which decreases the surface area and reduces the removal capacity. Stabilization of these particles can be achieved by surface coating or grafting with an organic layer (surfactant or polymer), coating with an inorganic layer (silica or carbon), and incorporating magnetic nanoparticles in polymer matrices (Scheme 2). In order to provide stabilize these particles, surface modification is required [1,51,139].

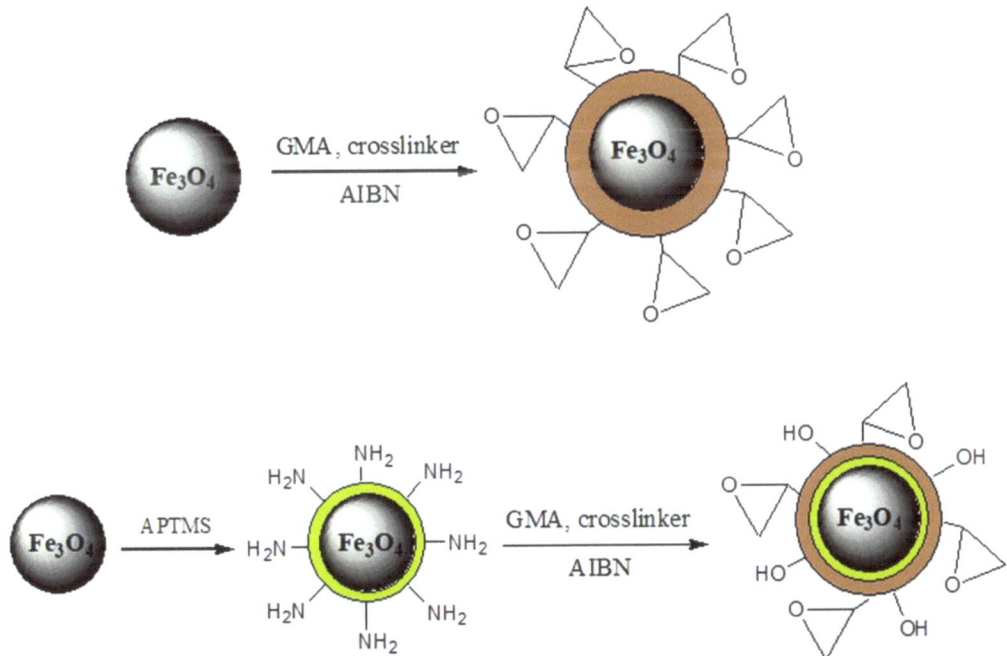

Scheme 2. Stabilization of Fe$_3$O$_4$ nanoparticles.

A wide range of different magnetic materials was used as metal sorbents. For example, Duranoğlu et al. used polyglycidyl methacrylate graft copolymer (PGMA) and polymer-supported magnetic nanoparticles (PGMAFe) as Cr(VI) sorbent [140]. Both adsorbents were useful for removing Cr(VI) from an aqueous solution over a wide pH range. The resulting graft copolymer and its Fe$_3$O$_4$ nanoparticles-coated form were highly effective for Cr(VI) sorption in column experiments, with a short contact time, up to 30 min. The column was efficiently regenerated with NaOH (10%, w/v) solution. However, PGMAFe had higher Cr(VI) adsorption capacity compared to PGMA, presumably as a consequence of the combined effects of Fe oxide and amine groups. Better correlation with experimental data was observed for both PGMA and PGMAFe sorbents. The maximum Cr(VI) adsorption capacities of PGMA and PGMAFe sorbents obtained at pH 4 were 132.5 and 162.9 mg/g, respectively. It was concluded that both samples were effective adsorbents for Cr(VI) in a wide pH range with relatively high adsorption capacity.

Atia et al. prepared magnetic methacrylate/divinylbenzene particles with a magnetite core and post-functionalized with ethylenediamine, diethylenetriamine, and tetraethylenepentamine. Synthesized magnetic core-shell polymer particles samples were tested as Hg(II) sorbent [141]. The Hg(II) sorption capacities of polymeric sorbent were found to be in the range 2.1–4.8 mmol/g.

Atia et al. prepared magnetic GMA sorbents, crosslinked with divinylbenzene (GMA/DVB-en) or N,N'-methylenebisacrylamide (GMA/MBA-en) and functionalized with tetraethylenepentamine [98]. The sorption behavior towards molybdate anions was studied. The Mo(VI) adsorption capacities of 4.24 and 6.18 mmol/g were obtained for GMA/DVB-en and GMA/MBA-en, respectively. The adsorption followed the pseudo-second-order model. Regeneration efficiency up to 90–96% was reached using an ammonia buffer.

Bayramoğlu et al. used magnetic terpolymer poly(glycidyl methacrylate–methyl methacrylate–ethylene glycol dimethacrylate) functionalized with ammonia for Hg(II) ions removal from aqueous solution in static conditions and in a magnetically stabilized fluidized bed (MFB) reactor [142]. The optimum removal of Hg(II) ions was observed at pH 5.5, with a maximum adsorption capacity of 124.8 mg/g.

Marković et al. studied the influence of different magnetite content on the porosity parameters, morphology, and magnetic properties of magnetic macroporous PGME copolymer (mPGME). The copolymer was post-functionalized by a ring-opening reaction with diethylene triamine. The amino-functionalized magnetic mPGME−deta, was studied as molybdenum, Mo(VI), and rhenium, Re(VII) sorbent from binary solutions [130]. The influence of pH, ionic strength, and coexisting cations (Ni^{2+}, Cd^{2+}, and Cu^{2+}) and anions (Cl^-, NO_3^- and SO_4^{2-}) on Mo(VI) and Re(VII) oxyanion sorption on mPGME-deta were investigated. Langmuir model was proven to be the most appropriate adsorption isotherm model, assuming monolayer adsorption at specific homogenous sites on the mPGME-deta surface. In addition, the selectivity of mPGME-deta for Re(VII) sorption was studied at a different contact time and Re/Mo ratio. High uptake of oxyanions was noted, i.e., 92% of Re(VII) and 98% of Mo(VI) were sorbed at pH 2.

Suručić et al. investigated the adsorption of vanadium (V) oxyanions from aqueous solutions onto diethylene triamine functionalized magnetic macroporous GMA-based copolymer prepared in the presence of magnetite nanoparticles coated with 3-aminopropyltrimethoxysilane, (m-Si-poly(GME)-deta) [104]. Vanadium (V) sorption was tested as a function of metal ions concentration, contact time, and pH. Sorption was rapid, with the sorption half time of 1 min and maximum sorption capacity of 28.7 μmol/g. The sorption process was best described by the pseudo-second-order model and Freundlich isotherm. The quantum chemical calculations were performed using the Gaussian09 software package (Gaussian, Inc., Wallingford, CT, USA). The sorption process is favorable in the pH range of 3–6 due to the strong electrostatic interactions between the absorption centers of copolymer and vanadium (V) oxyanions. In the investigated pH range, deta absorption centers with two and three protonated N atoms were in equilibrium as studied by quantum chemical modeling.

Perendija et al. tested magnetite (MG) modified cellulose membrane (Cell-MG), and diethylenetriaminepentaacetic acid dianhydride functionalized waste cell fibers (Cell-NH_2 and Cell-DTPA), and amino-modified diatomite in heavy metal ions [143]. The effects of sorption parameters on adsorption capacity and kinetics were studied. The capacities for nickel, lead, chromium, and arsenic (Ni(II), Pb(II), Cr(VI), and As(V)) ions were 88.2, 100.7, 95.8, and 78.2 mg/g, respectively.

Xie et al. observed that chitosan/organic rectorite-Fe_3O_4 composite magnetic adsorbent (CS/χOREC-Fe_3O_4), exhibited better adsorption capacity for removing Cd(II) and Cu(II) ions than magnetic organic-rectorite (OREC-Fe_3O_4) and chitosan [144]. The best fit for Cu(II) and Cd(II) uptake provided the Langmuir isotherm model and pseudo-second-order kinetic model. The XPS analysis indicated the adsorption of metal ions by -NH_2 on the adsorbent surface via physical and chemical adsorption. Recycling experiments showed that after four sorption/desorption circles with Na_2EDTA solutions, the adsorption capacity of CS/OREC-Fe_3O_4 was above 55%.

Shinozaki et al. used porous polymeric adsorbents obtained by suspension polymerization of styrene, divinylbenzene, and GMA and modified with diglycolamic acid ligands for the recovery of rare earth elements [63]. The adsorption isotherm was a Langmuir-type, with an adsorption capacity of 0.113 mmol/g.

9. Overview of Characterization Methods

9.1. Fourier Transform Infrared Spectroscopy (FTIR)

The metal ions adsorption mechanism may include physical and chemical adsorption. Chemical adsorption mainly involves mechanisms such as ion exchange, electrostatic attraction, surface complexation, and inner-sphere complexation, redox, and precipitation [35]. Therefore, different techniques could be used in order to understand the adsorption mechanism, such as Fourier Transform Infrared (FTIR) spectroscopy, X-ray photoelectron spectroscopy (XPS), 1H, and 13C solid-state nuclear magnetic resonance (NMR), etc.

The disappearance of bands characteristic for functional groups/ligands and the appearance of the bands that can be ascribed to the metal ions bonding in the FTIR spectra could be used as evidence of successful metal ions bonding. For example, Malović et al. used FTIR spectra of a copolymer of glycidyl methacrylate and ethylene glycol dimethacrylate, (PGME) modified with triethylene tetramine (PGME-teta) to prove the presence of -NH and -NH$_2$ groups as a result of successful functionalization [133]. Namely, a strong band occurs at 3500 cm^{-1}, where the valence vibrations for -NH, -NH$_2$, and -OH groups overlap.

Fourier transform infrared spectroscopy (FTIR) and X-ray photoelectron spectroscopy (XPS) were used for the analysis of the mechanism of Cu(II), Cd(II), and Pb(II) ions sorption from aqueous solutions by macroporous PGME with attached diethylene triamine, PGME-deta (sample PGME-10/12-deta) (Figures 2a and 3) [118].

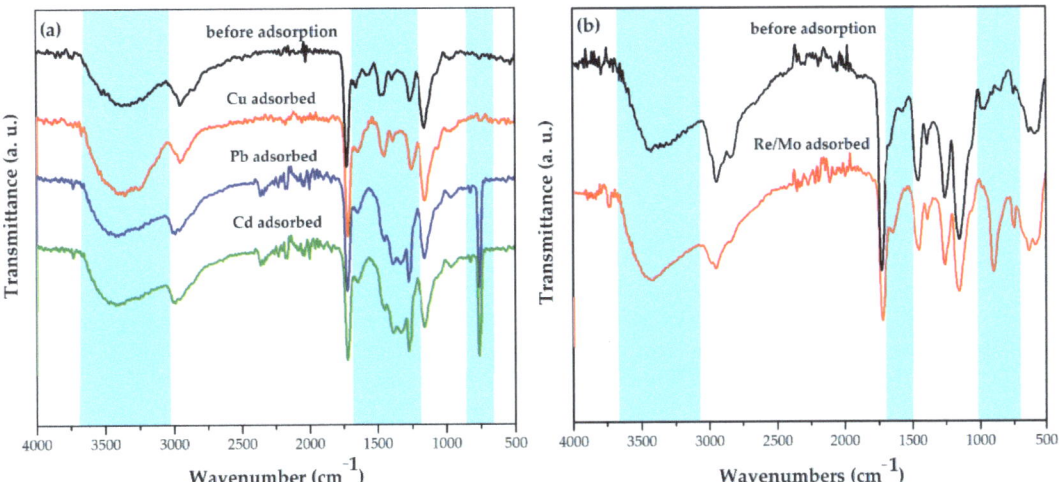

Figure 2. (**a**) FTIR-ATR spectra of PGME-deta before adsorption (black) and after adsorption: PGME-10/12-deta/Cu (red), PGME-10/12-deta/Pb (blue), and PGME-10/12-deta/Cd (green) [118], (**b**) FTIR-ATR spectra of mPGME-deta before adsorption (black) and after adsorption of Re(VII)/Mo(VI) (red) [130].

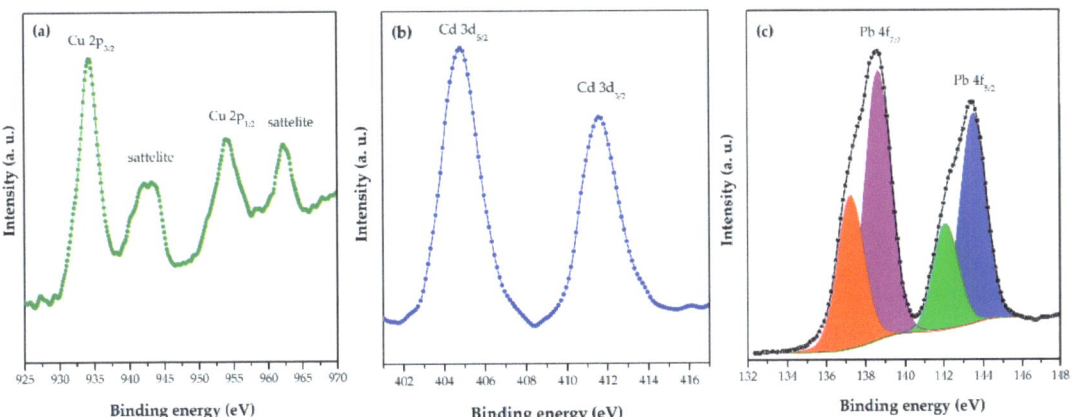

Figure 3. HRES (**a**) Cu 2p for PGME-10/12-deta/Cu, (**b**) Cd 3d for PGME-10/12-deta/Cd, and (**c**) Pb 4f for PGME-10/12-deta/Pb [118].

The disappearance of the peak for -NH at 1565 cm^{-1}, as well as the shift of the -NH$_2$ peak to ~1645 cm^{-1} in PGME-10/12-deta/Cu, PGME-10/12-deta/Cd, and PGME-10/12-deta/Pb spectra, clearly indicate the metal ions binding with amino-groups of PGME-10/12-deta. According to the literature data, the binding with the metal alters the hybridization type around nitrogen and causes the weakening of -NH bond [145]. However, from the standpoint of the metal ions sorption, the most significant part of the FTIR spectra was located in the 1000–700 cm^{-1} region The clear evidence of the Me(II) binding to PGME-10/12-deta was the appearance of new peaks at ~770, ~830 and ~970 cm^{-1} for PGME-10/12-deta/Cd and PGME-10/12-deta/Pb as well as the peaks at ~750 and ~970 cm^{-1} for PGME-10/12-deta/Cu, which can be ascribed to the formation of Me-O bond.

Similarly, Marković et al. observed the appearance of new bands in the region 700–1000 cm^{-1} (Me—O absorption bands) and a strong νCr—O band detected at 944 cm^{-1}, medium νCr—O band at 890 cm−1 and the band at ca 774 cm^{-1} in the FTIR spectra of PGME functionalized with hexamethylene diamine, PGME-HD, as the evidence of chromium binding [129].

Ekmeščić et al. considered the appearance of the wide band at 3060–3700 cm^{-1} (ν(NH) + ν(OH)), the bands at 1260 cm^{-1} ν(C-N), at 1560 cm^{-1} and 1650 cm^{-1} (δ (NH), δ (NH$_2$)), as well as the band at 1390 cm^{-1} (ν(NH)) in the FTIR spectra of PGME-deta with sorbed Re(VII)/Mo(VI) (Figure 2b) as confirmation of functionalization with diethylenetriamine [111]. In addition, the absence of characteristic bands for -NH and -NH$_2$ groups and the presence of the bands in the region of Mo-O absorption (1000–700 cm^{-1}) in the PGME-deta/Re(VII)/Mo(VI) spectra indicate that adsorption proceeds partially via coordination and electrostatic interactions.

Galhoum et al. also used FTIR analysis to prove lanthanum (La(III)) and yttrium (Y(III)) sorption onto polyaminophosphonic acid-functionalized polyglycidyl methacrylate (PGMA) [64]. The decrease of band intensity at 3357 cm^{-1} and 2966 cm^{-1} after La(III) and Y(III) sorption was related to the changes in the environment of -OH and -NH groups due to the metal binding. In addition, the main changes in FTIR spectra of Y(III)-loaded polymer sorbent were the shift of the band corresponding to Y-N bond at 503 cm^{-1} (to 531 cm^{-1}), the disappearance of the P-O-C band at 932 cm^{-1}, and the appearance of a new peak at 633 cm^{-1}. According to the literature, the latter can be ascribed to the formation of Me-O bonds [146].

Xiong et al. ascribed weakening of the C=N band at 1562 cm^{-1} in poly(glycidyl methacrylate) functionalized with 2-aminothiazole (A-PGMA) loaded with gold (A-PGMA-Au) to Au bonding to polymer sorbent [59]. According to the authors, Au adsorption proceeds via chelating and ion exchange between Au(III) and nitrogen groups on the surface of A-PGMA.

The appearance of new Cr-O bands at 944 cm^{-1} and 890 cm^{-1}, as well as Cr-N band at 420 cm^{-1} in the FTIR spectra of the chromium-loaded copolymer of glycidyl methacrylate and ethylene glycol dimethacrylate functionalized with hexamethylene diamine (PGME-HD), Marković et al. used as clear evidence of chromium bonding [147].

9.2. X-ray Photoelectron Spectroscopy (XPS)

X-ray photoelectron spectroscopy (XPS) is a quantitative technique for measuring the elemental composition of the surface of a material, and it also determines the binding states of the elements [148]. XPS normally probes to a depth of 10 nm. The energy and intensity of these peaks enable the identification and quantification of all surface elements present (except hydrogen).

This technique was used to elucidate the adsorption mechanism of metal binding to a new Cd(II) imprinted sorbent with interpenetrating polymer [149]. It was observed that after the Cd(II) adsorption, the N 1s bands shifted from 400.0 to 405.0 eV, indicating the formation of complexes, in which a pair of lone electrons from the N atoms was shared with the Cd(II), reducing the electron cloud density of the nitrogen atom, resulting in a higher BE peak observed.

The mechanism of Cu(II), Cd(II), and Pb(II) ions sorption from aqueous solutions by PGME-deta was studied by XPS and FTIR analysis [117]. Both techniques suggested complexation through the formation of Me-O and Me-N bonds with the OH, NH, and NH$_2$ groups as the possible mechanism of Cu(II), Cd(II), and Pb(II) sorption on PGME-deta.

The main Cu 2p peak (Figure 3a) for PGME-10/12-deta/Cu was positioned at 934.4 eV, which corresponds to Cu^{2+}. The presence of the well-known shake-up satellite found in Cu 2p spectra indicates the presence of Cu(II) species. The peak Cd3d$_{5/2}$ (Figure 3b) (which appears quite close to nitrogen N1 s peak) for sample PGME-10/12-deta/Cd was positioned at 404.8 eV, corresponding to Cd^{2+}. The Pb 4f doublet peak for sample PGME-10/12-deta/Pb (Figure 3c) was composed of two peaks with different oxidation state. The main peak Pb4f$_{7/2}$ can be thus fitted with two peaks positioned at 137.2 eV and 138.7 eV, corresponding to Pb^{4+} and Pb^{2+}, respectively. According to the authors, this indicates probable interaction between amino groups and Pb(II) ions due to chelation, electrostatic interaction with protonated amino groups, or formation of ternary complexes. In addition, FTIR and XPS analyses suggest complexation via Me-O and Me-N bonds with the -OH, -NH, and -NH$_2$ groups as the possible mechanism of Cu(II), Cd(II), and Pb(II) sorption.

XPS analysis was used for the investigation of the changes in the chemical composition and functional groups of the surface of magnetic macroporous crosslinked 10MAG-SGE60-deta prior and after sorption of Mo(VI) and Re(VII) ions [130]. The Mo 3d core-level spectrum (Figure 4a) of the sample was fitted into two components for Mo 3d$_{5/2}$ at 231.8 eV and 229.4 eV, which indicates molybdenum binding with reactive sites onto the 10MAG-SGE60-deta surface. The first peak was ascribed to Mo^{5+}, and the second one to the MoO$_2$ phase. The Re 4f narrow scan XPS spectra (Figure 4b) of 10MAG-SGE60-deta after adsorption show Re 4f$_{5/2}$ and Re 4f$_{7/2}$ doublet positioned at 45.7 eV and 39.2 eV indicating perrhenate binding with reactive sites onto the polymer surface. The more intense Re 4f$_{7/2}$ peak was deconvoluted into three components at 44.8, 46.6, and 48.1 eV, which signified the complexation and the existence different Re oxidation states in the sample.

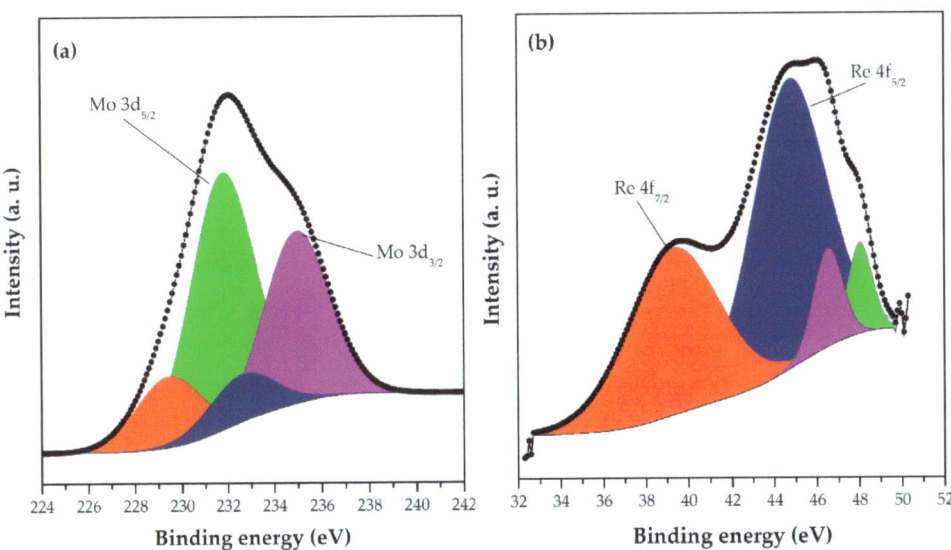

Figure 4. HRES spectra of (a) Mo 3d and (b) Re 4f for 10MAG-SGE60-deta after adsorption [130].

XPS spectroscopy was used to understand the chemical interactions between polyaminophosphonic acid-functionalized polyglycidyl methacrylate (PGMA) and lanthanum (La(III)) and yttrium (Y(III)) ions [64]. After the sorption from a binary La(III)/Y(III)solution, characteristic bands for La and Y appeared in the XPS spectra, i.e., La 3d$_{5/2}$ peaks with two couples of multiplet-splits at 836.2/839.9 eV, and 837.9/841.7 eV, as well as six Y 3d peaks (3 pairs for Y 3d$_{3/2}$ and Y 3d$_{5/2}$

bands) ascribed to chloride and phosphonate species. The La(III) and Y(III) sorption causes a shift toward lower BEs (binding energies) for P-O peaks, decreases the intensity of deprotonated phosphonate and increases of the intensity of protonated phosphonate peaks. XPS and FTIR analysis confirmed the contribution of phosphonate groups in metal binding with the co-existence of different complexes or different interactions with the neighboring reactive groups.

Xiong et al. also used analysis of XPS spectra for understanding the mechanism of adsorption of gold ions on poly(glycidyl methacrylate) functionalized with 2-aminothiazole (A-PGMA) [59]. As a result of gold ions adsorption, two new peaks at 82.6 eV and 86.3 eV appeared in the Au 4f spectra. The appearance of a new peak at 167.7 eV in S 2p spectra as well as a peak shift from 399.16 eV to 400.30 eV in the N 1s spectrum were ascribed to chelating between sulfur and nitrogen atoms with the gold ions. In conclusion, XPS analysis revealed that the gold ions adsorption on A-PGMA proceeds via ion exchange and chelation between the sulfur and nitrogen atoms on the surface of A-PGMA and $AuCl_4^-$–ions.

9.3. SEM/EDS and TEM

Scanning electron microscopy (SEM) can be applied to examine the shape, size, and morphology of the polymers. Additionally, SEM-EDX (energy-dispersive X-ray spectroscopy) analysis was used to identify the type of atoms present in the functionalized copolymers at a depth of 100–1000 nm from the surface. SEM-EDX provides information regarding the elemental distribution on the sorbent by elemental mapping of each component.

For example, Marković et al. examined the morphology of particle surface and cross-section for selected magneti10MAG-SGE60 and 10MAG-SGE60-deta samples by SEM analysis (Figure 5) [130]. The three-dimensional porous structure of the 10MAG-SGE60E-deta, composed of a large number of globules interconnected with channels and pores was visible on SEM image, which is consistent with reported values of porosity parameters, i.e., specific pore volume (0.99 cm^3/g), specific surface area (59 m^2/g) and pore diameter that corresponds to half of the pore volume (104 nm) [130].

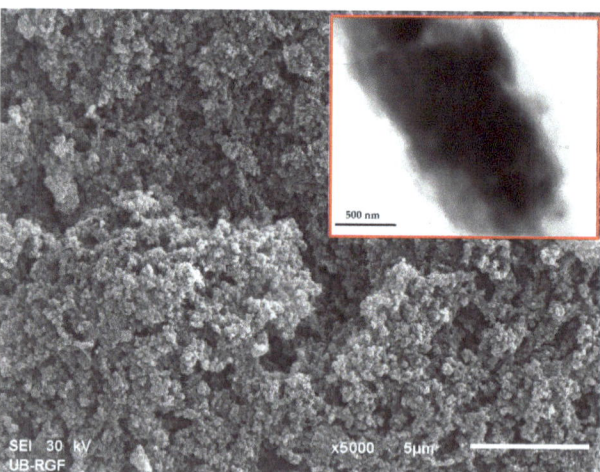

Figure 5. SEM micrograph of cross-section (magnification 5000×) and TEM micrograph (inset) of 10MAG-SGE60-deta.

In addition, SEM-EDS analysis confirmed the presence of N atoms at the particle surface, indicating that the reaction with diethylene triamine occurs mostly on the particle surface. The iron nanoparticles were also predominantly present at the particle surface and embedded in the bulk to a lesser extent. The distribution of dark magnetic nanoparticles throughout the gray copolymer matrix is visible from the TEM image, confirming magnetite incorporation in a macroporous polymer structure.

EDS analysis of PGME-10/12-deta with sorbed Cu(II), Cd(II), and Pb(II) ions performed by Nastasović et al., showed a significantly higher amount of Cu(II) and Pb(II) on the interior surface of the particles, supporting the significance of intra-particle diffusion as the controlling step of Cu(II), Cd(II), and Pb(II) sorption by PGME-deta [118].

9.4. Porosity Determination

Macroporous polymeric sorbents were synthesized in the shape of spherical particles by suspension copolymerization in the presence of a pore-forming agent (inert component, porogen), having a permanent well-developed porous structure even in the dry state [123]. The particles consist of smaller microspheres (10–20 nm), which are often fused. As a result of the mechanism of porous structure formation, a pore size distribution is obtained, i.e., micropores with diameters smaller than 2 nm, mesopores with diameters in the range 2–50 nm and macropores with diameters over 50 nm.

Porosity can be determined by two complementary methods-mercury porosimetry, and N_2 adsorption/desorption isotherms determination at 77 K. For the sake of illustration, pore size distribution (PSD) plots and nitrogen adsorption/desorption isotherm measured at 77 K for sample PGME-deta is presented in Figure 6a and 6b, respectively.

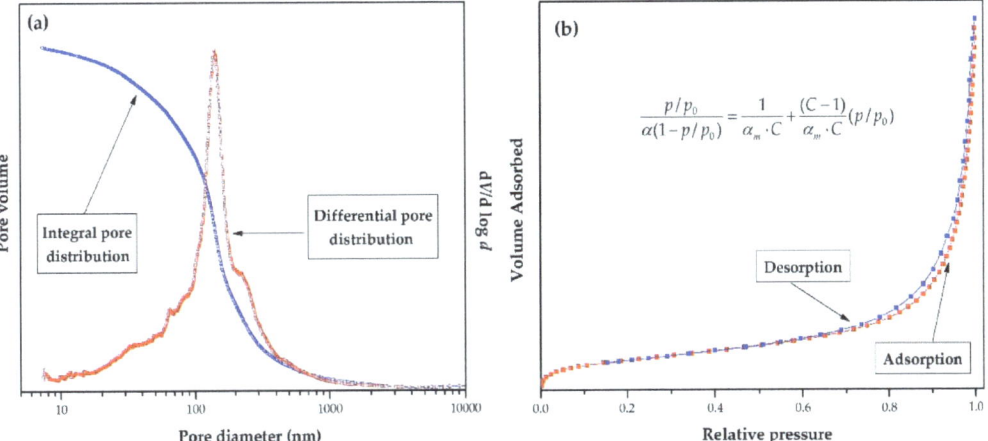

Figure 6. (**a**) Integral (cumulative) and differential pore size distribution curves and (**b**) adsorption/desorption isotherm measured at 77 K for sample PGME-deta.

Mercury porosimetry is based on the penetration of mercury into the pores as a function of the applied pressure and cover pores ranging from 7.5 nm (at the high pressure of 200 MPa) to 15 μm (at atmospheric pressure) [150]. Nitrogen sorption experiments were carried out at 77 K, in the relative pressure region of $p/p_0 = 0.05$ to $p/p_0 = 1.00$ and cover pores smaller than 7.5 nm. The mesopore distribution curve can be obtained from the adsorption branch of the N_2 isotherm by the Barrett–Joyner–Halenda (BJH) method. Specific surface area was then calculated from the well-known Brunauer–Emmett–Teller (BET) equation for multilayer adsorption, given in the inset of Figure 6b, where: p^0 is the adsorbate saturated vapor pressure, C is a constant related to the heat of adsorption. At the same time, α_m is the amount of adsorbate adsorbed in a monolayer. In the linear range of the adsorption isotherms ($0.05 \leq p/p_0 \leq 0.3$), α_m and C were estimated from the slope and the intercept of the straight line. The total pore volume was calculated as the volume of liquid adsorbate at a relative pressure of 0.99.

The porous structure of the macroporous polymeric sorbents can be described by porosity parameters: specific surface, S_{Hg}, specific pore volume, V_p, average pore diameter, d_p, and pore diameter that corresponds to half of the pore volume, $d_{V/2}$. The values of V_p, and $d_{V/2}$, of the copolymer samples, can be read from pore size distribution curves determined by mercury porosimetry. The specific surface area, S_{Hg}, can be calculated as the sum of incremental specific surface area from the pore size distribution curves, as described elsewhere [150].

Inverse gas chromatography under finite surface coverage, IGC-FC, can be used as an additional method, which enables the calculation of specific surface area by determining the adsorption isotherms of organic compounds [151]. The IGC method has been used for the investigation of polymer structure, the interactions of various liquids and gases with polymeric materials, and polymer-polymer miscibility [152–155]. The technique is especially advantageous for the investigation of macroporous crosslinked copolymers that conventional methods cannot characterize. A variety of polymer surface characteristics (dispersive component of surface free energies, enthalpy, and entropy of adsorption, acid/base constants), as well as interaction parameters and thermal transitions of polymers, can be calculated from the peak positions [156]. In the IGC-FC mode, measurable amounts of solutes were injected. From the peak shapes, adsorption isotherms, isosteric heat of adsorption, and adsorption energy distributions can be calculated.

Nastasović et al. used IGC-FC in order to determine specific surface area values for hexane, benzene, chloroform, and tetrahydrofuran sorption on macroporous PGME and PGME-deta [153]. The obtained S_a values were compared with the BET-specific surface areas measured by the nitrogen gas adsorption method. The deviations observed for the S_a values obtained by the BET method from the low-temperature nitrogen adsorption isotherms and hexane were attributed not only to the difference in molecule size but also to the specific polymer-adsorbate interactions.

10. Conclusions

As a result of rapid industrial development, growing demand for critical, precious, and rare earth metals, as well as environmental and health issues, intensive research on the new generation of polymeric and hybrid inorganic-organic sorbents with improved performances could be expected in the future. Furthermore, a deeper understanding of the sorbents structure and nature of interactions will be enabled by the development of modern techniques for structure analysis and polymer ligand-metal interactions. This review highlights the advantages of non-magnetic and magnetic porous glycidyl methacrylate copolymers, which can use as a potential alternative to low-cost but not recyclable materials. They can be adapted either by changing the porosity or by incorporating the appropriate functional groups and changing the surface chemistry, making them selective for targeted metal ions. This review is focused on methacrylate-based magnetic and non-magnetic sorbents with special attention to porous glycidyl methacrylates post-functionalized with amines and their applications in the removal of metal ions (cations and oxyanions) from aqueous solutions.

Author Contributions: Conceptualization, A.N., A.O.; methodology, A.N.; investigation, A.N.; writing—original draft preparation, A.N.; writing—review and editing, A.N., B.M., L.S. and A.O.; supervision, A.O. All authors have read and agreed to the published version of the manuscript.

Funding: This work was financially supported by the Ministry of Education, Science and Technological Development of the Republic of Serbia (Grant No. 451-03-68/2022-14/200026 and 451-03-68/2022-14/200135).

Institutional Review Board Statement: Not applicable.

Informed Consent Statement: Not applicable.

Data Availability Statement: The data presented in this study are available on request from thecorresponding author.

Conflicts of Interest: The authors declare no conflict of interest.

References

1. Liosis, C.; Papadopoulou, A.; Karvelas, E.; Karakasidis, T.E.; Sarris, I.E. Heavy Metal Adsorption Using Magnetic Nanoparticles for Water Purification: A Critical Review. *Materials* **2021**, *14*, 7500. [CrossRef] [PubMed]
2. Kazmierczak-Razna, J.; Zioła-Frankowska, A.; Nowicki, P.; Frankowski, M.; Wolski, R.; Pietrzak, R. Removal of Heavy Metal Ions from One- and Two-Component Solutions via Adsorption on N-Doped Activated Carbon. *Materials* **2021**, *14*, 7045. [CrossRef] [PubMed]
3. Lakherwal, D. Adsorption of Heavy Metals: A Review. *Int. J. Environ. Res. Dev.* **2014**, *4*, 41–48.
4. Fu, F.; Wang, Q. Removal of Heavy Metal Ions from Wastewaters: A Review. *J. Environ. Manag.* **2011**, *92*, 407–418. [CrossRef] [PubMed]
5. Tofan, L.; Wenkert, R. Chelating Polymers with Valuable Sorption Potential for Development of Precious Metal Recycling Technologies. *Rev. Chem. Eng.* **2022**, *38*, 167–183. [CrossRef]
6. Hu, Y.; Florek, J.; Larivière, D.; Fontaine, F.-G.; Kleitz, F. Recent Advances in the Separation of Rare Earth Elements Using Mesoporous Hybrid Materials. *Chem. Rec.* **2018**, *18*, 1261–1276. [CrossRef] [PubMed]
7. Xu, J.; Cao, Z.; Zhang, Y.; Yuan, Z.; Lou, Z.; Xu, X.; Wang, X. A Review of Functionalized Carbon Nanotubes and Graphene for Heavy Metal Adsorption from Water: Preparation, Application, and Mechanism. *Chemosphere* **2018**, *195*, 351–364. [CrossRef]
8. Ince, M.; Kaplan İnce, O. An Overview of Adsorption Technique for Heavy Metal Removal from Water/Wastewater: A Critical Review. *Int. J. Pure Appl. Sci.* **2017**, *3*, 10–19. [CrossRef]
9. Fei, Y.; Hu, Y.H. Design, Synthesis, and Performance of Adsorbents for Heavy Metal Removal from Wastewater: A Review. *J. Mater. Chem. A* **2022**, *10*, 1047–1085. [CrossRef]
10. Burakov, A.E.; Galunin, E.V.; Burakova, I.V.; Kucherova, A.E.; Agarwal, S.; Tkachev, A.G.; Gupta, V.K. Adsorption of Heavy Metals on Conventional and Nanostructured Materials for Wastewater Treatment Purposes: A Review. *Ecotoxicol. Environ. Saf.* **2018**, *148*, 702–712. [CrossRef]
11. Carlos, L.; Garcia Einschlag, F.S.; González, M.C.; Mártire, D.O. Applications of Magnetite Nanoparticles for Heavy Metal Removal from Wastewater. In *Waste Water—Treatment Technologies and Recent Analytical Developments*; Garca Einschlag, F.S., Ed.; InTech: London, UK, 2013; pp. 63–77.
12. Sud, D.; Mahajan, G.; Kaur, M. Agricultural Waste Material as Potential Adsorbent for Sequestering Heavy Metal Ions from Aqueous Solutions—A Review. *Bioresour. Technol.* **2008**, *99*, 6017–6027. [CrossRef] [PubMed]
13. Meseldzija, S.; Petrovic, J.; Onjia, A.; Volkov-Husovic, T.; Nesic, A.; Vukelic, N. Utilization of Agro-Industrial Waste for Removal of Copper Ions from Aqueous Solutions and Mining-Wastewater. *J. Ind. Eng. Chem.* **2019**, *75*, 246–252. [CrossRef]
14. Gómez Aguilar, D.L.; Rodríguez Miranda, J.P.; Astudillo Miller, M.X.; Maldonado Astudillo, R.I.; Esteban Muñoz, J.A. Removal of Zn(II) in Synthetic Wastewater Using Agricultural Wastes. *Metals* **2020**, *10*, 1465. [CrossRef]
15. Dzhardimalieva, G.I.; Uflyand, I.E. Synthetic Methodologies for Chelating Polymer Ligands: Recent Advances and Future Development. *ChemistrySelect* **2018**, *3*, 13234–13270. [CrossRef]
16. Radovanović, F.; Nastasović, A.; Tomković, T.; Vasiljević-Radović, D.; Nešić, A.; Veličković, S.; Onjia, A. Novel Membrane Adsorbers Incorporating Functionalized Polyglycidyl Methacrylate. *React. Funct. Polym.* **2014**, *77*, 1–10. [CrossRef]
17. Stajčić, A.; Nastasović, A.; Stajić-Trošić, J.; Marković, J.; Onjia, A.; Radovanović, F. Novel Membrane-Supported Hydrogel for Removal of Heavy Metals. *J. Environ. Chem. Eng.* **2015**, *3*, 453–461. [CrossRef]
18. Zheng, C.; He, C.; Yang, Y.; Fujita, T.; Wang, G.; Yang, W. Characterization of Waste Amidoxime Chelating Resin and Its Reutilization Performance in Adsorption of Pb(II), Cu(II), Cd(II) and Zn(II) Ions. *Metals* **2022**, *12*, 149. [CrossRef]
19. Samiey, B.; Cheng, C.-H.; Wu, J. Organic-Inorganic Hybrid Polymers as Adsorbents for Removal of Heavy Metal Ions from Solutions: A Review. *Materials* **2014**, *7*, 673–726. [CrossRef]
20. Dutta, K.; De, S. Aromatic Conjugated Polymers for Removal of Heavy Metal Ions from Wastewater: A Short Review. *Environ. Sci. Water Res. Technol.* **2017**, *3*, 793–805. [CrossRef]
21. Marjanovic, V.; Peric-Grujic, A.; Ristic, M.; Marinkovic, A.; Markovic, R.; Onjia, A.; Sljivic-Ivanovic, M. Selenate Adsorption from Water Using the Hydrous Iron Oxide-Impregnated Hybrid Polymer. *Metals* **2020**, *10*, 1630. [CrossRef]
22. Alcaraz, L.; Saquinga, D.N.; López, F.; Lima, L.D.; Alguacil, F.J.; Escudero, E.; López, F.A. Application of a Low-Cost Cellulose-Based Bioadsorbent for the Effective Recovery of Terbium Ions from Aqueous Solutions. *Metals* **2020**, *10*, 1641. [CrossRef]
23. Verma, M.; Lee, I.; Hong, Y.; Kumar, V.; Kim, H. Multifunctional β-Cyclodextrin-EDTA-Chitosan Polymer Adsorbent Synthesis for Simultaneous Removal of Heavy Metals and Organic Dyes from Wastewater. *Environ. Pollut.* **2022**, *292*, 118447. [CrossRef] [PubMed]
24. Haripriyan, U.; Gopinath, K.P.; Arun, J. Chitosan Based Nano Adsorbents and Its Types for Heavy Metal Removal: A Mini Review. *Mater. Lett.* **2022**, *312*, 131670. [CrossRef]
25. Shehzad, H.; Ahmed, E.; Sharif, A.; Farooqi, Z.H.; Din, M.I.; Begum, R.; Liu, Z.; Zhou, L.; Ouyang, J.; Irfan, A.; et al. Modified Alginate-Chitosan-TiO2 Composites for Adsorptive Removal of Ni(II) Ions from Aqueous Medium. *Int. J. Biol. Macromol.* **2022**, *194*, 117–127. [CrossRef] [PubMed]

26. Shehzad, H.; Farooqi, Z.H.; Ahmed, E.; Sharif, A.; Razzaq, S.; Mirza, F.N.; Irfan, A.; Begum, R. Synthesis of Hybrid Biosorbent Based on 1,2-Cyclohexylenedinitrilotetraacetic Acid Modified Crosslinked Chitosan and Organo-Functionalized Calcium Alginate for Adsorptive Removal of Cu(II). *Int. J. Biol. Macromol.* **2022**, *209*, 132–143. [CrossRef]
27. Castro, L.; Ayala, L.A.; Vardanyan, A.; Zhang, R.; Muñoz, J.Á. Arsenate and Arsenite Sorption Using Biogenic Iron Compounds: Treatment of Real Polluted Waters in Batch and Continuous Systems. *Metals* **2021**, *11*, 1608. [CrossRef]
28. Perumal, S.; Atchudan, R.; Edison, T.N.J.I.; Babu, R.S.; Karpagavinayagam, P.; Vedhi, C. A Short Review on Recent Advances of Hydrogel-Based Adsorbents for Heavy Metal Ions. *Metals* **2021**, *11*, 864. [CrossRef]
29. Antić, K.M.; Babić, M.M.; Vuković, J.J.J.; Vasiljević-Radović, D.G.; Onjia, A.E.; Filipović, J.M.; Tomić, S.L. Preparation and Characterization of Novel P(HEA/IA) Hydrogels for Cd^{2+} Ion Removal from Aqueous Solution. *Appl. Surf. Sci.* **2015**, *338*, 178–189. [CrossRef]
30. Naseem, K.; Farooqi, Z.H.; Begum, R.; Ur Rehman, M.Z.; Ghufran, M.; Wu, W.; Najeeb, J.; Irfan, A. Synthesis and Characterization of Poly(N-Isopropylmethacrylamide-Acrylic Acid) Smart Polymer Microgels for Adsorptive Extraction of Copper(II) and Cobalt(II) from Aqueous Medium: Kinetic and Thermodynamic Aspects. *Environ. Sci. Pollut. Res.* **2020**, *27*, 28169–28182. [CrossRef]
31. Shahid, M.; Farooqi, Z.H.; Begum, R.; Arif, M.; Irfan, A.; Azam, M. Extraction of Cobalt Ions from Aqueous Solution by Microgels for In-Situ Fabrication of Cobalt Nanoparticles to Degrade Toxic Dyes: A Two Fold-Environmental Application. *Chem. Phys. Lett.* **2020**, *754*, 137645. [CrossRef]
32. Ariffin, N.; Abdullah, M.M.A.B.; Mohd Arif Zainol, M.R.R.; Murshed, M.F.; Hariz-Zain; Faris, M.A.; Bayuaji, R. Review on Adsorption of Heavy Metal in Wastewater by Using Geopolymer. *MATEC Web Conf.* **2017**, *97*, 01023. [CrossRef]
33. Siyal, A.A.; Shamsuddin, M.R.; Khan, M.I.; Rabat, N.E.; Zulfiqar, M.; Man, Z.; Siame, J.; Azizli, K.A. A Review on Geopolymers as Emerging Materials for the Adsorption of Heavy Metals and Dyes. *J. Environ. Manag.* **2018**, *224*, 327–339. [CrossRef] [PubMed]
34. Da'na, E. Adsorption of Heavy Metals on Functionalized-Mesoporous Silica: A Review. *Microporous Mesoporous Mater.* **2017**, *247*, 145–157. [CrossRef]
35. Zhang, A.; Li, X.; Xing, J.; Xu, G. Adsorption of Potentially Toxic Elements in Water by Modified Biochar: A Review. *J. Environ. Chem. Eng.* **2020**, *8*, 104196. [CrossRef]
36. Lingamdinne, L.P.; Choi, J.-S.; Choi, Y.-L.; Yang, J.-K.; Koduru, J.R.; Chang, Y.-Y. Green Activated Magnetic Graphitic Carbon Oxide and Its Application for Hazardous Water Pollutants Removal. *Metals* **2019**, *9*, 935. [CrossRef]
37. Kegl, T.; Košak, A.; Lobnik, A.; Novak, Z.; Kralj, A.K.; Ban, I. Adsorption of Rare Earth Metals from Wastewater by Nanomaterials: A Review. *J. Hazard. Mater.* **2020**, *386*, 121632. [CrossRef]
38. Đolić, M.B.; Rajaković-Ognjanović, V.N.; Štrbac, S.B.; Dimitrijević, S.I.; Mitrić, M.N.; Onjia, A.E.; Rajaković, L.V. Natural Sorbents Modified by Divalent Cu^{2+}- and Zn^{2+}- Ions and Their Corresponding Antimicrobial Activity. *New Biotechnol.* **2017**, *39*, 150–159. [CrossRef]
39. Shoja Razavi, R.; Loghman-Estarki, M.R. Synthesis and Characterizations of Copper Oxide Nanoparticles Within Zeolite Y. *J. Clust. Sci.* **2012**, *23*, 1097–1106. [CrossRef]
40. Šljivić Ivanović, M.; Smičiklas, I.; Pejanović, S. Analysis and Comparison of Mass Transfer Phenomena Related to Cu^{2+} Sorption by Hydroxyapatite and Zeolite. *Chem. Eng. J.* **2013**, *223*, 833–843. [CrossRef]
41. Smičiklas, I.D.; Lazić, V.M.; Živković, L.S.; Porobić, S.J.; Ahrenkiel, S.P.; Nedeljković, J.M. Sorption of Divalent Heavy Metal Ions onto Functionalized Biogenic Hydroxyapatite with Caffeic Acid and 3,4-Dihydroxybenzoic Acid. *J. Environ. Sci. Health Part A* **2019**, *54*, 899–905. [CrossRef]
42. Pakade, V.; Chimuka, L. Polymeric Sorbents for Removal of Cr(VI) from Environmental Samples. *Pure Appl. Chem.* **2013**, *85*, 2145–2160. [CrossRef]
43. Sutirman, Z.A.; Sanagi, M.M.; Abd Karim, K.J.; Abu Naim, A.; Ibrahim, W.A.W. Chitosan-Based Adsorbents for the Removal of Metal Ions from Aqueous Solutions. *Malays. J. Anal. Sci.* **2018**, *22*, 839–850.
44. Guibal, E. Interactions of Metal Ions with Chitosan-Based Sorbents: A Review. *Sep. Purif. Technol.* **2004**, *38*, 43–74. [CrossRef]
45. Laus, R.; Costa, T.G.; Szpoganicz, B.; Fávere, V.T. Adsorption and Desorption of Cu(II), Cd(II) and Pb(II) Ions Using Chitosan Crosslinked with Epichlorohydrin-Triphosphate as the Adsorbent. *J. Hazard. Mater.* **2010**, *183*, 233–241. [CrossRef]
46. Ge, H.; Hua, T. Synthesis and Characterization of Poly(Maleic Acid)-Grafted Crosslinked Chitosan Nanomaterial with High Uptake and Selectivity for Hg(II) Sorption. *Carbohydr. Polym.* **2016**, *153*, 246–252. [CrossRef]
47. Lee, J.-Y.; Chen, C.-H.; Cheng, S.; Li, H.-Y. Adsorption of Pb(II) and Cu(II) Metal Ions on Functionalized Large-Pore Mesoporous Silica. *Int. J. Environ. Sci. Technol.* **2016**, *13*, 65–76. [CrossRef]
48. Mureseanu, M.; Reiss, A.; Stefanescu, I.; David, E.; Parvulescu, V.; Renard, G.; Hulea, V. Modified SBA-15 Mesoporous Silica for Heavy Metal Ions Remediation. *Chemosphere* **2008**, *73*, 1499–1504. [CrossRef]
49. Shiraishi, Y.; Nishimura, G.; Hirai, T.; Komasawa, I. Separation of Transition Metals Using Inorganic Adsorbents Modified with Chelating Ligands. *Ind. Eng. Chem. Res.* **2002**, *41*, 5065–5070. [CrossRef]
50. Asgari, M.; Zonouzi, A.; Rahimi, R.; Rabbani, M. Application of Porphyrin Modified SBA-15 in Adsorption of Lead Ions from Aqueous Media. *Orient. J. Chem.* **2015**, *31*, 1537–1544. [CrossRef]
51. Wang, Z.; Xu, W.; Jie, F.; Zhao, Z.; Zhou, K.; Liu, H. The Selective Adsorption Performance and Mechanism of Multiwall Magnetic Carbon Nanotubes for Heavy Metals in Wastewater. *Sci. Rep.* **2021**, *11*, 16878. [CrossRef]

52. Maleki, F.; Gholami, M.; Torkaman, R.; Torab-Mostaedi, M.; Asadollahzadeh, M. Multivariate Optimization of Removing of Cobalt(II) with an Efficient Aminated-GMA Polypropylene Adsorbent by Induced-Grafted Polymerization under Simultaneous Gamma-Ray Irradiation. *Sci. Rep.* **2021**, *11*, 18317. [CrossRef] [PubMed]
53. Elwakeel, K.Z.; Guibal, E. Potential Use of Magnetic Glycidyl Methacrylate Resin as a Mercury Sorbent: From Basic Study to the Application to Wastewater Treatment. *J. Environ. Chem. Eng.* **2016**, *4*, 3632–3645. [CrossRef]
54. Alexandratos, S.D. From Ion Exchange Resins to Polymer-Supported Reagents: An Evolution of Critical Variables: From Ion Exchange Resins to Polymer-Supported Reagents: An Evolution of Critical Variables. *J. Chem. Technol. Biotechnol.* **2018**, *93*, 20–27. [CrossRef]
55. Cyganowski, P. Synthesis of Adsorbents with Anion Exchange and Chelating Properties for Separation and Recovery of Precious Metals—A Review. *Solvent Extr. Ion Exch.* **2020**, *38*, 143–165. [CrossRef]
56. Pearson, R.G. Hard and Soft Acids and Bases. *J. Am. Chem. Soc.* **1963**, *85*, 3533–3539. [CrossRef]
57. Dharmapriya, T.N.; Lee, D.-Y.; Huang, P.-J. Novel Reusable Hydrogel Adsorbents for Precious Metal Recycle. *Sci. Rep.* **2021**, *11*, 19577. [CrossRef]
58. Pilśniak-Rabiega, M.; Wolska, J. Silver(I) Recovery on Sulfur-Containing Polymeric Sorbents from Chloride Solutions. *Physicochem. Probl. Miner. Process.* **2020**, *56*, 290–310. [CrossRef]
59. Xiong, C.; Wang, S.; Zhang, L.; Li, Y.; Zhou, Y.; Peng, J. Preparation of 2-Aminothiazole-Functionalized Poly(Glycidyl Methacrylate) Microspheres and Their Excellent Gold Ion Adsorption Properties. *Polymers* **2018**, *10*, 159. [CrossRef]
60. Kinemuchi, H.; Ochiai, B. Synthesis of Hydrophilic Sulfur-Containing Adsorbents for Noble Metals Having Thiocarbonyl Group Based on a Methacrylate Bearing Dithiocarbonate Moieties. *Adv. Mater. Sci. Eng.* **2018**, *2018*, 1–8. [CrossRef]
61. Nastasovic, A.; Jovanovic, S.; Jakovljevic, D.; Stankovic, S.; Onjia, A. Noble Metal Binding on Macroporous Poly(GMA-Co-EGDMA) Modified with Ethylenediamine. *J. Serbian Chem. Soc.* **2004**, *69*, 455–460. [CrossRef]
62. Nagarjuna, R.; Sharma, S.; Rajesh, N.; Ganesan, R. Effective Adsorption of Precious Metal Palladium over Polyethyleneimine-Functionalized Alumina Nanopowder and Its Reusability as a Catalyst for Energy and Environmental Applications. *ACS Omega* **2017**, *2*, 4494–4504. [CrossRef] [PubMed]
63. Shinozaki, T.; Ogata, T.; Kakinuma, R.; Narita, H.; Tokoro, C.; Tanaka, M. Preparation of Polymeric Adsorbents Bearing Diglycolamic Acid Ligands for Rare Earth Elements. *Ind. Eng. Chem. Res.* **2018**, *57*, 11424–11430. [CrossRef]
64. Galhoum, A.A.; Elshehy, E.A.; Tolan, D.A.; El-Nahas, A.M.; Taketsugu, T.; Nishikiori, K.; Akashi, T.; Morshedy, A.S.; Guibal, E. Synthesis of Polyaminophosphonic Acid-Functionalized Poly(Glycidyl Methacrylate) for the Efficient Sorption of La(III) and Y(III). *Chem. Eng. J.* **2019**, *375*, 121932. [CrossRef]
65. Yayayürük, A.E. The Use of Acrylic-Based Polymers in Environmental Remediation Studies. In *Acrylic Polymers in Healthcare*; Reddy, B.S.R., Ed.; InTech: London, UK, 2017; ISBN 978-953-51-3593-7.
66. Antic, K.; Babic, M.; Vukovic, J.; Onjia, A.; Filipovic, J.; Tomic, S. Removal of Pb2+ Ions from Aqueous Solution by P(HEA/IA) Hydrogels. *Hem. Ind.* **2016**, *70*, 695–705. [CrossRef]
67. Shakerian, F.; Kim, K.-H.; Kwon, E.; Szulejko, J.E.; Kumar, P.; Dadfarnia, S.; Haji Shabani, A.M. Advanced Polymeric Materials: Synthesis and Analytical Application of Ion Imprinted Polymers as Selective Sorbents for Solid Phase Extraction of Metal Ions. *TrAC Trends Anal. Chem.* **2016**, *83*, 55–69. [CrossRef]
68. Muzammill, E.M.; Khan, A.; Stuparu, M.C. Post-Polymerization Modification Reactions of Poly(Glycidyl Methacrylate)s. *RSC Adv.* **2017**, *7*, 55874–55884. [CrossRef]
69. Ambaye, T.G.; Vaccari, M.; van Hullebusch, E.D.; Amrane, A.; Rtimi, S. Mechanisms and Adsorption Capacities of Biochar for the Removal of Organic and Inorganic Pollutants from Industrial Wastewater. *Int. J. Environ. Sci. Technol.* **2021**, *18*, 3273–3294. [CrossRef]
70. Ayawei, N.; Ebelegi, A.N.; Wankasi, D. Modelling and Interpretation of Adsorption Isotherms. *J. Chem.* **2017**, *2017*, 1–11. [CrossRef]
71. Dada, A.O.; Adekola, F.A.; Odebunmi, E.O.; Ogunlaja, A.S.; Bello, O.S. Two–Three Parameters Isotherm Modeling, Kinetics with Statistical Validity, Desorption and Thermodynamic Studies of Adsorption of Cu(II) Ions onto Zerovalent Iron Nanoparticles. *Sci. Rep.* **2021**, *11*, 16454. [CrossRef]
72. Dada, A.O.; Adekola, F.A.; Odebunmi, E.O. Kinetics, Mechanism, Isotherm and Thermodynamic Studies of Liquid Phase Adsorption of Pb^{2+} onto Wood Activated Carbon Supported Zerovalent Iron (WAC-ZVI) Nanocomposite. *Cogent Chem.* **2017**, *3*, 1351653. [CrossRef]
73. Hashem, A.; Al-Anwar, A.; Nagy, N.M.; Hussein, D.M.; Eisa, S. Isotherms and Kinetic Studies on Adsorption of Hg(II) Ions onto Ziziphus Spina-Christi L. from Aqueous Solutions. *Green Process. Synth.* **2016**, *5*, 213–224. [CrossRef]
74. Nworie, F.S.; Nwabue, F.I.; Oti, W.; Mbam, E.; Nwali, B.U. Removal of methylene blue from aqueous solution using activated rice husk biochar: Adsorption isotherms, kinetics and error analysis. *J. Chil. Chem. Soc.* **2019**, *64*, 4365–4376. [CrossRef]
75. Ngakou, C.S.; Anagho, G.S.; Ngomo, H.M. Non-Linear Regression Analysis for the Adsorption Kinetics and Equilibrium Isotherm of Phenacetin onto Activated Carbons. *Curr. J. Appl. Sci. Technol.* **2019**, *36*, 1–18. [CrossRef]
76. Girish, C.R. Various Isotherm Models for Multicomponent Adsorption: A Review. *Int. J. Civ. Eng. Technol.* **2017**, *8*, 80–86.
77. Sahoo, T.R.; Prelot, B. Adsorption Processes for the Removal of Contaminants from Wastewater. In *Nanomaterials for the Detection and Removal of Wastewater Pollutants*; Bonelli, B., Freyria, F.S., Rossetti, I., Sethi, R., Eds.; Elsevier: Amsterdam, The Netherlands, 2020; pp. 161–222.

78. Wang, J.; Guo, X. Adsorption Kinetic Models: Physical Meanings, Applications, and Solving Methods. *J. Hazard. Mater.* **2020**, *390*, 122156. [CrossRef] [PubMed]
79. Diaz de Tuesta, J.L.; Silva, A.M.T.; Faria, J.L.; Gomes, H.T. Adsorption of Sudan-IV Contained in Oily Wastewater on Lipophilic Activated Carbons: Kinetic and Isotherm Modelling. *Environ. Sci. Pollut. Res.* **2020**, *27*, 20770–20785. [CrossRef]
80. Inyinbor, A.A.; Adekola, F.A.; Olatunji, G.A. Kinetics, Isotherms and Thermodynamic Modeling of Liquid Phase Adsorption of Rhodamine B Dye onto Raphia Hookerie Fruit Epicarp. *Water Resour. Ind.* **2016**, *15*, 14–27. [CrossRef]
81. Rojas, J.; Suarez, D.; Moreno, A.; Silva-Agredo, J.; Torres-Palma, R.A. Kinetics, Isotherms and Thermodynamic Modeling of Liquid Phase Adsorption of Crystal Violet Dye onto Shrimp-Waste in Its Raw, Pyrolyzed Material and Activated Charcoals. *Appl. Sci.* **2019**, *9*, 5337. [CrossRef]
82. Andrade, C.A.; Zambrano-Intriago, L.A.; Oliveira, N.S.; Vieira, J.S.; Quiroz-Fernández, L.S.; Rodríguez-Díaz, J.M. Adsorption Behavior and Mechanism of Oxytetracycline on Rice Husk Ash: Kinetics, Equilibrium, and Thermodynamics of the Process. *Water Air Soil Pollut.* **2020**, *231*, 103. [CrossRef]
83. Ebelegi, A.N.; Ayawei, N.; Wankasi, D. Interpretation of Adsorption Thermodynamics and Kinetics. *Open J. Phys. Chem.* **2020**, *10*, 166–182. [CrossRef]
84. Gupta, A.; Sharma, V.; Sharma, K.; Kumar, V.; Choudhary, S.; Mankotia, P.; Kumar, B.; Mishra, H.; Moulick, A.; Ekielski, A.; et al. A Review of Adsorbents for Heavy Metal Decontamination: Growing Approach to Wastewater Treatment. *Materials* **2021**, *14*, 4702. [CrossRef] [PubMed]
85. Liu, Y.; Liu, Y.-J. Biosorption Isotherms, Kinetics and Thermodynamics. *Sep. Purif. Technol.* **2008**, *61*, 229–242. [CrossRef]
86. Srivastava, S.; Goyal, P. *Novel Biomaterials: Decontamination of Toxic Metals from Wastewater*; Part of the Book Series of Environmental Science and Engineering, Environmental Engineering; Springer: Berlin/Heidelberg, Germany; New York, NY, USA, 2010; ISBN 978-3-642-11328-4.
87. Lata, S.; Singh, P.K.; Samadder, S.R. Regeneration of Adsorbents and Recovery of Heavy Metals: A Review. *Int. J. Environ. Sci. Technol.* **2015**, *12*, 1461–1478. [CrossRef]
88. Nastasović, A.; Jovanović, S.; Đorđević, D.; Onjia, A.; Jakovljević, D.; Novaković, T. Metal Sorption on Macroporous Poly(GMA-Co-EGDMA) Modified with Ethylene Diamine. *React. Funct. Polym.* **2004**, *58*, 139–147. [CrossRef]
89. Marković, B.M.; Stefanović, I.S.; Hercigonja, R.V.; Pergal, M.V.; Marković, J.P.; Onjia, A.E.; Nastasović, A.B. Novel Hexamethylene Diamine-Functionalized Macroporous Copolymer for Chromium Removal from Aqueous Solutions. *Polym. Int.* **2017**, *66*, 679–689. [CrossRef]
90. Wu, A.; Jia, J.; Luan, S. Amphiphilic PMMA/PEI Core–Shell Nanoparticles as Polymeric Adsorbents to Remove Heavy Metal Pollutants. *Colloids Surf. Physicochem. Eng. Asp.* **2011**, *384*, 180–185. [CrossRef]
91. Liu, X.; Chen, H.; Wang, C.; Qu, R.; Ji, C.; Sun, C.; Zhang, Y. Synthesis of Porous Acrylonitrile/Methyl Acrylate Copolymer Beads by Suspended Emulsion Polymerization and Their Adsorption Properties after Amidoximation. *J. Hazard. Mater.* **2010**, *175*, 1014–1021. [CrossRef]
92. Dinari, M.; Atabaki, F.; Pahnavar, Z.; Soltani, R. Adsorptive Removal Properties of Bivalent Cadmium from Aqueous Solution Using Porous Poly(N-2-Methyl-4-Nitrophenyl Maleimide-Maleic Anhydride-Methyl Methacrylate) Terpolymers. *J. Environ. Chem. Eng.* **2020**, *8*, 104560. [CrossRef]
93. Mohammadnezhad, G.; Dinari, M.; Soltani, R. The Preparation of Modified Boehmite/PMMA Nanocomposites by in Situ Polymerization and the Assessment of Their Capability for Cu^{2+} Ion Removal. *New J. Chem.* **2016**, *40*, 3612–3621. [CrossRef]
94. Moradi, O.; Mirza, B.; Norouzi, M.; Fakhri, A. Removal of Co(II), Cu(II) and Pb(II) Ions by Polymer Based 2-Hydroxyethyl Methacrylate: Thermodynamics and Desorption Studies. *Iran. J. Environ. Health Sci. Eng.* **2012**, *9*, 31. [CrossRef]
95. Shen, H.; Pan, S.; Zhang, Y.; Huang, X.; Gong, H. A New Insight on the Adsorption Mechanism of Amino-Functionalized Nano-Fe3O4 Magnetic Polymers in Cu(II), Cr(VI) Co-Existing Water System. *Chem. Eng. J.* **2012**, *183*, 180–191. [CrossRef]
96. Cifci, C.; Durmaz, O. Removal of Heavy Metal Ions from Aqueous Solutions by Poly(Methyl m Ethacrylate-Co-Ethyl Acrylate) and Poly(Methyl Methacrylate-Co-Buthyl m Ethacrylate) Membranes. *Desalination Water Treat.* **2011**, *28*, 255–259. [CrossRef]
97. Huš, S.; Kolar, M.; Krajnc, P. Separation of Heavy Metals from Water by Functionalized Glycidyl Methacrylate Poly (High Internal Phase Emulsions). *J. Chromatogr. A* **2016**, *1437*, 168–175. [CrossRef]
98. Atia, A.A.; Donia, A.M.; Awed, H.A. Synthesis of Magnetic Chelating Resins Functionalized with Tetraethylenepentamine for Adsorption of Molybdate Anions from Aqueous Solutions. *J. Hazard. Mater.* **2008**, *155*, 100–108. [CrossRef] [PubMed]
99. Zhao, J.; Wang, C.; Wang, S.; Zhou, Y.; Zhang, B. Experimental and DFT Studies on the Selective Adsorption of Pd(II) from Wastewater by Pyromellitic-Functionalized Poly(Glycidyl Methacrylate) Microsphere. *J. Mol. Liq.* **2020**, *300*, 112296. [CrossRef]
100. Liu, C.; Bai, R. Extended Study of DETA-Functionalized PGMA Adsorbent in the Selective Adsorption Behaviors and Mechanisms for Heavy Metal Ions of Cu, Co, Ni, Zn, and Cd. *J. Colloid Interface Sci.* **2010**, *350*, 282–289. [CrossRef] [PubMed]
101. Xiong, C.; Wang, S.; Zhang, L.; Li, Y.; Zhou, Y.; Peng, J. Selective Recovery of Silver from Aqueous Solutions by Poly (Glycidyl Methacrylate) Microsphere Modified with Trithiocyanuric Acid. *J. Mol. Liq.* **2018**, *254*, 340–348. [CrossRef]
102. Gupta, A.; Jain, R.; Gupta, D.C. Studies on Uptake Behavior of Hg(II) and Pb(II) by Amine Modified Glycidyl Methacrylate–Styrene–N,N′-Methylenebisacrylamide Terpolymer. *React. Funct. Polym.* **2015**, *93*, 22–29. [CrossRef]
103. Abd El-Magied, M.O.; Elshehy, E.A.; Manaa, E.-S.A.; Tolba, A.A.; Atia, A.A. Kinetics and Thermodynamics Studies on the Recovery of Thorium Ions Using Amino Resins with Magnetic Properties. *Ind. Eng. Chem. Res.* **2016**, *55*, 11338–11345. [CrossRef]

104. Suručić, L.; Tadić, T.; Janjić, G.; Marković, B.; Nastasović, A.; Onjia, A. Recovery of Vanadium (V) Oxyanions by a Magnetic Macroporous Copolymer Nanocomposite Sorbent. *Metals* **2021**, *11*, 1777. [CrossRef]
105. Chaipuang, A.; Phungpanya, C.; Thongpoon, C.; Watla-iad, K.; Inkaew, P.; Machan, T.; Suwantong, O. Synthesis of Copper(II) Ion-Imprinted Polymers via Suspension Polymerization. *Polym. Adv. Technol.* **2018**, *29*, 3134–3141. [CrossRef]
106. Bayramoglu, G.; Arica, M.Y. Polyethylenimine and Tris(2-Aminoethyl)Amine Modified p(GA–EGMA) Microbeads for Sorption of Uranium Ions: Equilibrium, Kinetic and Thermodynamic Studies. *J. Radioanal. Nucl. Chem.* **2017**, *312*, 293–303. [CrossRef]
107. Mafu, L.D.; Mamba, B.B.; Msagati, T.A.M. Synthesis and Characterization of Ion Imprinted Polymeric Adsorbents for the Selective Recognition and Removal of Arsenic and Selenium in Wastewater Samples. *J. Saudi Chem. Soc.* **2016**, *20*, 594–605. [CrossRef]
108. Bunina, Z.Y.; Bryleva, K.; Yurchenko, O.; Belikov, K. Sorption Materials Based on Ethylene Glycol Dimethacrylate and Methacrylic Acid Copolymers for Rare Earth Elements Extraction from Aqueous Solutions. *Adsorpt. Sci. Technol.* **2017**, *35*, 545–559. [CrossRef]
109. Zhang, B.; Wang, S.; Fu, L.; Zhang, L.; Zhao, J.; Wang, C. Selective High Capacity Adsorption of Au(III) from Aqueous Solution by Poly(Glycidyl Methacrylate) Functionalized with 2,6-Diaminopyridine. *Polym. Bull.* **2019**, *76*, 4017–4033. [CrossRef]
110. Han, L.; Peng, Y.; Ma, J.; Shi, Z.; Jia, Q. Construction of Hypercrosslinked Polymers with Styrene-Based Copolymer Precursor for Adsorption of Rare Earth Elements. *Sep. Purif. Technol.* **2022**, *285*, 120378. [CrossRef]
111. Ekmeščić, B.M.; Maksin, D.D.; Marković, J.P.; Vuković, Z.M.; Hercigonja, R.V.; Nastasović, A.B.; Onjia, A.E. Recovery of Molybdenum Oxyanions Using Macroporous Copolymer Grafted with Diethylenetriamine. *Arab. J. Chem.* **2019**, *12*, 3628–3638. [CrossRef]
112. Gao, B.; Zhang, Y.; Xu, Y. Study on Recognition and Separation of Rare Earth Ions at Picometre Scale by Using Efficient Ion-Surface Imprinted Polymer Materials. *Hydrometallurgy* **2014**, *150*, 83–91. [CrossRef]
113. Dong, T.; Yang, L.; Pan, F.; Xing, H.; Wang, L.; Yu, J.; Qu, H.; Rong, M.; Liu, H. Effect of Immobilized Amine Density on Cadmium(II) Adsorption Capacities for Ethanediamine-Modified Magnetic Poly-(Glycidyl Methacrylate) Microspheres. *J. Magn. Magn. Mater.* **2017**, *427*, 289–295. [CrossRef]
114. Liu, S.; Liu, L.; Su, G.; Zhao, L.; Peng, H.; Xue, J.; Tang, A. Enhanced Adsorption Performance, Separation, and Recyclability of Magnetic Core-Shell Fe3O4@PGMA-g-TETA-CSSNa Microspheres for Heavy Metal Removal. *React. Funct. Polym.* **2022**, *170*, 105127. [CrossRef]
115. Madrid, J.F.; Barba, B.J.D.; Pomicpic, J.C.; Cabalar, P.J.E. Immobilization of an Organophosphorus Compound on Polypropyl-Ene-g-Poly(Glycidyl Methacrylate) Polymer Support and Its Application in Scandium Recovery. *J. Appl. Polym. Sci.* **2022**, *139*, 51597. [CrossRef]
116. Donia, A.M.; Atia, A.A.; Moussa, E.M.M.; El-Sherif, A.M.; Abd El-Magied, M.O. Removal of Uranium(VI) from Aqueous Solutions Using Glycidyl Methacrylate Chelating Resins. *Hydrometallurgy* **2009**, *95*, 183–189. [CrossRef]
117. Sun, X.; Yang, L.; Xing, H.; Zhao, J.; Li, X.; Huang, Y.; Liu, H. Synthesis of Polyethylenimine-Functionalized Poly(Glycidyl Methacrylate) Magnetic Microspheres and Their Excellent Cr(VI) Ion Removal Properties. *Chem. Eng. J.* **2013**, *234*, 338–345. [CrossRef]
118. Nastasović, A.B.; Ekmeščić, B.M.; Sandić, Z.P.; Ranđelović, D.V.; Mozetič, M.; Vesel, A.; Onjia, A.E. Mechanism of Cu(II), Cd(II) and Pb(II) Ions Sorption from Aqueous Solutions by Macroporous Poly(Glycidyl Methacrylate-Co-Ethylene Glycol Dimethacrylate). *Appl. Surf. Sci.* **2016**, *385*, 605–615. [CrossRef]
119. Wang, Y.; Zhang, Y.; Hou, C.; He, X.; Liu, M. Preparation of a Novel TETA Functionalized Magnetic PGMA Nano-Absorbent by ATRP Method and Used for Highly Effective Adsorption of Hg(II). *J. Taiwan Inst. Chem. Eng.* **2016**, *58*, 283–289. [CrossRef]
120. Masoumi, A.; Ghaemy, M.; Bakht, A.N. Removal of Metal Ions from Water Using Poly(MMA-Co-MA)/Modified-Fe3O4 Magnetic Nanocomposite: Isotherm and Kinetic Study. *Ind. Eng. Chem. Res.* **2014**, *53*, 8188–8197. [CrossRef]
121. Jovanović, S.M.; Nastasović, A.; Jovanović, N.N.; Jeremić, K.; Savić, Z. The Influence of Inert Component Composition on the Porous Structure of Glycidyl Methacrylate/Ethylene Glycol Dimethacrylate Copolymers. *Angew. Makromol. Chem.* **1994**, *219*, 161–168. [CrossRef]
122. Kimmins, S.D.; Wyman, P.; Cameron, N.R. Amine-Functionalization of Glycidyl Methacrylate-Containing Emulsion-Templated Porous Polymers and Immobilization of Proteinase K for Biocatalysis. *Polymer* **2014**, *55*, 416–425. [CrossRef]
123. Okay, O. Macroporous Copolymer Networks. *Prog. Polym. Sci.* **2000**, *25*, 711–779. [CrossRef]
124. Suručić, L.T.; Janjić, G.V.; Rakić, A.A.; Nastasović, A.B.; Popović, A.R.; Milčić, M.K.; Onjia, A.E. Theoretical Modeling of Sorption of Metal Ions on Amino-Functionalized Macroporous Copolymer in Aqueous Solution. *J. Mol. Model.* **2019**, *25*, 177. [CrossRef]
125. Surucic, L.; Nastasovic, A.; Onjia, A.; Janjic, G.; Rakic, A. Design of Amino-Functionalized Chelated Macroporous Copolymer [Poly(GMA-EDGMA)] for the Sorption of Cu (II) Ions. *J. Serbian Chem. Soc.* **2019**, *84*, 1391–1404. [CrossRef]
126. Hercigonja, R.V.; Maksin, D.D.; Nastasović, A.B.; Trifunović, S.S.; Glodić, P.B.; Onjia, A.E. Adsorptive Removal of Technetium-99 Using Macroporous Poly(GMA-Co-EGDMA) Modified with Diethylene Triamine. *J. Appl. Polym. Sci.* **2012**, *123*, 1273–1282. [CrossRef]
127. Maksin, D.D.; Hercigonja, R.V.; Lazarević, M.Ž.; Žunić, M.J.; Nastasović, A.B. Modeling of Kinetics of Pertechnetate Removal by Amino-Functionalized Glycidyl Methacrylate Copolymer. *Polym. Bull.* **2012**, *68*, 507–528. [CrossRef]
128. Nastasović, A.; Sandić, Z.; Suručić, L.; Maksin, D.; Jakovljević, D.; Onjia, A. Kinetics of Hexavalent Chromium Sorption on Amino-Functionalized Macroporous Glycidyl Methacrylate Copolymer. *J. Hazard. Mater.* **2009**, *171*, 153–159. [CrossRef]

129. Maksin, D.D.; Nastasović, A.B.; Milutinović-Nikolić, A.D.; Suručić, L.T.; Sandić, Z.P.; Hercigonja, R.V.; Onjia, A.E. Equilibrium and Kinetics Study on Hexavalent Chromium Adsorption onto Diethylene Triamine Grafted Glycidyl Methacrylate Based Copolymers. *J. Hazard. Mater.* **2012**, *209–210*, 99–110. [CrossRef]
130. Marković, B.M.; Vuković, Z.M.; Spasojević, V.V.; Kusigerski, V.B.; Pavlović, V.B.; Onjia, A.E.; Nastasović, A.B. Selective Magnetic GMA Based Potential Sorbents for Molybdenum and Rhenium Sorption. *J. Alloys Compd.* **2017**, *705*, 38–50. [CrossRef]
131. Nastasović, A.; Jakovljević, D.; Sandić, Z.; Đorđević, D.; Malović, L.; Kljajević, S.; Marković, J.; Onjia, A. Amino-Functionalized Glycidyl Methacrylate Based Macroporous Copolymers as Metal Ion Sorbents. In *Reactive and Functional Polymers Research Advances*; Barroso, M.I., Ed.; Nova Science Publishers, Inc.: New York, NY, USA, 2007; pp. 79–112.
132. Jovanovic, S.; Nastasović, A. Macroporous Glycidyl Methacrylate Copolymers, Synthesis, Characterization and Application. In *Polymeric Materials*; Nastasović, A., Jovanović, S.M., Eds.; Transworld Research Network: Trivandrum, India, 2009; pp. 1–27.
133. Malović, L.; Nastasović, A.; Sandić, Z.; Marković, J.; Đorđević, D.; Vuković, Z. Surface Modification of Macroporous Glycidyl Methacrylate Based Copolymers for Selective Sorption of Heavy Metals. *J. Mater. Sci.* **2007**, *42*, 3326–3337. [CrossRef]
134. Jovanović, S.M.; Nastasović, A.; Jovanović, N.N.; Novaković, T.; Vuković, Z.; Jeremić, K. Synthesis, Properties and Applications of Crosslinked Macroporous Copolymers Based on Methacrylates. *Hem. Ind.* **2000**, *54*, 471–479.
135. Şenkal, B.F.; Yavuz, E. Crosslinked Poly(Glycidyl Methacrylate)-Based Resin for Removal of Mercury from Aqueous Solutions. *J. Appl. Polym. Sci.* **2006**, *101*, 348–352. [CrossRef]
136. Atia, A.A.; Donia, A.M.; Abou-El-Enein, S.A.; Yousif, A.M. Studies on Uptake Behaviour of Copper(II) and Lead(II) by Amine Chelating Resins with Different Textural Properties. *Sep. Purif. Technol.* **2003**, *33*, 295–301. [CrossRef]
137. Haratake, M.; Yasumoto, K.; Ono, M.; Akashi, M.; Nakayama, M. Synthesis of Hydrophilic Macroporous Chelating Polymers and Their Versatility in the Preconcentration of Metals in Seawater Samples. *Anal. Chim. Acta* **2006**, *561*, 183–190. [CrossRef]
138. Suručić, L.; Nastasović, A.; Rakić, A.; Janjić, G.; Onjia, A.; Popović, A. Comparative Study of W(VI) and Cr(VI) Oxyanions Binding Ability with Magnetic Amino-Functionalized Nanocomposite in Aqueous Solution. In Proceedings of the 5th World Congress on Mechanical, Chemical, and Material Engineering (MCM'19), Lisbon, Portugal, 17 August 2019.
139. Kalia, S.; Kango, S.; Kumar, A.; Haldorai, Y.; Kumari, B.; Kumar, R. Magnetic Polymer Nanocomposites for Environmental and Biomedical Applications. *Colloid Polym. Sci.* **2014**, *292*, 2025–2052. [CrossRef]
140. Duranoğlu, D.; Buyruklardan Kaya, İ.G.; Beker, U.; Şenkal, B.F. Synthesis and Adsorption Properties of Polymeric and Polymer-Based Hybrid Adsorbent for Hexavalent Chromium Removal. *Chem. Eng. J.* **2012**, *181–182*, 103–112. [CrossRef]
141. Atia, A.A.; Donia, A.M.; El-Enein, S.A.; Yousif, A.M. Effect of Chain Length of Aliphatic Amines Immobilized on a Magnetic Glycidyl Methacrylate Resin towards the Uptake Behavior of Hg(II) from Aqueous Solutions. *Sep. Sci. Technol.* **2007**, *42*, 403–420. [CrossRef]
142. Bayramoğlu, G.; Arica, M.Y. Kinetics of Mercury Ions Removal from Synthetic Aqueous Solutions Using by Novel Magnetic p(GMA-MMA-EGDMA) Beads. *J. Hazard. Mater.* **2007**, *144*, 449–457. [CrossRef]
143. Perendija, J.; Veličković, Z.S.; Cvijetić, I.; Rusmirović, J.D.; Ugrinović, V.; Marinković, A.D.; Onjia, A. Batch and Column Adsorption of Cations, Oxyanions and Dyes on a Magnetite Modified Cellulose-Based Membrane. *Cellulose* **2020**, *27*, 8215–8235. [CrossRef]
144. Xie, M.; Zeng, L.; Zhang, Q.; Kang, Y.; Xiao, H.; Peng, Y.; Chen, X.; Luo, J. Synthesis and Adsorption Behavior of Magnetic Microspheres Based on Chitosan/Organic Rectorite for Low-Concentration Heavy Metal Removal. *J. Alloys Compd.* **2015**, *647*, 892–905. [CrossRef]
145. Golcuk, K.; Altun, A.; Kumru, M. Thermal Studies and Vibrational Analyses of M-Methylaniline Complexes of Zn(II), Cd(II) and Hg(II) Bromides. *Spectrochim. Acta. A. Mol. Biomol. Spectrosc.* **2003**, *59*, 1841–1847. [CrossRef]
146. Coates, J. Interpretation of Infrared Spectra, A Practical Approach. In *Encyclopedia of Analytical Chemistry*; Meyers, R.A., Ed.; John Wiley & Sons, Ltd.: Chichester, UK, 2006; pp. 10815–10837. ISBN 978-0-470-02731-8.
147. Guibal, E.; Milot, C.; Eterradossi, O.; Gauffier, C.; Domard, A. Study of Molybdate Ion Sorption on Chitosan Gel Beads by Different Spectrometric Analyses. *Int. J. Biol. Macromol.* **1999**, *24*, 49–59. [CrossRef]
148. Mather, R.R. Surface Modification of Textiles by Plasma Treatments. In *Surface Modification of Textiles*; Wei, Q., Ed.; Elsevier: Amsterdam, The Netherlands, 2009; pp. 296–317.
149. Wang, J.; Liu, F. Synthesis and Application of Ion-Imprinted Interpenetrating Polymer Network Gel for Selective Solid Phase Extraction of Cd^{2+}. *Chem. Eng. J.* **2014**, *242*, 117–126. [CrossRef]
150. Rodríguez-Reinoso, F.; Sepúlveda-Escribano, A. Porous carbons in adsorption and catalysis. In *Handbook of Surfaces and Interfaces of Materials*; Nalwa, H., Ed.; Elsevier: Amsterdam, The Netherlands, 2001; Volume 5, pp. 309–355.
151. Nastasović, A.B.; Novaković, T.B.; Vuković, Z.M.; Ekmeščić, B.M.; Ranđelović, D.V.; Maksin, D.D.; Miladinović, Z.P. Polymer-Based Monolithic Porous Composite. In *Proceedings of the III Advanced Ceramics and Applications Conference*; Lee, W.E., Gadow, R., Mitic, V., Obradovic, N., Eds.; Atlantis Press: Paris, France, 2016; pp. 241–257.
152. Nastasović, A.B.; Onjia, A.E.; Milonjić, S.K.; Vuković, Z.M.; Jovanović, S.M. Characterization of Glycidyl Methacrylate Based Copolymers by Inverse Gas Chromatography under Finite Surface Coverage. *Macromol. Mater. Eng.* **2005**, *290*, 884–890. [CrossRef]
153. Nastasović, A.B.; Onjia, A.E.; Milonjić, S.K.; Jovanović, S.M. Determination of Thermodynamic Properties of Macroporous Glycidyl Methacrylate-Based Copolymers by Inverse Gas Chromatography. *J. Polym. Sci. Part B Polym. Phys.* **2005**, *43*, 2524–2533. [CrossRef]

154. Nastasović, A.B.; Onjia, A.E.; Milonjić, S.K.; Jovanović, S.M. Surface Characterization of Macroporous Glycidyl Methacrylate Based Copolymers by Inverse Gas Chromatography. *Eur. Polym. J.* **2005**, *41*, 1234–1242. [CrossRef]
155. Onjia, A.; Milonjić, S.K.; Jovanović, N.N.; Jovanović, S.M. An Inverse Gas Chromatography Study of Macroporous Copolymers Based on Methyl and Glycidyl Methacrylate. *React. Funct. Polym.* **2000**, *43*, 269–277. [CrossRef]
156. Nastasović, A.B.; Onjia, A.E. Determination of Glass Temperature of Polymers by Inverse Gas Chromatography. *J. Chromatogr. A* **2008**, *1195*, 1–15. [CrossRef]

Article

Dispersive Solid–Liquid Microextraction Based on the Poly(HDDA)/Graphene Sorbent Followed by ICP-MS for the Determination of Rare Earth Elements in Coal Fly Ash Leachate

Latinka Slavković-Beškoski [1], Ljubiša Ignjatović [2], Guido Bolognesi [3], Danijela Maksin [4], Aleksandra Savić [1], Goran Vladisavljević [3] and Antonije Onjia [5],*

[1] Anahem Laboratory, Mocartova 10, 11160 Belgrade, Serbia; latinka@anahem.org (L.S.-B.); stojanovic.aleksandra@anahem.org (A.S.)
[2] Faculty of Physical Chemistry, University of Belgrade, Studentski Trg 12-16, 11000 Belgrade, Serbia; ljignjatovic@ffh.bg.ac.rs
[3] Department of Chemical Engineering, Loughborough University, Leicestershire, Loughborough LE11 3TU, UK; g.bolognesi@lboro.ac.uk (G.B.); g.vladisavljevic@lboro.ac.uk (G.V.)
[4] Vinča Institute of Nuclear Sciences, University of Belgrade, 11001 Belgrade, Serbia; dmaksin@vin.bg.ac.rs
[5] Faculty of Technology and Metallurgy, University of Belgrade, Karnegieva 4, 11120 Belgrade, Serbia
* Correspondence: onjia@tmf.bg.ac.rs

Citation: Slavković-Beškoski, L.; Ignjatović, L.; Bolognesi, G.; Maksin, D.; Savić, A.; Vladisavljević, G.; Onjia, A. Dispersive Solid–Liquid Microextraction Based on the Poly(HDDA)/Graphene Sorbent Followed by ICP-MS for the Determination of Rare Earth Elements in Coal Fly Ash Leachate. *Metals* **2022**, *12*, 791. https://doi.org/10.3390/met12050791

Academic Editor: Antoni Roca

Received: 9 April 2022
Accepted: 29 April 2022
Published: 4 May 2022

Publisher's Note: MDPI stays neutral with regard to jurisdictional claims in published maps and institutional affiliations.

Copyright: © 2022 by the authors. Licensee MDPI, Basel, Switzerland. This article is an open access article distributed under the terms and conditions of the Creative Commons Attribution (CC BY) license (https://creativecommons.org/licenses/by/4.0/).

Abstract: A dispersive solid-phase microextraction (DSPME) sorbent consisting of poly(1,6-hexanediol diacrylate)-based polymer microspheres, with embedded graphene microparticles (poly(HDDA)/graphene), was synthesized by microfluidic emulsification/photopolymerization and characterized by optical microscopy and X-ray fluorescence spectrometry. This sorbent was applied for simple, fast, and sensitive vortex-assisted DSPME of rare earth elements (RREs) in coal fly ash (CFA) leachate, prior to their quantification by inductively coupled plasma mass spectrometry (ICP-MS). Among nine DSPME variables, the Plackett–Burman screening design (PBD), followed by the central composite optimization design (CCD) using the Derringer desirability function (D), identified the eluent type as the most influencing DSPME variable. The optimum conditions with maximum D (0.65) for the chelating agent di-(2-ethylhexyl) phosphoric acid (D2EHPA) amount, the sorbent amount, the eluting solvent, the extraction temperature, the centrifuge speed, the vortexing time, the elution time, the centrifugation time, and pH, were set to 60 µL, 30 mg, 2 M HNO$_3$, 25 °C, 6000 rpm, 1 min, 1 min, 5 min, and 4.2, respectively. Analytical validation of the DSPME method for 16 REEs (Sc, Y, La, Ce, Pr, Nd, Sm, Eu, Gd, Tb, Dy, Ho, Er, Tm, Yb, Lu) in CFA leachate samples estimated the detection limits at the low ppt level, the recovery range 43–112%, and relative standard deviation within ± 22%. This method was applied to a water extraction procedure (EP) and acetic acid toxicity characteristic leaching procedure (TCLP) for leachate of CFA, from five different coal-fired thermoelectric power plants. The most abundant REEs in leachate (20 ÷ 1 solid-to-liquid ratio) are Ce, Y, and La, which were found in the range of 22–194 ng/L, 35–105 ng/L, 48–95 ng/L, and 9.6–51 µg/L, 7.3–22 µg/L, 2.4–17 µg/L, for EP and TCLP leachate, respectively. The least present REE in TCLP leachate was Lu (42–125 ng/L), which was not detected in EP leachate.

Keywords: DSPME; REEs; Plackett–Burman; alkali-acid leaching; Derringer desirability; coal-fired power plant

1. Introduction

Rare earth elements (REEs) have recently gained an important role in various applications in hi-tech devices, specific catalysts, superconductors, telecommunications, laser technologies, etc. [1,2]. They are quite valuable due to their high conductivity and magnetism, which enable various engineering solutions.

In addition to ores, waste materials and by-products are increasingly being considered alternative sources for obtaining REEs [3,4]. Coal fly ash (CFA) is a promising source of REEs, whose potential as a source of REEs is being intensively studied [5]. REEs are found in CFA in various forms [6], so their recovery is very complex [7]. The major phase components of the CFA are quartz, mullite, hematite and amorphous glass. The rare earth elements are captured in this structure [8,9]. Therefore, the extraction of rare earth elements is very difficult.

The technology of obtaining REEs from CFA consists of several stages: mechanical grinding, magnetic separation, leaching, extraction, and refining [10]. Alkali-acid leaching is a common practice for conventional REEs recovery, while the chelating solvent extraction is usually used to separate REEs from CFA leachate. Finally, the refining stage involves electrolysis, zone melting, etc.

An important part of the REEs production from CFA is monitoring wastewater originating from CFA leaching, from both the recovery process and landfill. Two widely accepted standardized leaching testing procedures are the EN-12457-2 aqueous extraction procedure (EP) and US EPA 1311 toxicity characteristic leaching procedure (TCLP) [11].

The quantification of REEs in CFA leachate is quite challenging, consisting of several steps. Instrumental measurements for determining REEs somewhat converge towards ICP-MS [12,13]. However, even if ICP-MS is a powerful technique, it suffers from being not sensitive enough for some REEs present at a very low level [14,15] or interferences from high-matrix aqueous samples [16]. Therefore, the separation and preconcentration of REEs from a matrix solution are usually needed prior to an instrumental ICP-MS measurement [17].

Traditional methods for separating REEs include liquid–liquid extraction, ion-exchange, co-precipitation, and dry digestion [18–21]. Even if there have been tremendous advances in developing new solvents [22,23] and hybrid sorbents [24,25] for trace elements separations, these methods are somehow inconvenient, such as being time consuming, quite expensive, and not environmentally friendly [26].

One of the popular directions of research into improving the method of sample preparation is the introduction of microextraction in the field of analytical determination of trace elements [27,28]. Several microextraction sample preparation and preconcentration techniques for REEs prior to the instrumental measurements by ICP-OES and ICP-MS [29,30] have been investigated. REEs were subjected to preconcentration from groundwater by dispersive liquid–liquid microextraction (DLLME) followed by ICP-MS [31]. A few studies dealt with dispersive solid-phase microextraction (DSPME) of REEs, in which ICP-MS quantification was performed [16,32].

In most cases, the optimization of the microextraction procedure was conducted with the traditional "one-variable-at-a-time" (OVAT) approach. OVAT is an optimization technique in which one variable is changed while keeping all other variables constant. A more advanced chemometric approach using the design of experiments (DOE) enables optimization by changing all variables simultaneously. In addition to identifying the critical variables, it can also be used to achieve the desired response.

Although the chemometric optimization has been applied to the simultaneous preconcentration of several metals by microextraction prior to ICP-OES [33,34], only one study [16] has undertaken the optimization of DSPME of REEs in drinking water by using response surface methods. The chemometric approach can also be applied as a two-step optimization, consisting of a screening design followed by the response surface methodology. All cited works dealt with water samples or diluted aqueous solutions. However, microextraction from high-matrix CFA leachate could be much more difficult.

In this work, synthesized poly(HDDA)/graphene monodispersed particles were used as the sorbent in DSPME for the REEs separation from CFA leachate prior to their analysis by ICP-MS. Furthermore, since many variables in a DSPME process exist, a chemometric optimization of the experimental DSPME variables was conducted.

2. Materials and Methods

2.1. Chemicals and Reagents

The REEs' analytical standards were prepared from a mixed multi-element ICP-MS standard PE-MECAL2-ASL-1 (Accustandard Inc., New Haven, CT, USA) containing 10 µg/mL each of all REEs. This solution was also used to make the spiked samples. Single element stocks from Merck Co. (Darmstadt, Germany) for the elements Si, Al, Fe, Ca, Na, and Cl, were added to the spiked samples to give higher concentration levels for these major elements, similar to the CFA leachate matrix. The internal standard (ISTD) solution ICP-MS-IS-IN-1 (Accustandard Inc., New Haven, CT, USA) containing ^{115}In was used to control the instrument stability. Deionized Milli-Q water (Millipore, Burlington, USA) was used to prepare all solutions. All other chemicals used were purchased from Merck Co. (Darmstadt, Germany). TCLP extraction fluid consisted of 5.7 mL/L glacial acetic acid. Di-2-ethylhexylphosphoric acid (D2EHPA) was diluted to a concentration of 10% (v/v) in hexane. Composite polymer/graphene microspheres were produced using Darocur 1173 (2-hydroxy-2-methylpropiophenone) as a photo initiator, HDDA (1,6 hexanediol diacrylate) as a UV-curable monomer, and graphene oxide as a nanofiller used to increase the adsorption capacity of the particles, all from Sigma-Aldrich (Gillingham, UK).

2.2. Leaching of CFA Samples

A portion of 1.0 g CFA sample was mixed with a volume (20 mL) of EP or TCLP leaching fluid in a polyethylene bottle and rotary agitated at room temperature for 24 h. Then, the leachate was centrifuged at 6000 rpm for 10 min, and the supernatant was decanted. The decanted leachate was used for DSPME experiments. Since CFA is an alkaline solid, an unbuffered acetic acid (pH = 2.88) was used as the TCLP extraction fluid. Deionized water (18.2 MΩ·cm) was used to make the EP leachate.

2.3. Synthesis of Poly(HDDA)/Graphene

Graphene-embedded polymer microspheres were fabricated in a two-phase glass capillary microfluidic device. Emulsion droplets were first produced, followed by on-the-fly photopolymerization to solidify the droplets and form poly(HDDA)/graphene microspheres. The procedure of fabricating the microsphere used in this work is described in detail elsewhere [35]. Morphological investigation of these particles was performed by OMAX (Kent, WA, USA) model OM349P polarizing microscope, while chemical purity was checked by a Thermo Niton XL3t Goldd+ X-ray fluorescence spectrometer (Thermo Fisher Scientific, Waltham, MA, USA).

2.4. Factorial Design of DSPME

A Thermo mode Orion 3 pH-meter (Thermo Fisher Scientific, Chelmsford, MA, USA), Radwag analytical microbalance model MYA 5-3Y (Radwag, Radom, Poland), Centrifuge model LACE16 (Colo lab Expert, Novo Mesto, Slovenia), Lauda model RM-6 water bath (Lauda-Brinkmann, Delran, NJ, USA), and Vortex model IKA MS2 (IKA-Werke, Staufen, Germany) were used in the DSPME experiments.

The following DSPME procedure was used: 25 mL of a spiked sample or CFA leachate was taken in a 50 mL centrifuge tube, and its pH was adjusted with HNO_3 or NaOH. Then, an accurately weighted mass of poly(HDDA)/graphene sorbent and a volume of D2EHPA solution were added to form a chelating complex with REEs in the solution. Then, an emulsion was produced by vortexation. Next, the DSPME sorbent containing chelated REEs complexes was separated using centrifugation. Afterwards, nitric acid was added to the residue to release REEs. The final volume was made up to 2.5 mL with deionized water and further diluted prior to ICP-MS measurement. The experimental DSPME variables that were optimized are listed in Table 1.

Table 1. The variables and their coded values (−1, +1) for the Plackett–Burman design.

No.	Variable	Symbol	Level −1	Level +1
1.	poly(HDDA)/graphene adsorbent amount (mg)	m_a	10	50
2.	pH value	pH	3	11
3.	D2EHPA chelating (10%) agent content (μL)	Che	20	200
4.	Vortex time (min)	tv	1	5
5.	Extraction temperature (°C)	T	10	40
6.	Centrifuge time (min)	tc	1	5
7.	Centrifuge speed (rpm)	w_R	2000	10000
8.	Eluent type *	E	E_A	E_B
9.	Eluent time (min)	te	1	5

* Eluent type: E_A—2M HNO_3; E_B—2M HNO_3 in methanol/acetone.

2.5. ICP-MS Measurements

A Thermo Scientific ICP-MS instrument (Thermo Fisher Scientific, Waltham, MA, USA) model iCAP Q, equipped with a Cetac ASX-520 autosampler and controlled via Qtegra software was used in this work to measure the content of REEs. The sample introduction system includes a standard Peltier-cooled quartz vortex spray chamber, PFA nebulizer with removable quartz rectangular central tube (0.25 mm id) and standard nickel sampling and skimmer cones. The instrument runs in single kinetic energy discrimination (KED) collision cell mode, using pure helium as the collision gas.

Table 2 presents the ICP-MS instrument parameters and isotopes with potentially interfering masses for each REE.

Table 2. ICP-MS instrument setup and isotopes (interference) of each REE.

ICP-MS Parameter	Value	Isotope (Interference)
Plasma power	1550 W	Analytes:
Cool flow (Ar)	13.8 L/min	^{45}Sc (COO, COOH)
Auxiliary flow (Ar)	0.82 L/min	^{89}Y
Nebulizer flow (Ar)	0.97 L/min	^{139}La
KED mode gas flow (He)	5 mL/min	^{140}Ce
Peristaltic pump speed	35 rpm	^{141}Pr
Injector	Quartz, 2.5 mm ID	^{146}Nd
Interface cones	Nickel	^{147}Sm
Sweeps/reading	20	^{153}Eu (BaO)
Replicates	3	^{157}Gd (CeOH, PrO)
Points per peak	3	^{159}Tb (NdO)
Dwell times	10–40 ms	^{163}Dy (SmO)
Scan mode	Peak hopping	^{165}Ho (SmO)
Sweps	30	^{166}Er (SmO, NdO)
Sample flush time	4 s	^{169}Tm (SmO, EuO)
Read delay time	20 s	^{172}Yb (GdO)
Wash time	60 s	^{175}Lu (GdO, TbO)
Calibration type	Matrix-matched, external	^{115}In (I.S.)

A mixed-matrix-matched ICP-MS standard solution of 16 REEs was diluted with 1% nitric acid to a concentration of 0.1 ng/L to 50 μg/L. Each standard solution was spiked to contain 10 mg/L Si, 5.0 mg/L Al, 2.0 mg/L Fe, and 1.0 mg/L Ca. These standard spiked solutions were used to test the linearity and recovery.

3. Results and Discussion

3.1. Characterization of Poly(HDDA)/Graphene Particles

Optical microscopy measurements clearly indicated spherical poly(HDDA)/graphene particles (Figure 1). Four groups of particles with different diameters (A, B, C, D) are identified. It is also noticeable that there is an agglomeration of particles, in which a single particle is attracted to a neighboring one. This property is beneficial in the DSPME process, in which an aqueous solution is to be separated from particles.

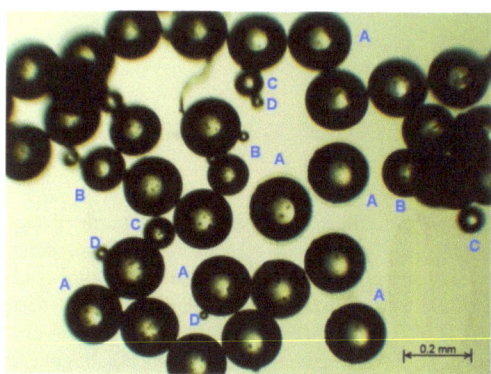

Figure 1. Optical microscopy image of the fabricated poly(HDDA)/graphene microspheres.

A critical characteristic of a DSPME sorbent is chemical purity. In particular, the absence of trace elements in DSPME sorbent is a must when determining trace elements. Thus, the presence of elements in the sorbent, even if they are not analytes, can lead to various isobaric and polyatomic interferences in an ICP-MS measurement. Therefore, the DSPME sorbent used was checked for the presence of trace elements by x-ray fluorescence spectrometry (XRF) before use. Figure 2 shows an XRF spectrum of poly(HDDA)/graphene particles. It is obvious that no metal elements were detected. Just to note, the peaks in the spectrum belong to the instrumental blank.

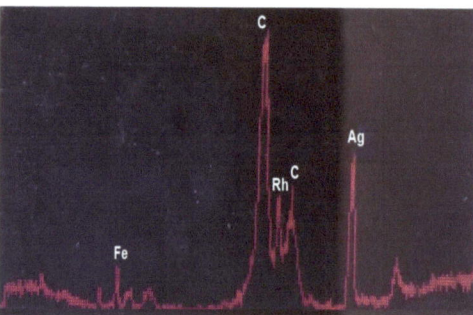

Figure 2. EDXRF spectrum of poly(HDDA)/graphene microspheres (C—Compton peaks; Rh—Kα + Kβ Rayleigh).

3.2. Factorial Optimization of DSPME

3.2.1. Plackett–Burman Screening

PBD design was used to screen nine independent variables. The Derringer [36] desirability function (D) derived from recoveries was used as a response variable. D is obtained from individual desirabilities, i.e., recoveries, using the geometric mean and is calculated according to the following equation:

$$D = (d_1^{r_1} \cdot d_2^{r_2} \cdot d_3^{r_3} \cdot d_4^{r_4} \ldots d_n^{r_n})^{1/\Sigma r_i} \quad (1)$$

where d_i represents individual desirabilities, n is the number of REEs, and r_i is the coefficient of the importance of the variable compared to other variables. This coefficient can vary, but in this case, it was assumed that all coefficients are of equal importance, so that no weights were assigned to different REEs. The result of the PBD design analysis is shown in the form of a Pareto plot in Figure 3.

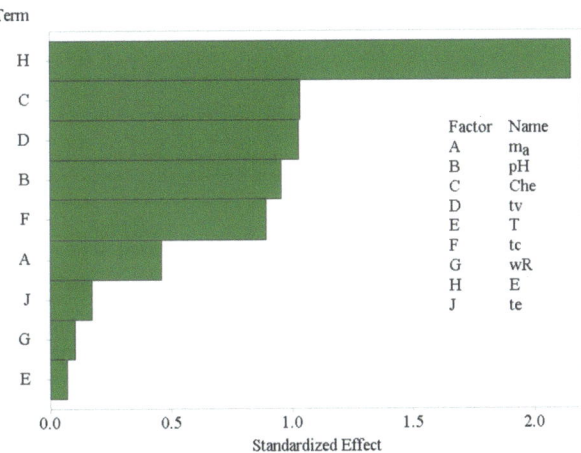

Figure 3. Pareto plot for the Plackett–Burman screening experiments.

It is obvious that the eluting solvent used to extract REEs from the DSPME sorbent was the most influencing variable. The addition of organic solvents, such as methanol or acetone, was found to have a strong negative effect on the REEs recoveries. The variation in the amount of D2EHPA chelating agent, the vortexation time, pH, and centrifugation time were also found to be significant variables. An increase in the D2EHPA amount and pH negatively affect the DSPME process, while the decrease has an opposite effect. The remaining variables are negligible. Thus, the poly(HDDA)/graphene amount, the extraction temperature, and the centrifuge speed were set to their middle values in the experimental domain of 30 mg, 25 °C, and 6000 rpm, respectively. Vortexing and elution time were minimized to 1 min, but centrifugation time was set to the maximum (5 min). Two variables, pH and the D2EHPA amount, were selected for the subsequent step in the DSPME optimization by response surface methodology.

3.2.2. Central Composite Design Optimization

According to the CCD experiments, 13 runs were carried out, and the results of the response procedures in five different levels of the two independent variables are summarized. The effect of the amount of chelating agent D2EHPA, in the range of 20–110 µL and the pH values from 1.0 to 6.0, was investigated by unblocked CCD with axial points. Derringer aggregate response for all 16 REEs was maximized during the optimization. The response surface plot is shown in Figure 4. These data were fitted by a second-order polynomial expression model, including linear, polynomial, and cross terms, in Equation (2) for the pH values and the D2EHPA amounts.

$$D = -1.30 + 0.368 \cdot pH + 0.0357 \cdot Che - 0.0367 \cdot pH^2 - 0.000251 \cdot Che^2 - 0.00085 \cdot pH \cdot Che \quad (2)$$

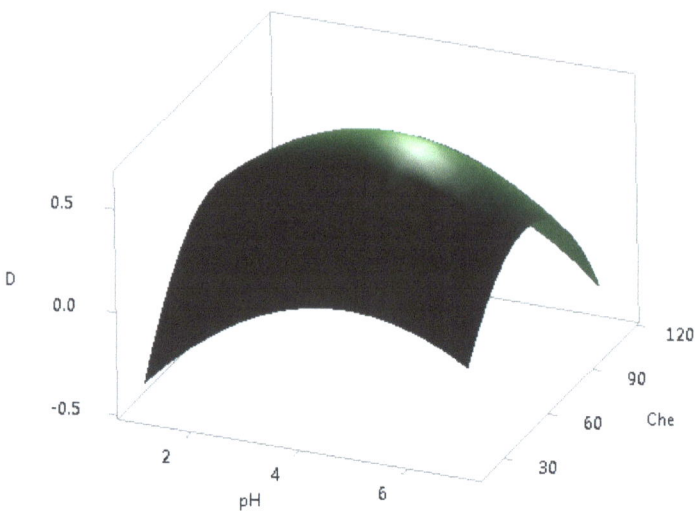

Figure 4. Response surface plot for DSPME optimization.

The maximum D value of 0.65 was reached for the D2EHPA amount of 60 µL and pH = 4.2. Finally, the overall optimum for the DSPME process was obtained at 30 mg, 25 °C, and 6000 rpm, 1 min, 1 min, 5 min, 60 µL, 2 M HNO_3, and 4.2, for the sorbent amount, the extraction temperature, the centrifuge speed, the vortexing time, the elution time, the centrifugation time, the D2EHPA amount, the eluting solvent, and pH, respectively. These optimized values were further used for validation study and the application of the DSPME method to real CFA leachate samples.

3.3. Analytical Characteristics

In order to assess the validity of the developed method, spiked aqueous solutions with increased Si, Al, Fe, and Ca content were examined by determining the limit of detection (LOD), the linear correlation coefficients (R^2), average recovery (R), and relative standard deviation (RSD). Each standard, spike, and CFA leachate sample was spiked with an internal standard used to correct for shifts in signal intensity.

Good linearity in the method was proved in the range of 0.1 ng/L–50 µg/L of REEs in the diluted spiked solutions. This range covers the LOD levels for all REEs studied at which they can be found in CFA leachates. LOD, average recovery, and RSD for each RRE are all presented in Table 3. One can see that LODs are between 0.6 and 83 ng/L, the recovery ranges from 43 and 112%, while RSD values are within ±22%. Thus, the method based on a combination of DSPME with ICP-MS may be considered acceptable to determine the REEs concentrations in these high-matrix aqueous samples.

Table 3. Analytical characteristics of DSPME-ICP-MS of REEs.

No.	REE	CAS No.	LOD (ng/L)	Linear Equation (a + b·x)	R^2	Recovery (%)	RSD (%)
1.	Scandium (Sc)	7440-45-1	83	6.55 + 7707·x	0.9965	59	17
2.	Yttrium (Y)	7429-91-6	27	730 + 76,388·x	0.9992	67	8.7
3.	Lanthanum (La)	7440-52-0	2.4	46.6 + 263,993·x	0.9997	96	9.2
4.	Cerium (Ce)	7440-53-1	1.8	230 + 328,932·x	0.9996	85	8.1
5.	Praseodymium (Pr)	7440-54-2	1.5	6.66 + 352,101·x	0.9998	90	7.9
6.	Neodymium (Nd)	7440-60-0	3.1	10.0 + 72,372·x	0.9996	112	9.8
7.	Samarium (Sm)	7439-91-0	3.3	3.33 + 64,019·x	0.9997	92	6.6
8.	Europium (Eu)	7439-94-3	2.7	3.38 + 226,836·x	0.9997	93	8.7
9.	Gadolinium (Gd)	7440-00-8	2.7	3.37 + 116,690·x	0.9997	94	12
10.	Terbium (Tb)	7440-10-0	1.9	3.31 + 603,632·x	0.9999	87	13
11.	Dysprosium (Dy)	7440-20-2	2.1	3.37 + 153,391·x	0.9998	79	11
12.	Holmium (Ho)	7440-19-9	2.3	6.67 + 641,315·x	0.9999	68	13

Table 3. Cont.

No.	REE	CAS No.	LOD (ng/L)	Linear Equation (a + b·x)	R^2	Recovery (%)	RSD (%)
13.	Erbium (Er)	7440-27-9	2.4	16.7 + 222,524·x	0.9997	67	14
14.	Thulium (Tm)	7440-29-1	3.6	1.21 + 724,602·x	0.9999	56	16
15.	Ytterbium (Yb)	7440-30-4	4.5	1.03 + 177,804·x	0.9997	44	19
16.	Lutetium (Lu)	7440-65-5	6.7	1.12 + 430,574·x	0.9999	43	22

3.4. Analytical Applications

A recent study on the extraction of REEs from CFA by sequential extraction [37] showed that most REEs are in the residual fraction, so it is necessary to use strong mineral acids for the efficient leaching of REEs. On the other hand, a significant portion of the REEs is trapped in alumina matrices, which may be more easily leached by alkaline agents [38]. Therefore, the CFA leachate from the REEs recovery process, produced by alkaline roasting [39] and followed by acid leaching [40], is likely to contain a high level of matrix elements.

The next step in the REEs recovery from CFA leachate includes removing the matrix elements (Al, Si, and Fe) with some of the separation techniques, and finally, REEs separation. These separations rely on precipitation, adsorption, ion exchange, and chelating extraction [41]. Unfortunately, these processes are characterized by high consumption of energy and reagents. Therefore, continued research is underway to make the REE recovery more economical.

From the ecological point of view, a significant amount of the CFA recovery process leachate, accompanied by the CFA landfill leachate, ending in wastewater streams, causes serious concern and needs to be monitored. In this study, the proposed DSPME-ICP-MS method was used to analyze the EP and TCLP leachates of CFA from five different coal-fired thermal power plants in Serbia (Power plants: A—Tent A; B—Tent B; C—Kolubara; D—Morava; E—Kostolac). Table 4 shows the REEs content in CFA leachates of 20 ÷ 1 liquid-to-solid (L/S) ratio. The following order of REEs in the decreasing content was observed: Ce > Y > La > Nd > Dy > Gd > Sm > Er > Pr > Tb > Eu > Ho > Yb > Tm > Sc,Lu (n.d.) for aqueous leachate, and Ce > La > Y > Nd > Er > Gd > Sm > Dy > Pr > Tm > Ho > Tb > Eu > Sc > Yb > Lu for acetic acid leachate.

The most abundant REE in studied CFA aqueous leachates was Ce (22–194 ng/L), followed by Y (35–105 ng/L) or La (48–95 ng/L). In contrast, the lowest concentrations were found for Lu (0.048–0.084 ng/L).

It is seen that the samples (C_{EP} and C_{TCLP}) from the Kolubara power plant have a higher REE content, while REEs' concentrations in the Kostolac power plant samples are at the lowest level. These differences can be attributed to different coals used in the power plants.

Note that the present method detected no Sc and Lu in aqueous leachate. On the other hand, the content of REEs in TCLP leachate, by two orders of magnitude, is higher compared to the aqueous leachate. In this case, the concentrations range from 42 ng/L (Lu) to 51 µg/L (Ce). In this study, the ratio of the concentrations of REEs in the TCLP leachate to aqueous extracts ranged from 104 to 352.

Table 4. REEs content (ng/L, except for ΣREEs is µg/L) in CFA leachate (20 ÷ 1 L/S). Leaching agents: EP—water; TCLP—acetic acid. Samples from coal-fired power plants: A—Tent A; B—Tent B; C—Kolubara; D—Morava; E—Kostolac.

No.	REE	A_{EP}	B_{EP}	C_{EP}	D_{EP}	E_{EP}	A_{TCLP}	B_{TCLP}	C_{TCLP}	D_{TCLP}	E_{TCLP}
1.	Sc	n.d.	n.d.	n.d.	n.d.	n.d.	211	186	172	198	249
2.	Y	105	56	430	35	37	10,136	18,986	22,716	8958	7343
3.	La	92	72	101	48	95	6199	11,264	17,285	4235	2465
4.	Ce	194	102	331	22	39	23,413	39,866	51,249	16,625	9660
5.	Pr	25	28	42	2.8	5.2	2977	5181	6338	2168	1206

Table 4. Cont.

No.	REE	A_{EP}	B_{EP}	C_{EP}	D_{EP}	E_{EP}	A_{TCLP}	B_{TCLP}	C_{TCLP}	D_{TCLP}	E_{TCLP}
6.	Nd	97	72	179	11	20	12,305	21,778	25,608	9189	5402
7.	Sm	26	26	55	6.4	8.3	2838	4832	5361	2096	1323
8.	Eu	14	17	15	6.1	n.d.	642	1088	1204	512	324
9.	Gd	30	25	61	4.3	6.6	2823	4854	5437	2090	1403
10.	Tb	16	13	13	n.d.	n.d.	388	664	764	291	219
11.	Dy	34	22	84	4.8	7.6	2129	3704	4234	1632	1339
12.	Ho	17	12	17	2.9	n.d.	407	728	844	327	278
13.	Er	22	14	47	n.d.	6.2	1094	2016	2374	918	819
14.	Tm	12	6.0	5.1	n.d.	n.d.	135	253	311	113	117
15.	Yb	11	6.3	18	n.d.	n.d.	598	1405	1829	604	704
16.	Lu	n.d.	n.d.	n.d.	n.d.	n.d.	168	125	98	42	102
17.	ΣREEs	0.70	0.47	1.40	0.14	0.22	66	117	146	50	33

4. Conclusions

A new DSPME sorbent, consisting of spherical particles of poly(HDDA) and graphene, was synthesized by microfluidic emulsification, characterized, and applied in the DSPME of REEs prior to ICP-MS. The proposed DSPME-ICP-MS method is fast, has a low-consuming sorbent, and is specifically green. The main advantage of the DSPME technique is that it provides an extensive interface between poly(HDDA)/graphene particles and the aqueous phase after a cloudy solution formation. The separation factors for REEs were efficiently maximized by applying a two-step optimization using Plackett–Burman design, central composite designs, and Derringer desirability aggregate response function. Analytical characteristics and the method robustness are acceptable for most of the studied REEs for the analysis of coal fly ash leachate for REEs. Several leachate samples from CFA from different coal-fired power plants were analyzed by the proposed method. Cerium, La, and Y were found to be the most abundant REEs in CFA leachates. A significant difference between CFA leachate samples, in terms of the REEs content, was attributed to the coal properties.

Author Contributions: Investigation, writing—original draft preparation, L.S.-B.; formal analysis, project administration, L.I.; validation, resources, G.B.; data curation, software, A.S.; visualization, D.M.; conceptualization, methodology, G.V.; supervision, writing—review and editing, A.O. All authors have read and agreed to the published version of the manuscript.

Funding: This research was supported by the Science Fund of the Republic of Serbia (Grant No. 7743343 SIW4SE).

Institutional Review Board Statement: Not applicable.

Informed Consent Statement: Not applicable.

Data Availability Statement: The data presented in this study are available on request from the corresponding author.

Conflicts of Interest: The authors declare no conflict of interest.

References

1. Stopic, S.; Friedrich, B. Advances in Understanding of the Application of Unit Operations in Metallurgy of Rare Earth Elements. *Metals* **2021**, *11*, 978. [CrossRef]
2. Balaram, V. Rare Earth Elements: A Review of Applications, Occurrence, Exploration, Analysis, Recycling, and Environmental Impact. *Geosci. Front.* **2019**, *10*, 1285–1303. [CrossRef]
3. Jyothi, R.K.; Thenepalli, T.; Ahn, J.W.; Parhi, P.K.; Chung, K.W.; Lee, J.-Y. Review of Rare Earth Elements Recovery from Secondary Resources for Clean Energy Technologies: Grand Opportunities to Create Wealth from Waste. *J. Clean. Prod.* **2020**, *267*, 122048. [CrossRef]
4. Palaparthi, J.; Chakrabarti, R.; Banerjee, S.; Guin, R.; Ghosal, S.; Agrahari, S.; Sengupta, D. Economically Viable Rare Earth Element Deposits along Beach Placers of Andhra Pradesh, Eastern Coast of India. *Arab. J. Geosci.* **2017**, *10*, 201. [CrossRef]

5. Vilakazi, A.Q.; Ndlovu, S.; Chipise, L.; Shemi, A. The Recycling of Coal Fly Ash: A Review on Sustainable Developments and Economic Considerations. *Sustainability* **2022**, *14*, 1958. [CrossRef]
6. Fu, B.; Hower, J.C.; Zhang, W.; Luo, G.; Hu, H.; Yao, H. A Review of Rare Earth Elements and Yttrium in Coal Ash: Content, Modes of Occurrences, Combustion Behavior, and Extraction Methods. *Prog. Energy Combust. Sci.* **2022**, *88*, 100954. [CrossRef]
7. Zhang, W.; Noble, A.; Yang, X.; Honaker, R. A Comprehensive Review of Rare Earth Elements Recovery from Coal-Related Materials. *Minerals* **2020**, *10*, 451. [CrossRef]
8. Keller, V.; Stopić, S.; Xakalashe, B.; Ma, Y.; Ndlovu, S.; Mwewa, B.; Simate, G.S.; Friedrich, B. Effectiveness of Fly Ash and Red Mud as Strategies for Sustainable Acid Mine Drainage Management. *Minerals* **2020**, *10*, 707. [CrossRef]
9. Ma, Y.; Stopic, S.; Xakalashe, B.; Ndlovu, S.; Forsberg, K.; Friedrich, B. A Cleaner Approach for Recovering Al and Ti from Coal Fly Ash via Microwave-Assisted Baking, Leaching, and Precipitation. *Hydrometallurgy* **2021**, *206*, 105754. [CrossRef]
10. Wen, Z.; Zhou, C.; Pan, J.; Cao, S.; Hu, T.; Ji, W.; Nie, T. Recovery of Rare-Earth Elements from Coal Fly Ash via Enhanced Leaching. *Int. J. Coal Prep. Util.* **2020**, *284*, 124725. [CrossRef]
11. Tsiridis, V.; Samaras, P.; Kungolos, A.; Sakellaropoulos, G.P. Application of Leaching Tests for Toxicity Evaluation of Coal Fly Ash. *Environ. Toxicol.* **2006**, *21*, 409–416. [CrossRef] [PubMed]
12. Wu, S.; Hong, W.; Zhang, B.; Yang, C.; Wang, J.; Gao, J.; Mi, F.; Zhang, H.; Zhao, X.; Li, Q. Study on the Determination of Rare Earth Elements in Coal Ash by ICP-MS. *Integr. Ferroelectr.* **2019**, *198*, 116–121. [CrossRef]
13. Wysocka, I. Determination of Rare Earth Elements Concentrations in Natural Waters—A Review of ICP-MS Measurement Approaches. *Talanta* **2021**, *221*, 121636. [CrossRef] [PubMed]
14. Hann, S.; Boeck, K.; Koellensperger, G. Immunoaffinity Assisted LC-ICP-MS—a Versatile Tool in Biomedical Research. *J. Anal. Spectrom.* **2010**, *25*, 18–20. [CrossRef]
15. Fisher, A.; Kara, D. Determination of Rare Earth Elements in Natural Water Samples—A Review of Sample Separation, Preconcentration and Direct Methodologies. *Anal. Chim. Acta* **2016**, *935*, 1–29. [CrossRef]
16. Manousi, N.; Gomez-Gomez, B.; Madrid, Y.; Deliyanni, E.A.; Zachariadis, G.A. Determination of Rare Earth Elements by Inductively Coupled Plasma-Mass Spectrometry after Dispersive Solid Phase Extraction with Novel Oxidized Graphene Oxide and Optimization with Response Surface Methodology and Central Composite Design. *Microchem. J.* **2020**, *152*, 104428. [CrossRef]
17. Ebihara, M.; Hayano, K.; Shirai, N. Determination of Trace Rare Earth Elements in Rock Samples Including Meteorites by ICP-MS Coupled with Isotope Dilution and Comparison Methods. *Anal. Chim. Acta* **2020**, *1101*, 81–89. [CrossRef]
18. Milicic, L.; Terzic, A.; Pezo, L.; Mijatovic, N.; Brceski, I.; Vukelic, N. Assessment of Efficiency of Rare Earth Elements Recovery from Lignite Coal Combustion Ash via Five-Stage Extraction. *Sci. Sinter.* **2021**, *53*, 169–185. [CrossRef]
19. Rubinos, D.A.; Barral, M.T. Sorptive Removal of Hg^{II} by Red Mud (Bauxite Residue) in Contaminated Landfill Leachate. *J. Environ. Sci. Health Part A* **2017**, *52*, 84–98. [CrossRef]
20. Balaram, V.; Subramanyam, K.S.V. Sample Preparation for Geochemical Analysis: Strategies and Significance. *Adv. Sample Prep.* **2022**, *1*, 100010. [CrossRef]
21. Ma, Y.; Stopic, S.; Gronen, L.; Milivojevic, M.; Obradovic, S.; Friedrich, B. Neural Network Modeling for the Extraction of Rare Earth Elements from Eudialyte Concentrate by Dry Digestion and Leaching. *Metals* **2018**, *8*, 267. [CrossRef]
22. Li, Y.; Peng, G.; He, Q.; Zhu, H.; Al-Hamadani, S.M.Z.F. Dispersive Liquid–Liquid Microextraction Based on the Solidification of Floating Organic Drop Followed by ICP-MS for the Simultaneous Determination of Heavy Metals in Wastewaters. *Spectrochim. Acta. A Mol. Biomol. Spectrosc.* **2015**, *140*, 156–161. [CrossRef] [PubMed]
23. Smith, R.C.; Taggart, R.K.; Hower, J.C.; Wiesner, M.R.; Hsu-Kim, H. Selective Recovery of Rare Earth Elements from Coal Fly Ash Leachates Using Liquid Membrane Processes. *Environ. Sci. Technol.* **2019**, *53*, 4490–4499. [CrossRef] [PubMed]
24. Marjanovic, V.; Peric-Grujic, A.; Ristic, M.; Marinkovic, A.; Markovic, R.; Onjia, A.; Sljivic-Ivanovic, M. Selenate Adsorption from Water Using the Hydrous Iron Oxide-Impregnated Hybrid Polymer. *Metals* **2020**, *10*, 1630. [CrossRef]
25. Suručić, L.; Tadić, T.; Janjić, G.; Marković, B.; Nastasović, A.; Onjia, A. Recovery of Vanadium (V) Oxyanions by a Magnetic Macroporous Copolymer Nanocomposite Sorbent. *Metals* **2021**, *11*, 1777. [CrossRef]
26. Peiravi, M.; Ackah, L.; Guru, R.; Mohanty, M.; Liu, J.; Xu, B.; Zhu, X.; Chen, L. Chemical Extraction of Rare Earth Elements from Coal Ash. *Miner. Metall. Process.* **2017**, *34*, 170–177. [CrossRef]
27. Aguirre, M.Á.; Baile, P.; Vidal, L.; Canals, A. Metal Applications of Liquid-Phase Microextraction. *TrAC Trends Anal. Chem.* **2019**, *112*, 241–247. [CrossRef]
28. Rajakovic, L.; Todorovic, Z.; Rajakovic-Ognjanovic, V.; Onjia, A. Analytical Methods for Arsenic Speciation Analysis. *J. Serb. Chem. Soc.* **2013**, *78*, 1461–1479. [CrossRef]
29. Sajid, M.; Asif, M.; Ihsanullah, I. Dispersive Liquid–Liquid Microextraction of Multi-Elements in Seawater Followed by Inductively Coupled Plasma-Mass Spectrometric Analysis and Evaluation of Its Greenness. *Microchem. J.* **2021**, *169*, 106565. [CrossRef]
30. Labutin, T.A.; Lednev, V.N.; Ilyin, A.A.; Popov, A.M. Femtosecond Laser-Induced Breakdown Spectroscopy. *J. Anal. At. Spectrom.* **2016**, *31*, 90–118. [CrossRef]
31. Krishnan Chandrasekaran, S.; Dheram Karunasagar, K.; Jayaraman Arunachalam, G. Dispersive Liquid-Liquid Micro-Extraction for Simultaneous Preconcentration of 14 Lanthanides at Parts per Trillion Levels from Groundwater and Determination Using a Micro-Flow Nebulizer in Inductively Coupled Plasma-Quadrupole Mass Spectrometry. *J. Anal. Spectrom.* **2010**, *25*, 18–20. [CrossRef]

32. Chen, S.; Yan, J.; Li, J.; Lu, D. Magnetic $ZnFe_2O_4$ Nanotubes for Dispersive Micro Solid-Phase Extraction of Trace Rare Earth Elements Prior to Their Determination by ICP-MS. *Microchim. Acta* **2019**, *186*, 228. [CrossRef] [PubMed]
33. Sereshti, H.; Khojeh, V.; Samadi, S. Optimization of Dispersive Liquid–Liquid Microextraction Coupled with Inductively Coupled Plasma-Optical Emission Spectrometry with the Aid of Experimental Design for Simultaneous Determination of Heavy Metals in Natural Waters. *Talanta* **2011**, *83*, 885–890. [CrossRef]
34. Pinheiro, F.C.; Aguirre, M.Á.; Nóbrega, J.A.; González-Gallardo, N.; Ramón, D.J.; Canals, A. Dispersive Liquid-Liquid Microextraction Based on Deep Eutectic Solvent for Elemental Impurities Determination in Oral and Parenteral Drugs by Inductively Coupled Plasma Optical Emission Spectrometry. *Anal. Chim. Acta* **2021**, *1185*, 339052. [CrossRef] [PubMed]
35. Zhao, Y.; Moshtaghibana, S.; Zhu, T.; Fayemiwo, K.A.; Price, A.; Vladisavljević, G. Microfluidic Fabrication of Novel Polymeric Core-shell Microcapsules for Storage of CO2 Solvents and Organic Chelating Agents. *J. Polym. Sci.* **2022**, pol.20210959. [CrossRef]
36. Derringer, G.; Suich, R. Simultaneous Optimization of Several Response Variables. *J. Qual. Technol.* **1980**, *12*, 214–219. [CrossRef]
37. Park, S.; Kim, M.; Lim, Y.; Yu, J.; Chen, S.; Woo, S.W.; Yoon, S.; Bae, S.; Kim, H.S. Characterization of Rare Earth Elements Present in Coal Ash by Sequential Extraction. *J. Hazard. Mater.* **2021**, *402*, 123760. [CrossRef]
38. Wang, Z.; Dai, S.; Zou, J.; French, D.; Graham, I.T. Rare Earth Elements and Yttrium in Coal Ash from the Luzhou Power Plant in Sichuan, Southwest China: Concentration, Characterization and Optimized Extraction. *Int. J. Coal Geol.* **2019**, *203*, 1–14. [CrossRef]
39. Tang, M.; Zhou, C.; Pan, J.; Zhang, N.; Liu, C.; Cao, S.; Hu, T.; Ji, W. Study on Extraction of Rare Earth Elements from Coal Fly Ash through Alkali Fusion—Acid Leaching. *Miner. Eng.* **2019**, *136*, 36–42. [CrossRef]
40. King, J.F.; Taggart, R.K.; Smith, R.C.; Hower, J.C.; Hsu-Kim, H. Aqueous Acid and Alkaline Extraction of Rare Earth Elements from Coal Combustion Ash. *Int. J. Coal Geol.* **2018**, *195*, 75–83. [CrossRef]
41. Rybak, A.; Rybak, A. Characteristics of Some Selected Methods of Rare Earth Elements Recovery from Coal Fly Ashes. *Metals* **2021**, *11*, 142. [CrossRef]

Article

Extraction of Cu(II), Fe(III), Zn(II), and Mn(II) from Aqueous Solutions with Ionic Liquid R₄NCy

Jonathan Castillo [1], Norman Toro [2], Pía Hernández [3], Patricio Navarro [4], Cristian Vargas [4], Edelmira Gálvez [5] and Rossana Sepúlveda [1,*]

1. Departamento de Ingeniería en Metalurgia, Universidad de Atacama, Copiapó 1531772, Chile; jonathan.castillo@uda.cl
2. Faculty of Engineering and Architecture, Universidad Arturo Prat, Iquique 1100000, Chile; notoro@unap.cl
3. Departamento de Ingeniería Química y Procesos de Minerales, Universidad de Antofagasta, Avenida Angamos 601, Antofagasta 1270300, Chile; pia.hernandez@uantof.cl
4. Departamento de Ingeniería Metalúrgica, Facultad de Ingeniería, Universidad de Santiago de Chile (USACH), Santiago 9170124, Chile; patricio.navarro@usach.cl (P.N.); cristian.vargas@usach.cl (C.V.)
5. Department of Metallurgical and Mining Engineering, Universidad Católica del Norte, Avenida Angamos 0610, Antofagasta 1270709, Chile; egalvez@ucn.cl
* Correspondence: rossana.sepulveda@uda.cl

Abstract: The leaching of copper ores produces a rich solution with metal interferences. In this context, Fe(III), Zn(II), and Mn(II) are three metals contained in industrial copper-rich solutions in high quantities and eventually can be co-extracted with the copper. The purpose of the current study was to determine the feasibly of solvent extraction with the use of ionic liquid methyltri-octyl/decylammonium bis (2,4,4-trimethylpentyl)phosphinate (R₄NCy) as an extractant of Cu(II) in the presence of Fe(III), Zn(II), and Mn(II). In general terms, the results showed a high single extraction efficiency of all the metals under study. In the case of Fe(III) and Zn(II), the extraction was close to 100%. On the contrary, the stripping efficiency was poor to Fe(III) and discrete to Zn(II), but very high to Cu(II) and Mn(II). Finally, the findings of this study suggest that the ionic liquid R₄NCy is feasible for the pre-treatment of the copper solvent extraction process to remove metal impurities such as Fe(III) and Zn(II).

Keywords: ionic liquids; solvent extraction; green chemistry; mining

Citation: Castillo, J.; Toro, N.; Hernández, P.; Navarro, P.; Vargas, C.; Gálvez, E.; Sepúlveda, R. Extraction of Cu(II), Fe(III), Zn(II), and Mn(II) from Aqueous Solutions with Ionic Liquid R₄NCy. *Metals* **2021**, *11*, 1585. https://doi.org/10.3390/met11101585

Academic Editors: Antoni Roca and Antonije Onjia

Received: 12 August 2021
Accepted: 1 October 2021
Published: 5 October 2021

Publisher's Note: MDPI stays neutral with regard to jurisdictional claims in published maps and institutional affiliations.

Copyright: © 2021 by the authors. Licensee MDPI, Basel, Switzerland. This article is an open access article distributed under the terms and conditions of the Creative Commons Attribution (CC BY) license (https://creativecommons.org/licenses/by/4.0/).

1. Introduction

Hydrometallurgy is an important technique used in mineral processing to obtain pure metal from ores. The hydrometallurgical route commonly involves ore leaching, solvent extraction (SX), and pure metals' electrowinning [1–3]. SX is a technique with several advantages, it is a continuous operation, efficient, high production, low cost, and employs of simple equipment to recover valuable elements from impurities [4,5]. Opposite, SX has industrial-scale severe problems, such as emulsification, solvent volatilization, or degradation of extractant. The negative effect of SX results in decreased production, an increase of production cost, low quality of final products, and environmental implications, mainly by loss of volatile compounds. Hydroxyoxime type extractants are one of the large-scale copper extractants. Currently, a variety of hydroxy oximes are used for the copper industry, either ketoxime or aldoxime or a combination of both [6,7].

The recent advance in metal extraction with SX finds to improve the separation, extraction efficiencies, and selectivity, mainly involving the use of synergistic extractants, ionic liquids, and both, that is, a mixture of extractant and ionic liquids [8–10].

Ionic liquids (ILs) are a green kind of reagents, comprising organic cations and organic/inorganic anions, with a high potential to reduce or replace the hazardous organic solvents [11–13]. The properties of ILs have globally captured the attention, negligible vapor pressure, high solvating power, thermal and electrochemical stability, good conductivity, and recyclability. The application of ILs is extensive, ranging from refrigeration, lubricants to solvents [14–16]. ILs are proposed in several industrial applications, for example, metal dissolution, metal and organic extraction, electrolytes for batteries, water treatment, and environmental processes [17–22].

In the case of metal extraction with ILs, the list is too long; several elements are studied with these green reagents, for example, molybdenum, rare earth, precious metals, chromium, iron, copper, cobalt, lead, and zinc [23–33].

Some ILs (e.g., based on imidazolium cations) have solubility and decomposition problems in some solvent extraction systems. To resolve the problems described above, ILs based on quaternary ammonium and phosphonium ions are proposed [34–36]. In this context, Swain et al. [37] used trihexyl(tetradecyl)phosphonium chloride (Cyphos IL 101) and trioctylmethylammonium chloride (Aliquat 336) for cadmium recovery. The results have shown 99% of cadmium extraction using 0.4 M of Cyphos IL 101, and 0.2 M of Aliquat 336. The stripping performance was 70 and 75% from Cyphos IL 101, and Aliquat 336, respectively, used 0.1 M of EDTA. In 2020, Deferm and coworkers reported the separation of In(III) and Zn(II) using Cyphos IL 101. The elements were extracted from ethylene glycol since the amount of IL in the chloride solution was inefficient [38]. Padhan and Sarangi extract Nb and Pr from the magnet with conventional extractant and ionic liquid-based of this same extractant. The efficiency of different extractants Nd and Pr extraction followed the order $R_4NCy>R_4ND>$Cyanex 272>D2EHPA>Aliquat 336 [39]. Jing and coworkers recovered cobalt selectively from lithium batteries using the ionic liquid $[C_8H_{17}NH_2]$ [Cyanex 727]. The system has a good performance and produces a final product $CoSO_4 \times 7H_2O$, with high purity (99.7%) [24]. Rybka and Regel-Rosocka report Ni(II) and Co(II) extraction with the use of Cyphos IL 101 and Cyphos IL 104 in chloride media. Both ILs are only very efficient to extract Co(II), with an extraction rate over 95% with Cyphos IL 104, Ni(II) extraction up 20% as maximum [40].

In the present investigation, ionic liquids were synthesized by reacting the commercial extractants Aliquat 336 and Cyanex 272 to investigate the extraction of Cu, Fe, Zn, and Mn from sulfate solutions. The mono and multielement extraction were studied under the following experimental conditions: IL concentration, O/A ratio, and pH. As a final step, the elements loaded in the IL were stripped, varying the sulfuric acid solution and O/A ratio.

2. Materials and Methods

2.1. Reagents and Synthesis of Ionic Liquids

The SX experiments were carried out by a synthetic pregnant leach solution (PLS). The PLS was prepared according to the information of Chilean mining reported by Quijada-Maldonado et al. [41]. The concentrations of Cu(II), Zn(II), Mn(II), and Fe(III) of the PLS are summarized in Table 1. The reagents used in the aqueous phases were copper(II) sulfate pentahydrate ($CuSO_4 \cdot 5H_2O$), Fe(III) sulfate hydrate ($Fe_2(SO_4)_3 \cdot nH_2O$), Mn(II) sulfate monohydrate ($MnSO_4 \cdot H_2O$), Zn sulfate heptahydrate ($ZnSO_4 \cdot 7H_2O$), sulfuric acid, and distilled water. All reagents used in aqueous phases were analytical grade purchased from Merck.

Table 1. Concentration of metals in PLS.

Metal	Concentration (mg/L)	
	Literature Report [41]	This Study
Cu(II)	3360	3360
Fe(III)	1570	1570
Mn(II)	>1000	1000
Zn(II)	107	110

The reagents used in the organic phase were the commercial extractants trialkyl methylammonium chloride (Aliquat 336), bis 2,4,4-trimethylpentyphosphinic acid (Cyanex 272), kerosene, and sodium hydrogen carbonate. All the reagents were provided by Sigma-Aldrich (St. Louis, MO, USA), except Cyanex 272, provided by Solvay (Santiago, Chile). Aliquat 336 and Cyanex 272 (commercial grade) were used without further purification; all other used compounds were analytical grade.

Ionic liquid methyltrioctyl/decylammonium bis 2,4,4-(trimethylpentyl)phosphinate denoted as R_4NCy was synthesized according to procedures published in the literature [35,36], mixing equimolar ratio of Cyanex 272 (HCy) and Aliquat 336 (R_4NCl), and dissolved the mixture in kerosene to obtain the desired IL concentration. The mixture of Cyanex 272 and Aliquat 336 was washed twice with sodium bicarbonate 0.5 M to remove the chloride anion and the proton from the organic phase. The general reaction of ionic liquid formation is shown in the following equation [39]:

$$R_4NCl + HCy + NaHCO_3 = R_4NCy + NaCl + CO_2 + H_2O \quad (1)$$

Schematic structures of the ionic liquids and reagent used in this work are depicted in Figure 1. To identify the formation of R4NCy, the FT-IR spectra was performed on the IL synthesized. The concentration of IL in kerosene was 0.1 M. The frequencies of the char-acteristic stretching vibrational bands of this extractant in kerosene are presented in Figure 2.

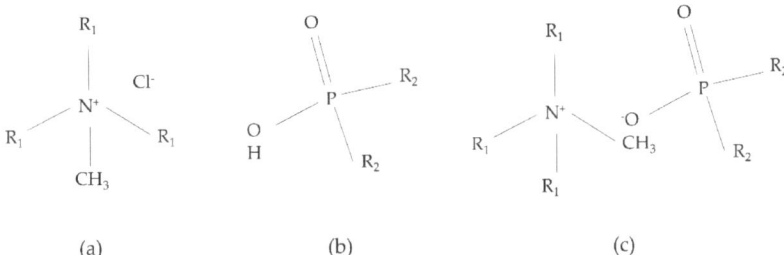

Figure 1. Molecular structure of the reagents and IL used in this work (**a**) Aliquat 336 (**b**) Cyanex 272, and (**c**) R_4NCl.

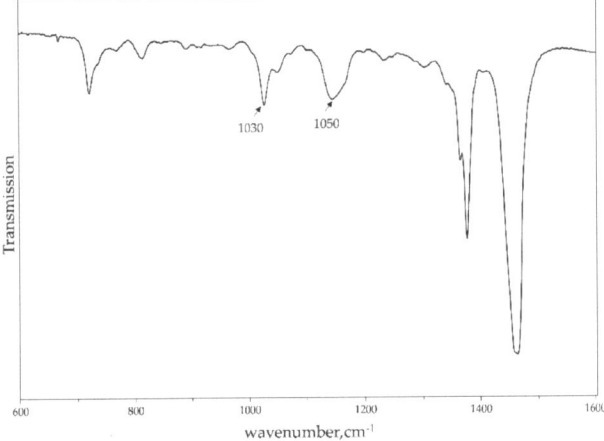

Figure 2. FT−IR of R_4NCy LiCY in kerosene. IL concentration 0.1 M.

2.2. Liquid/Liquid Extraction

The solvent extraction tests were performed in shake flasks (50 mL), at ambient temperature (22 ± 1 °C), with a magnetic stirring set to 200 rev/min for 20 min. The elemental analysis was measured in an aqueous solution by atomic absorption spectrometry (PinAAcle 900F, Perkin Elmer, Waltham, MA, USA). For extraction tests, 10 mL of aqueous phase containing metal was contacted with a volume of the organic phase to obtain the organic/aqueous phase ratio (O/A). After extraction, two phases were centrifuged at 3000 rpm for 5 min for total phase separation. The aqueous sample was analyzed for metal concentration. The liquid-liquid extraction tests were carried out in duplicate, and a standard deviation of ±2% was obtained for all the tests. The performance of the liquid-liquid extractions was determined by Equations (2)–(4) as the extraction percentage, the stripping percentage, and distribution ratio of the specie (D). The stripping procedure was similar to that of extraction tests. The mass balance was confirmed by measuring the concentration of the metals in the stripped aqueous phase:

$$\text{Extraction (\%)} = ([M]_i - [M]_{aq})/[M]_i \times 100 \qquad (2)$$

$$\% \text{ Stripping} = ([M]^*_{aq}/[M]_{org}) \times 100 \qquad (3)$$

$$D = ([M]_i - [M]_{aq})/[M]_{aq} \qquad (4)$$

where $[M]_i$ and $[M]_{aq}$ are the initial and final concentrations of metal in the aqueous solution, and $[M]_{org}$ and $[M]^*_{aq}$ are the concentrations of a metal ion in the loaded organic phase before stripping and in aqueous phase after stripping, respectively. The concentration of metals in the loaded organic phase was determined by mass balance ($[M]_{org} = [M]_i - [M]_{aq}$). The equations shown above are valid for a 1/1 phase relationship. Otherwise, the mass balances incorporate the phase volumes.

3. Results and Discussion

3.1. Effect of the R_4NCy Concentration in Metal Extraction

Table 2 shows the single metals extraction as a function of the different molar concentrations of R_4NCy in the organic phase. It is observed that R_4NCy strongly extracts Fe (III) and Zn(II) in all ranges of concentration of IL. In the case of Cu(II) and Mn(II), the extraction increases when the concentration of R_4NCy increases, from 44.14% and 30.63% with 0.1 M of extractant to achieving 97.64% and 93.22% respectively with 0.54 M of extractant. These results are totally surprising, considering that the purpose of the current study was to determine the feasibly of solvent extraction with the use of ionic liquid R_4NCy as an extractant of Cu(II) in the presence of Fe(III), Zn(II), and Mn(II). Therefore, the R_4NCy would not replace the commercial extractants, for example, LIX and ACORGA type, widely used in the hydrometallurgical copper industry. Anyway, the more efficient extraction of Fe and Zn could be positive, considering that are potential impurities that could affect the process SX.

The trend to the high efficiency of extraction of Fe(III) and Zn(II) over the Cu(II) and Mn(II), is reported in literature using similar ILs. Devi [42], despite working with lower copper and ionic liquid concentrations, reported that some ILs showed a similar tendency for Cu extraction. In the case of Fe(III) and Zn(II), Regel-Rosocka et al. [43] report high extraction percentage in chloride media using phosphonium ILs (Cyphos IL 101 and Cyphos IL 104). Another study similar to the previous one, Baczyńska et al. [44], uses three phosphonium ILs (Cyphos IL 101, Cyphos IL 104, and Cyphos IL 167) to extract Zn(II) and Fe(III) in membrane process, the result showed a good extraction efficiency. In the case of extraction of Mn(II) and Fe(III), Ola et al. [45] report a similar tendency, the Fe(III) is extracted more efficiently than Mn(II). Ola et al. used Cyphos IL 101 in chloride media. Finally, Nguyen and Lee report good extraction of Mn(II) with R_4NCy from leaching solution containing Co(II), Ni(II), Mn(II), and Li(I) [46].

Table 2. Effect concentration of R$_4$NCy on extraction efficiency and distribution ratio of the species. Experimental conditions: initial pH = 2, O/A = 1, 400 min^{-1}, and 20 min of stirring time.

Metal	IL Concentration [M]	E(%)	D
Cu(II)	0.1	44.14	0.79
	0.27	84.73	5.54
	0.54	97.64	41.4
Fe(III)	0.1	99.77	435
	0.27	99.82	559
	0.54	99.73	372
Mn(II)	0.1	30.63	0.44
	0.27	71.58	2.51
	0.54	93.22	13.7
Zn(II)	0.1	95.55	21.4
	0.27	96.09	24.6
	0.54	92.82	12.9

The mechanism of metal extraction with R$_4$NCy involves both the metal and the medium in which the tests are carried out. Previous work [35] reported of Cu(II) extraction mechanism with R$_4$NCy as shown below:

$$2R_4NCy + Cu^{2+} + SO_4^{2-} = CuCy_2 \cdot (R_4N)_2SO_4 \quad (5)$$

This mechanism can be generalized to the rest of the metals in this study, as shown below:

$$nR_4NCy + M^{n+} + SO_4^{2-} = MCy_n \cdot (R_4N)_nSO_4 \quad (6)$$

3.2. Effect of Initial pH in Metal Extraction

The results shown in Figure 3 indicate two hugely different single metal extraction performance tendencies with R$_4$NCy. In the case of Fe(III) and Zn(II), the efficiency of extraction is not influenced by initial pH in the range 1 to 3; this result confirms the affinity of R$_4$NCy for Fe(III) and Zn(II).

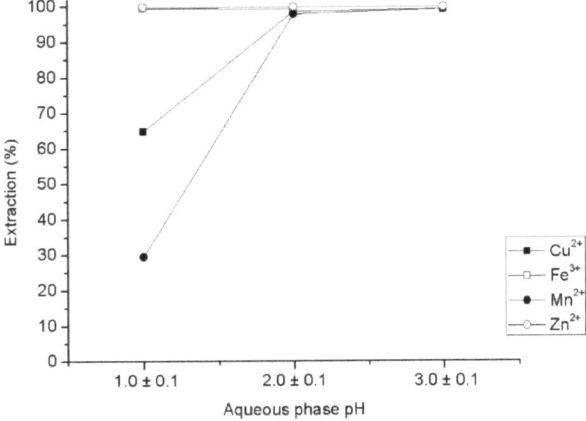

Figure 3. Effect of initial pH on the single extraction percentage of metals from aqueous solution. Experimental conditions: [R$_4$NCy] = 0.54 M, O/A = 1, 400 min^{-1}, and 20 min of stirring time.

On the contrary, the efficiency of extraction of Cu(II) and Mn(II) is clearly influenced by the initial pH of the aqueous solution, decrease the percentage from close 97% at pH 2 and 3 to 65% and 30% at pH 1, respectively. This makes it possible to carry out a multi-element extraction at pH values from 2 or, failing that, selective extraction of Fe(III) and Zn(II) at pH 1, like a pretreatment of Cu PLs or recovery of Fe and Zn.

Previous work [35] demonstrates the influence of pH on the Cu(II) extraction using R_4NCy. A similar tendency of pH effect on Cu extraction is reported by Devi [42]. This research uses the same IL for Cu(II) extraction in a diluted system with low IL concentration. The results are very similar in this study; the efficiency of Cu(II) extraction decrease at low pH. This fact is because the solutions with acid pH and high sulfate concentrations produce the undissociated neutral molecule $CuSO_{4aq}$. In the case of Zn(II) and Mn(II), Zhu et al. [47] report a higher Zn(II) extraction over Mn(II) as a function of pH. This research used Cyphos IL 101 as extractant.

3.3. Effect of O/A Ratio in Single Metal Extraction

The results show in Table 3 show a total extraction of Fe(III) and Zn(II). This result is impressive because the 1/6 is a high O/A ratio, and the extraction is not affected, again the R_4NCy demonstrates the high affinity for these cations. On the other hand, Cu(II) and Mn(II) extraction efficiency decreases significantly with the increment of the O/A ratio. The results of these sets of tests show a clear tendency to the high efficiency of extraction of Fe(III) and Zn(II) over the Cu(II) and Mn(II). In this context, to study the O/A ratio, Foltova et al. [48] used quaternary ammonium ionic liquids to extract Co and Sm. The results of Foltova show a similar tendency of this study, the Cu extraction efficiency decay when the O/A relationship is under 0.25 (or 1/4).

Table 3. Effect of O/A ratio on extraction efficiency and distribution ratio of the species. Experimental conditions: [R_4NCy] = 0.27 M, initial pH = 2, 400 min^{-1}, and 20 min of stirring time.

Metal	O/A Rate	E(%)	D
Cu(II)	1:1	84.73	5.54
	1:3	28.60	0.40
	1:6	9.08	0.09
Fe(III)	1:1	99.82	559
	1:3	99.82	540
	1:6	91.40	10.6
Mn(II)	1:1	71.58	2.51
	1:3	22.70	0.29
	1:6	11.90	0.13
Zn(II)	1:1	96.09	24.58
	1:3	96.73	29.55
	1:6	82.73	4.78

3.4. Effect of Sulfuric Acid in Metal Stripping

Each loaded organic phase contained 3.25 g/L Cu(II), 1.55 g/L Fe(III), 0.96 g/L Mn(II), and 0.106 g/L Zn(II). Figure 4 shows the stripping performance as a function of the sulfuric acid. It is observed that the sulfuric acid concentration of 2 M all metals are removed over 90% of organic phases. If the acid concentration is reduced to 1 and 0.5 M, the performance decrease significantly in the case of Fe(III); the other metals maintain the best stripping efficiency.

The most interesting aspect of this graph is sensitivity to acid concentration for the stripping Fe(III) loaded IL and the contrast with the stripping performance of the other metals in the study. Cu(II), Zn(II), and Mn(II) were easily stripped from loaded ILs for all acid concentrations studied. In the case of Fe(III), the stripping efficiency is near 95% for 2 M of H_2SO_4 but decreases significantly at 1 and 0.5 M of acid concentration (65% and 20% respectively). This result is relevant for subsequent studies since it is possible to perform a selective stripping of iron concerning the remaining metals extracted in the IL but at higher sulfuric acid concentrations.

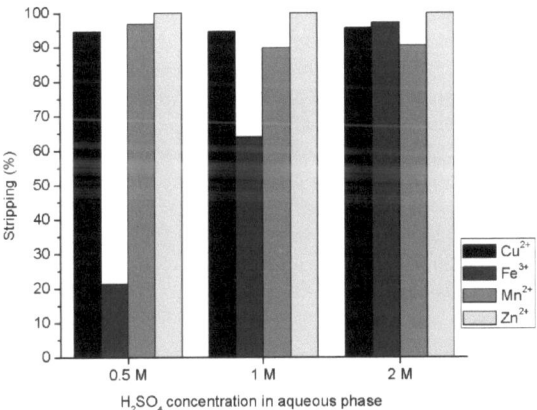

Figure 4. Effect of sulfuric acid on the stripping percentage of metals from the loaded organic phases. Experimental conditions: [R_4NCy] = 0.54 M, O/A = 1, 400 min^{-1}, and 20 min of stirring time.

In the literature it has been reported that HNO_3, HCl, and H_2SO_4 solutions can strip the metals in this study from the loaded ionic liquids. In this context, Nguyen and Lee report the stripping of Mn(II) from ALiCY with HCl, and the best performance of 1 M of acid [46]. Tran et al. reported the stripping tests for Fe(III) loaded ILs using 3 M of H_2SO_4 and HCl. The ILs used by the researchers were [C4min][N88SA], [C4Py][N88SA], ALiCy, ALiD2, and ALiPC. The results of the stripping tests are very heterogeneous, but in general terms, the best performance of stripping from Fe(III) loaded ILs was using H_2SO_4 [49].

3.5. Effect of O/A in Metal Stripping

The loaded organic phases were similar to the previous point. To concentrate the metals in stripping solution, the O/A ratio was investigated from 1/1 to 6/1 (Figure 5). The tests showed a different high performance for metals in the study. It is possible to obtain relatively high stripping efficiency for Mn(II) and Cu(II), even at high O/A ratios (6/1), which reach 60% and 30%, respectively. For Zn(II) and Fe(III), the stripping at 3/1 of O/A ratio were 20% and 2%, respectively, and for 6/1 ratio, both metals were not stripped.

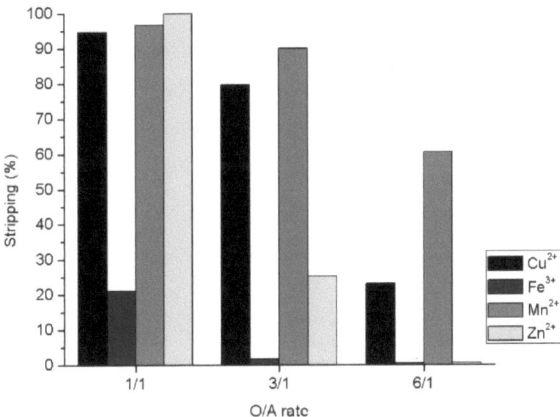

Figure 5. Effect of O/A ratio on the stripping percentage of metals from the organic loaded phases. Experimental conditions: [R_4NCy] = 0.54 M, [H_2SO_4] = 1 M, 400 min^{-1}, and 20 min of stirring time.

To evaluate the effect of the O/A ratio in the stripping efficiency, the relation of organic and aqueous phases was variate from 1/1 to 6/1. Figure 5 shows a clear trend of decreasing stripping efficiency when increasing the O/A ratio from 1/1 to 6/1, but at the same time, the aqueous phase is concentrated with the discharged metals. The Cu(II) and especially the Mn(II) maintain relative ease for being stripped of the organic phase. On the other hand, Zn(II) and Fe(III) have a poor stripping efficiency; both metals practically are not stripped from load IL over 3/1 O/A ratio.

This result is not entirely evident in solvent extraction systems. For example, Quijada et al. [41] reported a stripping efficiency of Fe(III) close to 50% for similar operational conditions, but in this case, the researchers used TFA as extractant and a Bmim-based ionic liquid as diluent. Finally, these results may be useful to design a selective process of separation of metals contained in a PLS of copper, or a pretreatment of this solution to remove Fe(III) and Zn(II).

4. Conclusions

The present research aimed to evaluate the extraction and stripping of ionic liquid R_4NCy in the presence of a simulated commercial PLS of copper industry containing Cu(II), Mn(II), Zn(II), and Fe(III). The main conclusions are the following:

(1) The extraction tests showed a strongly selective extraction of Fe(III) and Zn(II) over Cu(II) and Mn(II). A practically total extraction of Fe(III) and Zn(II) was obtained for all the conditions under study. In the case of Cu(II) and Mn(II), the extraction efficiency is strongly influenced by the R_4NCy concentration, the PLS pH, and the O/A ratio.

(2) The use of 2 M of H_2SO_4 produces a very efficient stripping of all metals loaded into the ionic liquid. The efficiency does not decline as acid concentration decreases, except for Fe(III). The stripping efficiency of iron loaded into the ionic liquid is close to 20% for 0.5 M of sulfuric acid. This result is very promising because it would allow developing a very selective process concerning Fe(III).

(3) Despite its exploratory nature, this study offers a clear insight into the strong affinity of ionic liquid R_4NCy for Fe(III) and Zn(II) over Cu(II) and Mn(II). This fact opens the possibility of using the ionic liquid R_4NCy in the pre-treatment of copper-rich solutions, removing the impurities as Fe(III) and Zn(II).

Author Contributions: Conceptualization, P.N. and C.V.; methodology, P.H. and E.G.; validation, N.T.; formal analysis, P.H. and E.G.; investigation, J.C. and R.S.; data curation, N.T.; writing—original draft preparation, J.C. and R.S.; writing—review and editing, J.C. and R.S.; visualization, C.V.; supervision, P.N. All authors have read and agreed to the published version of the manuscript.

Funding: This research received no external funding.

Institutional Review Board Statement: Not applicable.

Informed Consent Statement: Not applicable.

Acknowledgments: This work was supported by the Universidad de Atacama (project DIUDA 22317). The proofreading was supported by the project "FIUDA 2030". The FTIR spectra collections was supported by project "FT-IR Thermo Nicolet iS50, DIUDAQUIP-2017 UDA".

Conflicts of Interest: The authors declare no conflict of interest.

References

1. Cao, X.; Zhang, T.A.; Zhang, W.; Lv, G. Solvent Extraction of Sc(III) by D2EHPA/TBP from the Leaching Solution of Vanadium Slag. *Metals* **2020**, *10*, 790. [CrossRef]
2. Sekisov, A.; Rasskazova, A. Assessment of the Possibility of Hydrometallurgical Processing of Low-Grade Ores in the Oxidation Zone of the Malmyzh Cu-Au Porphyry Deposit. *Minerals* **2021**, *11*, 69. [CrossRef]
3. Mahmoudi, A.; Shakibania, S.; Rezaee, S.; Mokmeli, M. Effect of the Chloride Content of Seawater on the Copper Solvent Extraction Using Acorga M5774 and LIX 984N Extractants. *Sep. Purif. Technol.* **2020**, *251*, 117394. [CrossRef]
4. Rao, M.; Zhang, T.; Li, G.; Zhou, Q.; Luo, J.; Zhang, X.; Zhu, Z.; Peng, Z.; Jiang, T. Solvent Extraction of Ni and Co from the Phosphoric Acid Leaching Solution of Laterite Ore by P204 and P507. *Metals* **2020**, *10*, 545. [CrossRef]

5. Ying, Z.; Ren, X.; Li, J.; Wu, G.; Wei, Q. Recovery of Chromium(VI) in Wastewater Using Solvent Extraction with Amide. *Hydrometallurgy* **2020**, *196*, 105440. [CrossRef]
6. Zhang, W.; Xie, X.; Tong, X.; Du, Y.; Song, Q.; Feng, D. Study on the Effect and Mechanism of Impurity Aluminum on the Solvent Extraction of Rare Earth Elements (Nd, Pr, La) by P204-P350 in Chloride Solution. *Minerals* **2021**, *11*, 61. [CrossRef]
7. Shakibania, S.; Mahmoudi, A.; Mokmeli, M.; Rashchi, F. The Effect of the Chloride Ion on Chemical Degradation of LIX 984N Extractant. *Miner. Eng.* **2020**, *159*, 106628. [CrossRef]
8. Wang, L.Y.; Guo, Q.J.; Lee, M.S. Recent Advances in Metal Extraction Improvement: Mixture Systems Consisting of Ionic Liquid and Molecular Extractant. *Sep. Purif. Technol.* **2019**, *210*, 292–303. [CrossRef]
9. Hong, T.; Liu, M.; Ma, J.; Yang, G.; Li, L.; Mumford, K.A.; Stevens, G.W. Selective Recovery of Rhenium from Industrial Leach Solutions by Synergistic Solvent Extraction. *Sep. Purif. Technol.* **2020**, *236*, 116281. [CrossRef]
10. Hidayah, N.N.; Abidin, S.Z. The Evolution of Mineral Processing in Extraction of Rare Earth Elements Using Liquid-Liquid Extraction: A Review. *Miner. Eng.* **2018**, *121*, 146–157. [CrossRef]
11. Keskin, S.; Kayrak-Talay, D.; Akman, U.; Hortaçsu, Ö. A Review of Ionic Liquids towards Supercritical Fluid Applications. *J. Supercrit. Fluids* **2007**, *43*, 150–180. [CrossRef]
12. Park, J.; Jung, Y.; Kusumah, P.; Lee, J.; Kwon, K.; Lee, C.K. Application of Ionic Liquids in Hydrometallurgy. *Int. J. Mol. Sci.* **2014**, *15*, 15320–15343. [CrossRef] [PubMed]
13. Weng, J.; Wang, C.; Li, H.; Wang, Y. Novel Quaternary Ammonium Ionic Liquids and Their Use as Dual Solvent-Catalysts in the Hydrolytic Reaction. *Green Chem.* **2006**, *8*, 96. [CrossRef]
14. Quijada-maldonado, E.; Olea, F.; Sepúlveda, R.; Cabezas, R.; Merlet, G.; Romero, J. Possibilities and Challenges for Ionic Liquids in Hydrometallurgy. *Sep. Purif. Technol.* **2020**, *251*, 117289. [CrossRef]
15. Lee, J.; Yeo, C.-D.; Hu, Z.; Thalangama-Arachchige, V.D.; Kaur, J.; Quitevis, E.L.; Kumar, G.; Koh, Y.P.; Simon, S. Friction and Wear of Pd-Rich Amorphous Alloy (Pd43Cu27Ni10P20) with Ionic Liquid (IL) as Lubricant at High Temperatures. *Metals* **2019**, *9*, 1180. [CrossRef]
16. Alguacil, F.J.; López, F.A. Permeation of $AuCl_4^-$ Across a Liquid Membrane Impregnated with $A324H^+Cl^-$ Ionic Liquid. *Metals* **2020**, *10*, 363. [CrossRef]
17. Diabate, P.D.; Dupont, L.; Boudesocque, S.; Mohamadou, A. Novel Task Specific Ionic Liquids to Remove Heavy Metals from Aqueous Effluents. *Metals* **2018**, *8*, 412. [CrossRef]
18. Rodríguez, M.; Ayala, L.; Robles, P.; Sepúlveda, R.; Torres, D.; Carrillo-Pedroza, F.R.; Jeldres, R.I.; Toro, N. Leaching Chalcopyrite with an Imidazolium-Based Ionic Liquid and Bromide. *Metals* **2020**, *10*, 183. [CrossRef]
19. Salar-García, M.J.; Ortiz-Martínez, V.M.; Hernández-Fernández, F.J.; de los Ríos, A.P.; Quesada-Medina, J. Ionic Liquid Technology to Recover Volatile Organic Compounds (VOCs). *J. Hazard. Mater.* **2017**, *321*, 484–499. [CrossRef]
20. Carlesi, C.; Cortes, E.; Dibernardi, G.; Morales, J.; Muñoz, E. Ionic Liquids as Additives for Acid Leaching of Copper from Sulfidic Ores. *Hydrometallurgy* **2016**, *161*, 29–33. [CrossRef]
21. Jin, Y.; Zhang, J.; Song, J.; Zhang, Z.; Fang, S.; Yang, L.; Hirano, S.I. Functionalized Ionic Liquids Based on Quaternary Ammonium Cations with Two Ether Groups as New Electrolytes for $Li/LiFePO_4$ Secondary Battery. *J. Power Sources* **2014**, *254*, 137–147. [CrossRef]
22. Egashira, M.; Okada, S.; Yamaki, J.I.; Yoshimoto, N.; Morita, M. Effect of Small Cation Addition on the Conductivity of Quaternary Ammonium Ionic Liquids. *Electrochim. Acta* **2005**, *50*, 3708–3712. [CrossRef]
23. Matsumoto, M.; Yamaguchi, T.; Tahara, Y. Extraction of Rare Earth Metal Ions with an Undiluted Hydrophobic Pseudoprotic Ionic Liquid. *Metals* **2020**, *10*, 502. [CrossRef]
24. Jing, X.; Wu, Z.; Zhao, D.; Li, S.; Kong, F.; Chu, Y. Environmentally Friendly Extraction and Recovery of Cobalt from Simulated Solution of Spent Ternary Lithium Batteries Using the Novel Ionic Liquids of $[C_8H_{17}NH_2][Cyanex\ 272]$. *ACS Sustain. Chem. Eng.* **2021**, *9*, 2475–2485. [CrossRef]
25. Łukomska, A.; Wiśniewska, A.; Dąbrowski, Z.; Domańska, U. Liquid-Liquid Extraction of Cobalt(II) and Zinc(II) from Aqueous Solutions Using Novel Ionic Liquids as an Extractants. *J. Mol. Liq.* **2020**, *307*, 112955. [CrossRef]
26. Tran, T.T.; Liu, Y.; Lee, M.S. Recovery of Pure Molybdenum and Vanadium Compounds from Spent Petroleum Catalysts by Treatment with Ionic Liquid Solution in the Presence of Oxidizing Agent. *Sep. Purif. Technol.* **2021**, *255*, 117734. [CrossRef]
27. Binnemans, K. Lanthanides and Actinides in Ionic Liquids. *Compr. Inorg. Chem. II Second. Ed. Elem. Appl.* **2013**, *2*, 641–673. [CrossRef]
28. Maria, L.; Cruz, A.; Carretas, J.M.; Monteiro, B.; Galinha, C.; Gomes, S.S.; Araújo, M.F.; Paiva, I.; Marçalo, J.; Leal, J.P. Improving the Selective Extraction of Lanthanides by Using Functionalised Ionic Liquids. *Sep. Purif. Technol.* **2020**, *237*, 116354. [CrossRef]
29. Devi, N.; Sukla, L.B. Studies on Liquid-Liquid Extraction of Yttrium and Separation from Other Rare Earth Elements Using Bifunctional Ionic Liquids. *Miner. Process. Extr. Metall. Rev.* **2019**, *40*, 46–55. [CrossRef]
30. Rzelewska-Piekut, M.; Regel-Rosocka, M. Separation of Pt(IV), Pd(II), Ru(III) and Rh(III) from Model Chloride Solutions by Liquid-Liquid Extraction with Phosphonium Ionic Liquids. *Sep. Purif. Technol.* **2019**, *212*, 791–801. [CrossRef]
31. Sepúlveda, R.; Romero, J.; Sánchez, J. Copper Removal from Aqueous Solutions by Means of Ionic Liquids Containing a B-Diketone and the Recovery of Metal Complexes by Supercritical Fluid Extraction. *J. Chem. Technol. Biotechnol.* **2014**, *89*, 899–908. [CrossRef]

32. Kim, B.K.; Lee, E.J.; Kang, Y.; Lee, J.J. Application of Ionic Liquids for Metal Dissolution and Extraction. *J. Ind. Eng. Chem.* **2018**, *61*, 388–397. [CrossRef]
33. Valdés Vergara, M.A.; Lijanova, I.V.; Likhanova, N.V.; Olivares Xometl, O.; Jaramillo Vigueras, D.; Morales Ramirez, A.J. Recycling and Recovery of Ammonium-Based Ionic Liquids after Extraction of Metal Cations from Aqueous Solutions. *Sep. Purif. Technol.* **2014**, *155*, 110–117. [CrossRef]
34. Nguyen, V.N.H.; Le, M.N.; Lee, M.S. Comparison of Extraction Ability between a Mixture of Alamine 336/Aliquat 336 and D2EHPA and Ionic Liquid ALi-D2 from Weak Hydrochloric Acid Solution. *Metals* **2020**, *10*, 1678. [CrossRef]
35. Castillo, J.; Coll, M.T.; Fortuny, A.; Navarro Donoso, P.; Sepúlveda, R.; Sastre, A.M. Cu(II) Extraction Using Quaternary Ammonium and Quaternary Phosphonium Based Ionic Liquid. *Hydrometallurgy* **2014**, *141*, 89–96. [CrossRef]
36. Fortuny, A.; Coll, M.T.; Sastre, A.M. Use of Methyltrioctyl/Decylammonium Bis 2,4,4-(Trimethylpentyl)Phosphinate Ionic Liquid (ALiCY IL) on the Boron Extraction in Chloride Media. *Sep. Purif. Technol.* **2012**, *97*, 137–141. [CrossRef]
37. Swain, S.S.; Nayak, B.; Devi, N.; Das, S.; Swain, N. Liquid-Liquid Extraction of Cadmium(II) from Sulfate Medium Using Phosphonium and Ammonium Based Ionic Liquids Diluted in Kerosene. *Hydrometallurgy* **2016**, *162*, 63–70. [CrossRef]
38. Deferm, C.; Onghena, B.; Nguyen, V.T.; Banerjee, D.; Fransaer, J.; Binnemans, K. Non-Aqueous Solvent Extraction of Indium from an Ethylene Glycol Feed Solution by the Ionic Liquid Cyphos IL 101: Speciation Study and Continuous Counter-Current Process in Mixer–Settlers. *RSC Adv.* **2020**, *10*, 24595–24612. [CrossRef]
39. Padhan, E.; Sarangi, K. Recovery of Nd and Pr from NdFeB Magnet Leachates with Bi-Functional Ionic Liquids Based on Aliquat 336 and Cyanex 272. *Hydrometallurgy* **2017**, *167*, 134–140. [CrossRef]
40. Rybka, P.; Regel-Rosocka, M. Nickel(II) and Cobalt(II) Extraction from Chloride Solutions with Quaternary Phosphonium Salts. *Sep. Sci. Technol.* **2012**, *47*, 1296–1302. [CrossRef]
41. Quijada-Maldonado, E.; Romero, J.; Osorio, I. Selective Removal of Iron(III) from Synthetic Copper(II) Pregnant Leach Solutions Using [Bmim][Tf2N] as Diluent and TFA as Extracting Agent. *Hydrometallurgy* **2016**, *159*, 54–59. [CrossRef]
42. Devi, N. Solvent Extraction and Separation of Copper from Base Metals Using Bifunctional Ionic Liquid from Sulfate Medium. *Trans. Nonferrous Met. Soc. China* **2016**, *26*, 874–881. [CrossRef]
43. Regel-Rosocka, M.; Nowak, Ł.; Wiśniewski, M. Removal of Zinc(II) and Iron Ions from Chloride Solutions with Phosphonium Ionic Liquids. *Sep. Purif. Technol.* **2012**, *97*, 158–163. [CrossRef]
44. Baczyńska, M.; Regel-Rosocka, M.; Coll, M.T.; Fortuny, A.; Sastre, A.M.; Wiśniewski, M. Transport of Zn(II), Fe(II), Fe(III) across Polymer Inclusion Membranes (PIM) and Flat Sheet Supported Liquid Membranes (SLM) Containing Phosphonium Ionic Liquids as Metal Ion Carriers. *Sep. Sci. Technol.* **2016**, *51*, 2639–2648. [CrossRef]
45. Ola, P.D.; Kurobe, Y.; Matsumoto, M. Solvent Extraction and Stripping of Fe and Mn from Aqueous Solution Using Ionic Liquids as Extractants. *Chem. Eng. Trans.* **2017**, *57*, 1135–1140. [CrossRef]
46. Nguyen, V.N.H.; Lee, M.S. Separation of Co(II), Ni(II), Mn(II) and Li(I) from Synthetic Sulfuric Acid Leaching Solution of Spent Lithium Ion Batteries by Solvent Extraction. *J. Chem. Technol. Biotechnol.* **2021**, *96*, 1205–1217. [CrossRef]
47. Zhu, Z.; Yoko, P.; Cheng, C.Y. Recovery of Cobalt and Manganese from Nickel Laterite Leach Solutions Containing Chloride by Solvent Extraction Using Cyphos IL 101. *Hydrometallurgy* **2017**, *169*, 213–218. [CrossRef]
48. Sobekova Foltova, S.; vander Hoogerstraete, T.; Banerjee, D.; Binnemans, K. Samarium/Cobalt Separation by Solvent Extraction with Undiluted Quaternary Ammonium Ionic Liquids. *Sep. Purif. Technol.* **2019**, *210*, 209–218. [CrossRef]
49. Tran, T.T.; Iqbal, M.; Lee, M.S. Comparison of the Extraction and Stripping Behavior of Iron (III) from Weak Acidic Solution between Ionic Liquids and Commercial Extractants. *J. Korean Inst. Met. Mater.* **2019**, *57*, 787–794. [CrossRef]

Article

Influence of Electrolyte Impurities from E-Waste Electrorefining on Copper Extraction Recovery

Jovana Djokić [1,*], Dragana Radovanović [2], Zlatko Nikolovski [3], Zoran Andjić [1] and Željko Kamberović [4]

1. Innovation Centre of Faculty of Chemistry in Belgrade Ltd., University of Belgrade, Studentski Trg 12-16, 11000 Belgrade, Serbia; zoranandjic@yahoo.com
2. Innovation Centre of Faculty of Technology and Metallurgy in Belgrade Ltd., University of Belgrade, Karnegijeva 4, 11000 Belgrade, Serbia; divsic@tmf.bg.ac.rs
3. Institute Mol, Nikole Tesle 15, 22300 Stara Pazova, Serbia; zlatko_nbg37@hotmail.com
4. Faculty of Technology and Metallurgy, University of Belgrade, Karnegijeva 4, 11000 Belgrade, Serbia; kamber@tmf.bg.ac.rs
* Correspondence: djokic@chem.bg.ac.rs

Abstract: In order to reflect possible issues in future sole e-waste processing, an electrolyte of complex chemical composition reflecting system of sole e-waste processing was obtained by following a specially designed pyro-electrometallurgical method. The obtained non-standard electrolyte was further used for the purpose of comprehensive metal interference evaluation on the copper solvent extraction (SX) process. Optimization of the process included a variation of several process parameters, allowing determination of the effect of the most abundant and potentially the most influential impurities (Ni, Sn, Fe, and Zn) and 14 other trace elements. Moreover, comparing three commercial extractants of different active chelating groups, it was determined that branched aldoxime reagent is favorable for Cu extraction from the chemically complex system, as can be expected in future e-waste recycling. The results of this study showed that, under optimal conditions of 20 vol.% extractant concentration, feed pH 1.5, O/A ratio 3, and 10-min phase contact time, 88.1% of one stage Cu extraction was achieved. Co-extraction of the Fe, Zn, Ni, and Sn was under 8%, while Pb and trace elements were negligible. Optimal conditions (H_2SO_4 180 g/L, O/A = 2, and contact time 5 min) enabled 95.3% Cu stripping and under 6% of the most influential impurities. In addition, an impurity monitoring and distribution methodology enabled a better understanding and design of the process for the more efficient valorization of metals from e-waste.

Keywords: e-waste; electrolyte recycling; solvent extraction; chelating extractants; copper; impurities influence; metal distribution

Highlights:

- Electrolyte of highly complex chemical composition resulting from specially designed pyro-electrometallurgical e-waste process, aiming to reflect future obstacles considering sole e-waste recycling
- A one-factor solvent extraction methodology for the comparison of three commercial extractants in terms of efficiency, distribution coefficients, selectivity, and influence of impurities on Cu extraction from experimentally obtained electrolyte
- Feed pH, extractant, and stripping agent concentration affect solvent extraction most significantly through reaction equilibrium shifts and active centers availability
- Optimized process conditions enable selective Cu extraction among highly abundant Fe, Zn, Pb, Ni, and Sn and trace elements (i.e., Al, Co, Cr, Mg, Na, Sb, Ga, Ge)
- Transfer monitoring, distribution, and methodology for additional valorization of metals

1. Introduction

Resource shortage, environmental and human health issues, and economic incentives have promoted recycling to become mandatory in the modern world. Efforts to reduce natural exploitation, preserve the environment, and promote the reuse and valorization of previously discarded materials, thereby minimizing process waste, have become one of the most important goals. This approach leads to closing up the material flow loop and achieving a circular economy [1]. The same concept is applied to electrical and electronic waste (e-waste) also. As the waste stream increases at an ever-growing rate, recycling has become of great concern in recent decades. In terms of metal content, printed circuit boards (PCB) are the most valuable part of e-waste, containing more than 60 elements of the periodic table, from 28% [2] to 40% [3] of metals, including base and precious metals, rare, scarce elements, but also hazardous, toxic, metals. Besides metals, this material contains various organic compounds. This extremely heterogeneous composition as well as chemical complexity represent the main obstacle for efficient valorization, making e-waste and PCB recycling quite challenging [4]. At the same time, metal value makes it a major secondary resource and an attractive material for recycling [5]. Considering the abundance of copper and precious metals in e-waste, which is higher than in ore deposits [6], the usual valorization methods are analogous to those in primary metal production, built on pyro-electro and/or hydrometallurgical bases [7]. The application and significance of the electrometallurgical approach in the production of metals from solution are given in a detailed review by Rai et al., emphasizing that there are limitations in e-waste processing due to the complexity of the material itself [8].

If the pyrometallurgical route (PM) is employed, the smelting feed is usually a synergetic mixture of ore concentrate and other valuable scrap materials. Several plants for smelting and refinery, such as Umicore [9], Boliden [10], and Naoshima (Mitsubishi Materials Corporation) [11], have integrated e-waste in processing, accounting for 15–30 wt.% of the feed. Yet, considering the rising trend of e-waste, it is inevitable that a mass fraction of this material in the feed will also grow. So, if smelting feed is predominantly or solely e-waste, it is expected to negatively affect the traditional production process due to inlet deviating in terms of compounds and concentration. Due to similar physical-chemical properties, selective separation of metals between smelting phases is hindered [12]. In general, obtained blister copper is further refined through an anode oxidation process to decrease the abundance of the impurities, such as Al, Ba, Be, Ca, Cr, Fe, Mg, Mn, Na, and Zn. At the same time, precious metals, as well as Bi, Pb, Ni, Cd, Sb, and Sn, remain in molten blister copper [13]. Further, the PM route is followed by electrochemical deposition where, besides cathode Cu, electrolyte and anode slime are generated as by-products [14]. If feed material is sole e-waste, a non-standard tin precipitate is also generated during electrorafination (ER) due to a high Sn concentration in the starting material. This non-standard by-product requires additional treatment steps [15]. Due to upstream recycling chemistry and complex electrolyte composition, the subsequent electrowinning (EW) step is also challenging. Besides Ni, Co, As, and Sb, the main obstacle for EW is Fe content [16], whose concentration above 1 g/L causes current efficiency losses due to Fe cathode reduction as a competitor reaction, resulting in low cathode deposition of Cu with poor quality [17].

Likewise, in the case of pyro-electro, in a hydrometallurgical route, where leaching by various lixiviants can be conducted directly on target e-waste components or take place after smelting, EW is an established metal separation method [18]. Moreover, the hydrometallurgical route is a prospective alternative to pyrometallurgical processing, especially for small and medium-sized enterprises [19]. Yet again, with all upstream recycling chemistry issues inherent, another issue is the buildup of metal impurities in the cyclic treatment set-up, where impurities severely impact the EW process [20].

Accordingly, no matter which route is chosen, efficiency and selectivity are low due to the material composition complexity mentioned earlier. Besides target elements, usually Cu, the obtained electrolyte or leaching solution contains simultaneously dissolved elements that are considered impurities, negatively affecting the recycling process.

The incorporation of e-waste into established metal production routes is challenging as e-waste composition constantly changes. Even more challenging is how to effectively process this feed material with additional value without prior dilution by primary and other scrap materials, in response to today's circular economy demands [21].

Therefore, in order to achieve selective metal valorization and obtain high-purity products, generated solutions require downstream purification and concentration processes. Besides the aforementioned EW, other purification/concentration methods include solvent extraction, ion exchange, precipitation, and their combinations [22,23]. A choice depends on the lixiviant nature, target metal concentration, effect of impurities, and process economics [24]. Among them, solvent extraction is an established Cu purification technique commonly applied in industry and small-scale operations [25].

Solvent extraction (SX) is a traditional separation method that has been used for decades in hydrometallurgy of primary metal production for the purpos of separation, purification and concentration of metals in the solution. Now, the goal is to assimilate this technology into e-waste recycling. SX is characterized by easy operation, low capital and economic costs, and flexibility [26]. The latter is very important considering the chemical composition diversity of e-waste when rigid treatment methods can hardly follow composition fluctuation. Extractants for SX are under constant development and improvements to ensure selectivity, increase extraction efficiency, and facilitate re-extraction. Several extraction types and mixtures, with or without modifiers, have been used to extract Cu, including hydroxyoximes, hydroxy benzophenone oximes, aldoxime, quinolines, organic acids, and organophosphorus based compounds [27]. As e-waste recycling solutions are non-standard and burdened by various elements, the efficiency of different extraction agents has been studied. Due to low selectivity, many have found that removing metal impurities, especially Fe, is necessary before SX [28,29], meaning an additional technological step is required, representing a new point of metal values loss.

In this study, aldoxime and ketoxime/aldoxime blend with or without modifier were used to determine selectivity and efficiency in metal separation. Accordingly, the results of the literature survey data are presented as follows. It is reported that Acorga® aldoxime reagents exhibit high selectivity and recovery toward Cu even when the concentration of Fe is predominant in the solution [30]. By optimizing the solvent extraction conditions (feed pH 1.1, M5640 16 vol.%, O/A = 1, contact time 3 min), the efficiency of copper extraction from the e-waste sulfate leaching solution was more than 85%. It is also concluded that the presence of iron negatively affected Cu extraction efficiency and that the strength of H_2SO_4 (254 g/L as the optimal one), as the stripping agent, highly influences Fe stripping percentage [3]. Deep [31] and Ferreira [32] used Acorga® M5640 modified by isotridecanol to selectively extract Cu from a sulfate solution dominated by Fe and Zn and found that the modifier hindered Cu loading but facilitated stripping. The results considering Fe co-extraction deviate, but the percentages are undoubtedly low and influenced by the pH of the feed solution. Ochromowicz and Chmielewski compared commercial LIX® reagents (984N and 612N-LV) with the Acorga® M5640 for the Cu extraction in the presence of the equal Fe concentration (both about 30 g/L). The latter extractant proved to be more efficient, yet it was determined that Fe co-extraction is highly influenced by feed pH and extractant concentration. Optimal process conditions were M5640 30 vol.%, and 0.75 pH [26]. Linden et al. came to a similar conclusion in their study. They determined that, even with a low metal concentration (Cu, Fe, Cr, Zn, Pb, Ca, and Ni, each of 500 ppm), co-extraction of Fe from synthetic feed could reach almost 40% and 12% by using LIX® 984N and Acorga® M5640, respectively [33]. Even if Cu is predominant in the solution, selectivity is not achieved, neither by LIX® series or Acorga® OPT, which is extractant modified to facilitate extraction and stripping [34], and Fe co-extraction is challenging to avoid [35]. On the contrary, Banza concluded that LIX® 984N proves to be Cu/Fe selective if pH is above 2.5 [36]. Yet, the precipitation of Fe during Cu extraction at the mentioned pH was not discussed.

The aforementioned discussion showed quite extensive research on copper SX. When it comes to the influence of metal on the process efficiency, the focus is usually on Fe, with less attention paid to the interference of other metals. Having in mind the traditional circuit set-up as a coupled hydro-electro treatment, the buildup of impurities in specific phases of the process stages is unavoidable.

According to the literature review, recycling is based on an e-waste dilution methodology in primary pyro-processing feed. Moreover, published studies considering solution purification of e-waste processing mainly address the issue of the influence of base metals on Cu extraction in simplified systems.

This paper addresses the knowledge gap of sole e-waste processing and the comprehensive determination of impurities interferences. Hence, at the core of this study was the aim to determine the effect of the most abundant and potentially the most influential impurities (Ni, Sn, Fe, and Zn, of which Ni and Sn are the most challenging since they cannot be removed by pyro-processing) as well as 14 other trace elements (i.e., Al, Bi, Ca, Co, Mg, Na, Sb, Ga, etc.) on copper SX. In addition, metal distribution through electro-hydrometallurgical e-waste processing was determined in order to understand interferences in a chemically complex system.

For that purpose, various e-waste categories as feed material have been intentionally selected to provide the generation of an electrolyte of non-standard chemical composition, further used as feed for SX optimization. The effect of extractant concentration, aqueous to liquid ratios, phase contact time, and feed pH was determined and compared for each of the three investigated extractants. Stripping optimization also included one-factor parameters variations. According to extraction/stripping efficiency and co-extraction levels, proper SX conditions and metal distributions were defined. These approaches allow to understand future obstacles in the processing of sole e-waste and contribute to the design of a process that will enable an efficient flow of materials.

2. Experimental

2.1. Material

A total of 400 kg of various e-waste categories, which consisted of 60 wt.% metallic fractions granulate, 35 wt.% waste PCBs fraction, of which 5 wt.% were RAM cards, and 5 wt.% CPU fractions, were received from local suppliers. E-waste fractions, all but metallic granulate, went through the preparation step.

2.2. Sole E-Waste Preparation Process

The vacuum pyrolysis depolymerization method was conducted on PCBs and CPUs fractions of received e-waste. Organic compounds that were removed by this step (8.9 wt.% of gases and 17.5 wt.% of liquids) would have otherwise jeopardized further processing.

The depolymerized sample (30 wt.%), altogether with metallic granulate (65 wt.%) and flux material (slag, 5 wt.%), was the feed for reducing smelting in a DC electric arc furnace. Besides slag and filter dust, the main smelting product was metal phase as non-standard Cu alloy weighing 75% by weight of the feed, and this was used for anode casting. The traditional step of anode oxidative refining was excluded with the intention of preserving a high content of Fe, Zn, Na, Cr, Mn, Ti, Al, Ba, Be, Mg, and Ca in the cast anodes, which would otherwise be mostly removed from the molten copper. This approach makes it possible to obtain a particularly non-standard complex electrolyte, which was the intention for this study. Each casted anode weighted between 23 and 24 kg, and 4 anodes, with a total mass of 92.4 kg, were used in the next step, electrorefining (ER). The ER products were cathode copper (85 kg of 99.83% purity), as the main standard product, anode slime, and electrolyte as standard by-products (13 kg and 200 L, respectively) and tin precipitate as a non-standard by-product (4 kg of fine suspended particles periodically extracted by coagulation from electrolyte during ER).

Depolymerisation, smelting, ER, and tin precipitate extraction processes are described by Djokić et al. [15].

All experimentally obtained products and by-products were analyzed, while the electrolyte of complex chemical composition was used as the aqueous phase (feed) for solvent extraction optimization and for monitoring the distribution of impurities.

2.3. Reagents and Procedure

Three extractants were used in the current study: Acorga® series M5640 (branched, C9 modified aldoxime reagent with active 5-nonyl-2-hydroxy-benzaldoxime), OPT 5510 (C9 ester modified aldoxime-ketoxime reagent), and LIX® series 984N reagent (non-modified aldoxime-ketoxime blend containing 5-nonylsalicylaldoxime and 2-hydroxy-5-nonylacetophenone oxime). Extractants were not additionally purified. Other used chemicals were p.a. grade.

The organic phase (OP) was prepared by diluting extraction agents in kerosene to achieve desired concentrations. This diluent was selected based on its physical-chemical properties (i.e., hydrophobic, low dielectric constant favorable for cation extraction because it does not promote the polymerization of chelating extractants over hydrogen bridges), proper safety, health, and environmental characteristics, and applicability in industry.

The solvent extraction (SX) experiments were carried out by varying extractant concentration in the organic phase (M5640, 5–30 vol.%), organic to aqueous phase (O/A) ratio (1/3, 1/2, 1/1, 2/1, 3/1, and 4/1), contact time (2.5, 5, 7.5, 10, 12.5, and 15 min), and pH of feed solution (0, 1, 1.5, and 2). NaOH (Fisher Chemical) was used for pH adjustment by adding the appropriate amount to the electrolyte until the desired pH was reached. Experiments were conducted in separating funnels at room temperature by contacting corresponding volumes of aqueous and organic phases. After complete phase disengagement, which required 3–5 min, raffinate was sampled for analysis, and the loaded organic phase was further used for stripping. This step optimization included a variation of stripping agent concentration (H_2SO_4(Carlo Erba reagents), 100–200 g/L), O/A ratio (1/2, 1, 2, and 4), and contact time (2.5 to 15 min). Appropriate volumes of phases are contacted for desired time. After separation by gravitation, analysis of aqueous phases (raffinate and loaded strip) followed.

In the subsequent experimental set, the extraction behavior of metals using LIX® 984N and OPT 5510 was investigated under experimentally determined optimal parameters for the M5640 extractant. All experiments were conducted as one-stage extraction/stripping.

Determined parameters were used to define optimal process conditions for SX which takes place according to Equation (1), [26]:

$$M^{n+} + nHA = MA_n + nH^+ \qquad (1)$$

where M^{n+} is a metal cation and HA is an extractant.

This process is reversible, and according to stoichiometry, for each mole of extracted metal ion, it is expected to increase raffinate acidity (i.e., one mol of divalent metal ion generates two moles of H^+), which shifts the equilibrium towards a stripping reaction.

2.4. Analytical Methods

The chemical characterization of casted anodes was determined by analysis of the elemental composition (Elementar™ Vario EL III, CHNS/O elemental analyzer), X-ray fluorescence spectrometry (XRF, Thermo Scientific Niton™ XL3t, calibrated to operate in alloy mode), and optical emission spectrometry (OES, SPECTROMAXx™ arc/spark, calibrated for Cu alloy composition determination). A representative sample of casted anodes was digested in HNO_3/HF (3/1 volume ratio, 15 min at 180 °C) mixture to perform a more comprehensive analysis. The chemical composition of all obtained solutions was determined via the inductively coupled plasma method with an optical emission spectrometer (ICP-OES, SPECTROBLUE™ TI, Smart Analyzer Vision data processing software, in relation to the AccuStandard® analytical reference standards). Measurements of pH were conducted by using an INO-LAB® pH-720 pH meter, calibrated by LLG™ buffer solutions to operate in acid environment.

The concentrations of metals in aqueous phases were determined by direct instrumental analysis as triplicates, including the standard deviation of the measurements and the calibration of the instrument, while mass balance calculations determined the concentrations of metals in the loaded organic phases. Parameters such as the distribution coefficient (D), the extraction/stripping efficiency (%E and %S, respectively), and the separation factor ($\beta_{A/B}$) were calculated for each experiment according to Equations (2)–(5).

$$D = ([M]_0 - [M]_{aq})/[M]_{aq} \quad (2)$$

$$\%E = (D/(D + (V_{aq}/V_{org}))) \times 100 \quad (3)$$

$$\beta_{A/B} = D_A/D_B \quad (4)$$

$$\%S = ([M]_{aq\text{-}s}/[M]_{org}) \times 100 \quad (5)$$

where $[M]_0$ and $[M]_{aq}$ represent the initial and metal concentration in the aqueous phase (raffinate), V_{aq} and V_{org} represent the volumes of aqueous and organic phases, A and B represent different metals whose separation factor is the subject of determination, and $[M]_{aq\text{-}s}$ and $[M]_{org}$ are concentrations of metal ions in aqueous stripping solution and initial organic phase. All experimentally obtained data were processed as described.

Measured and calculated values are further used to construct McCabe–Thiele diagrams to determine the theoretical number of extraction/stripping stages.

In addition, the Sankey diagram was constructed to present the mass balance of the SX process under defined optimal conditions and to perceive the metal distribution.

3. Results and Discussion

3.1. Feed Composition

Casted anodes and electrolyte obtained by anodes electrorefining were chemically characterized in detail, and their composition is shown in Table 1. The chemical composition of minor components of smelting feed as well as of the process by-products are not presented.

Table 1. Chemical composition of pyro-electro and solvent extraction feed (n.a., not analyzed; <DL, below detection limit).

Element	Metallic Granulate	Anodes		Electrolyte	
	%	%	ppm	mg/L	g/L
Cu	69.12	83.40	-	-	41.37
Fe	5.90	3.90	-	-	20.67
Zn	12.71	5.40	-	-	26.55
Ni	1.68	2.32	-	-	9.89
Sn	5.89	7.27	-	453.1	-
Pb	3.30	3.52	-	3.1	-
Bi	n.a.	0.04	-	112.9	-
Ca	n.a.	0.10	-	346.3	-
Co	0.01	0.05	-	163.4	-
Cr	0.12	0.03	-	95.8	-
Mg	n.a.	0.03	-	61.1	-
Ag	0.64	0.64	-	<DL	-
Au	30 ppm	0.11	-	<DL	-
Al	0.12	-	63	19.3	-
Cd	0.04	-	62	21.7	-
Mn	0.10	-	84	29.7	-
B	n.a.	-	32	8.7	-
Na	n.a.	-	120	41.8	-
Sb	n.a.	-	23	5.8	-
Ga	n.a.	-	47	12.5	-
Ge	n.a.	-	4	1.2	-

As can be seen from Table 1, obtained products are characterized by extremely complex chemical composition, with a high concentration of Cu, Fe, Zn, Ni, and Sn, and burdened with trace elements. This non-standard chemical composition and abundance of base and precious metals reflect the applied treatments' feed material composition and chemistry mechanisms. More specifically, during smelting, base (Co, Cu, Ni, Sn) and precious metals (Au and Ag) are concentrated in blister copper. In addition, due to excluded step of oxidative anode refining prior to casting, impurities share in casted anodes is increased. Half of the Zn and a small share of Fe are lost by dissipation among smelting by-products (filter dust and slag). During electrorefining, metals from anodes are dissolved to the electrolyte, while most Sn, Pb, and precious metals are precipitated (as anode slime and suspended tin precipitate). The obtained values differ from the current electrolytes generated by traditional recycling by the dilution method of smelting feed. However, they are expected in the future if the smelting feed is predominantly or solely e-waste.

3.2. Solvent Extraction (SX) Optimization

3.2.1. Effect of Extractant Concentrations and O/A Ratio

A series of experiments were carried out to determine the effect of extractant concentration and O/A ratio on the copper SX efficiency. The aqueous phase was raw electrolyte of pH 0, and contact time was 10 min to ensure maximal mass transfer. The M5640 concentration vs. copper extraction efficiency plot was constructed (Figure 1). As can be seen, Cu extraction efficiency almost linearly increases with an increase of extractant concentration from 5% to 20%, after which efficiency rises insignificantly. Moreover, it can be seen that, in the experiments with a phase ratio O/A = 1 or less, Cu loading is below 14%. It is presumed that, for O/A below 1, the predominance of raw electrolyte and its high acidity negatively affect extraction. Similar results are obtained for the co-extraction of the impurities (not shown).

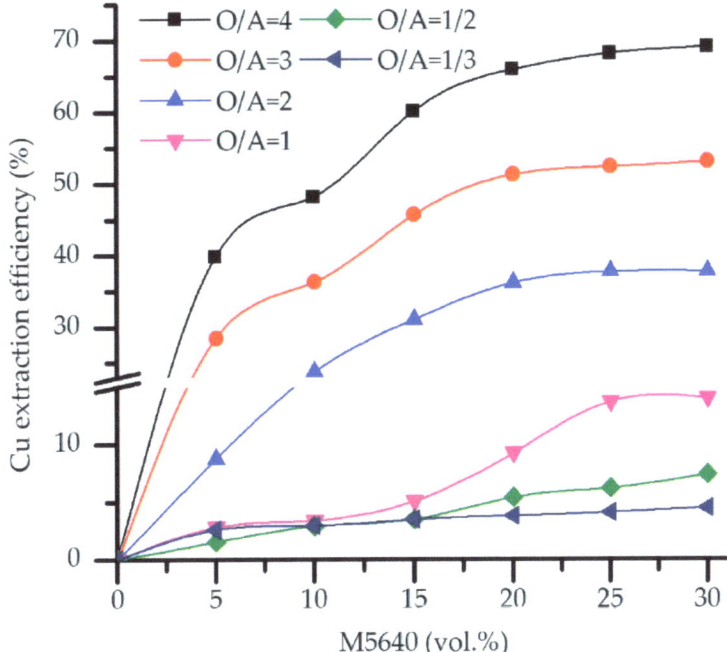

Figure 1. Copper extraction efficiency as a function of M5640 concentration and O/A phase ratio for the feed pH 0, temperature 20 °C, and contact time 10 min.

According to what is stated above, a detailed analysis of extraction efficiencies and separation coefficients was performed only for O/A ratios above 2.

Figures 2 and 3 show the significant influence of extractant concentration and O/A ratio on extraction efficiency and Cu vs. impurities separation. Yet, the extraction of B, Ca, Na, Mg, Pb, Ga, and Ge do not occur at O/A ratios above 2 in the whole extractant concentration range. Mentioned elements are co-extracted (below 3%) only at O/A = 1/3 when their feed input is the highest (not shown). As shown in Figures 2a and 3a, at O/A = 2, both extraction efficiency and separation coefficient are low so this phase ratio is excluded from subsequent optimization experiments. Further, at O/A = 3 and 4, extraction of the most abundant impurities slightly decreases with extractant concentration while Cu plateau is reached at M5640 20 vol.%. Although at O/A = 4, Cu extraction efficiency is the highest, it is also the general co-extraction of accompanying metals, predominantly trace elements (Figure 2c). Therefore, Fe, Zn, Ni, and Sn loading, for M5640 20 vol.%, is 6.9, 6.1, 4.1, and 7.4 at O/A = 3 (Figure 2b), and significantly higher at O/A = 4: 14.4, 9.6, 10.05, and 19.80, respectively (Figure 2c).

The separation coefficient (Figure 3b,c) shows significant differences comparing O/A = 3 and 4 ratios, but only at the upper M5640 concentration limit, which, as previously established, has not contributed to Cu extraction efficiency. For M5640 up to 20 vol.%, better separation is achieved for O/A = 3 ratio.

Extraction and selectivity results, as shown in Figures 2 and 3, respectively, could be explained by M5640 high affinity towards Cu versus other metals present in high concentrations, e.g., Ni, Sn, Fe, and Zn. However, when copper extraction approaches saturation, due to the excess of the organic phase (at O/A = 4), the number of complexing centers available to extract accompanying (trace) metals increase, resulting in the co-extraction of elevated impurities. Namely, according to Equation (1), the dimerization of extractant is necessary for Cu^{2+} extraction (Equation (6)):

$$Cu^{2+} + 2HA = CuA_2 + 2H^+ \tag{6}$$

This process is facilitated under high extractant concentration in the organic phase, which also allows the formation of the complexes with ions of higher valence than of Cu (i.e., Bi^{3+}, Sb^{4+}, and Sn^{4+}). In addition, although a high co-extraction of $Fe^{2+/3+}$, Ni^{2+}, and Zn^{2+} could be expected based on their valence and extraction mechanism, this is not the case, probably due to their ionic radius and steric interference during di- and trimerization of the extractant. In the case of Zn^{2+}, whose ionic radius is close to Cu^{2+} (74 and 73 pm, respectively), the extraction of Zn is hindered, possibly due to the salting-out effect in a high sulfate concentration environment and the formation of $Zn(SO_4)_2^{2-}$ instead of Zn^{2+}. Considering that oximes do not extract anionic species at low pH, Zn co-extraction is minimal. Thus, when the concentration of M5640 is 20 vol.%, and the O/A ratio is 3, the Fe, Zn, and Ni concentrations in the loaded organic phase were 1.4, 1.6, and 0.4 g/L. At the same time, Sn and Bi are measured in 34 and 41 ppm, respectively. The total concentration of trace metals was 25 ppm.

Comparing extraction efficiency and selectivity, considering impurities behavior, extractant (M5640) concentration of 20 vol.%, and O/A ratio of 3 was selected as optimal. In addition, the determined optimal concentration of M5640 in the organic phase is also beneficial considering that, for concentrations of M5640 higher than 25 vol.%, an increase in the viscosity of the organic phase was observed, leading to mixing difficulties and prolonged phase disengagement.

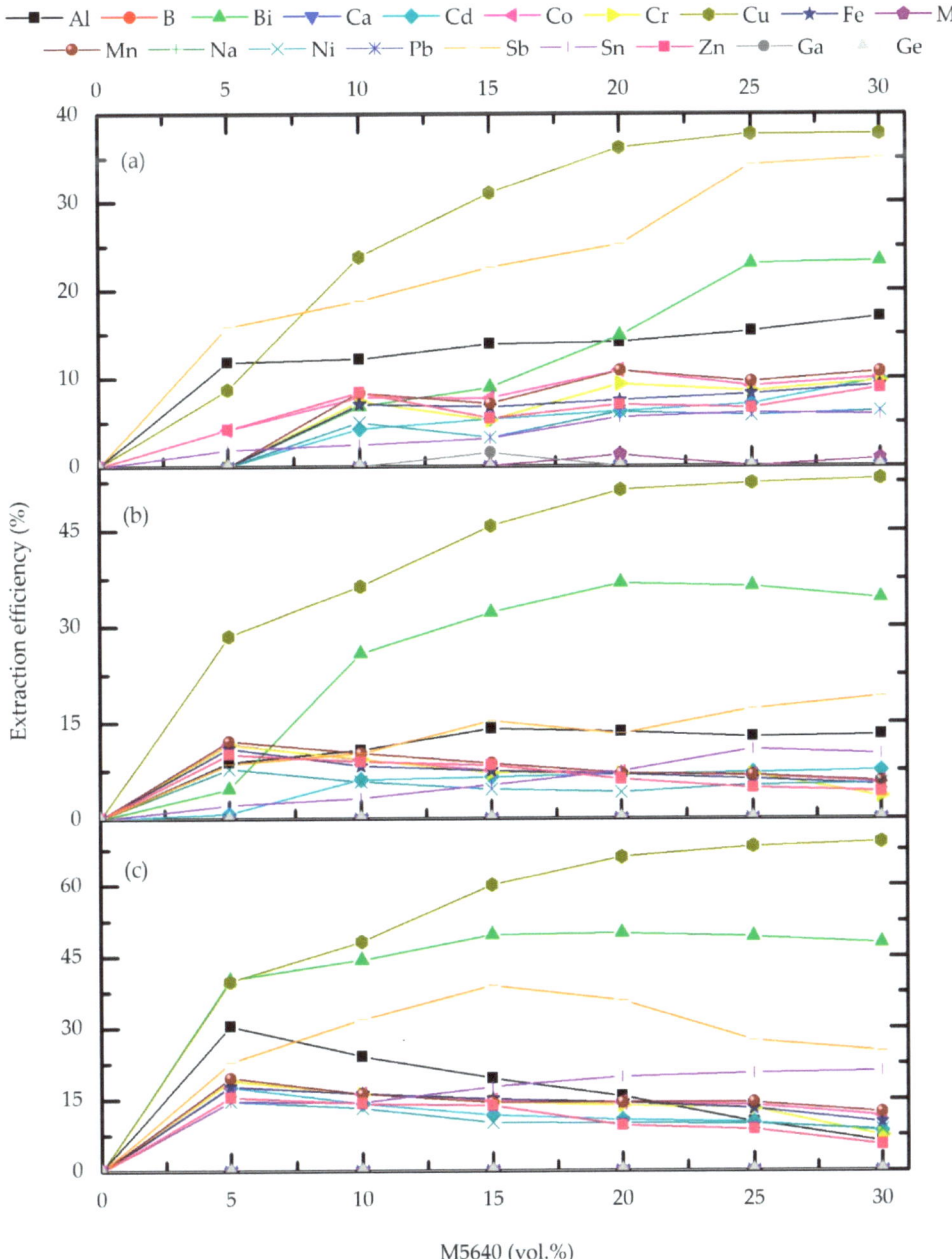

Figure 2. Effect of M5640 concentration and O/A phase ratio 4 (**a**), 3 (**b**), and 2 (**c**) on extraction efficiency for the feed pH 0, temperature 20 °C, and contact time 10 min.

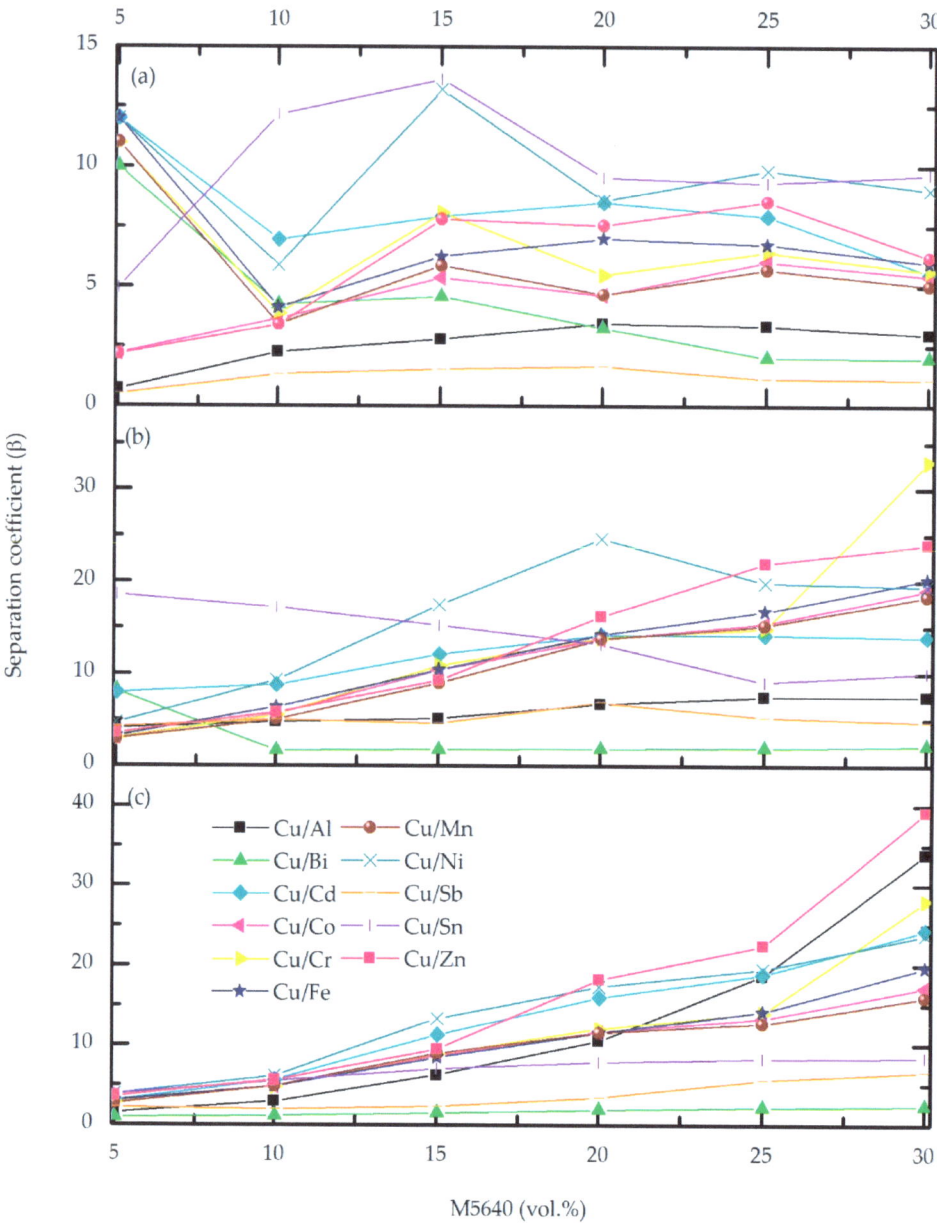

Figure 3. Effect of M5640 concentration and O/A phase ratio 4 (**a**), 3 (**b**), and 2 (**c**) on Cu-metal separation factor for the feed pH 0, temperature 20 °C, and contact time 10 min.

3.2.2. Effect of Phase Contact Time

In order to investigate the phase contact time effect (Figure 3), a set of experiments was conducted at previously optimized extractant concentration (20 vol.%) and phase ratio (O/A = 3). As in previous experiments, the pH of the feed electrolyte was constant: 0. The extraction efficiency of Cu and co-extraction of metal impurities were taken into account to determine the phase contact time parameter.

As can be seen from Figure 4a, the extraction percentage generally rises with contact time. For the majority of trace metals, as well as for the most abundant impurities (Ni, Sn, Fe, and Zn), the change in extraction efficiency is negligible. Moreover, prolonged contact time does not affect the extraction of B, Ca, Na, Mg, Pb, Ga, and Ge, which remain in the raffinate. The extraction percentage of Sb sharply rises after 10 min of phase contact from 13.2% to 21.65%, indicating slow extraction kinetics, while at this time point, Cu and Bi reach the plateau. These results can be explained by the strong binding capacity of Cu-M5640 comparing to the other metals, leading to increased extraction until equilibrium is reached. After that point, the extraction of impurities rises. As shown in Figure 4b, the separation coefficient is influenced by phase contact time until the equilibrium of each metal is achieved, which for most of them is around 10 min. Considering the contact time effect on the metal's behavior during extraction, the mentioned time point of 10 min makes a breaking point, and thus is it is chosen as the optimal one.

3.2.3. Effect of Feed pH

In order to determine the influence of the acidity of the aqueous feed on the efficiency of Cu extraction, a series of experiments were conducted by varying the pH while the other process conditions were constant. According to Figure 5a, raising the feed pH from 0 to 2, the effect on the co-extraction of the Fe, Zn, Ni, and Sn was negligible, and overall extraction was below 8%. Moreover, trace elements loading is slightly affected by feed pH. The most drastic change in the extraction, affected by feed acidity, was observed for Cu loading, which increased from 51.30% to 98.87% in the experimental pH range. High Cu extraction efficiency at elevated feed pH can be expected, having in the mind extraction mechanism and H^+ influence on the reaction equilibrium (Equation (6)). Thus, in the extremely high acidity conditions, during the extraction step, simultaneous re-extraction occurs, leading to stripping of Cu from the organic phase. This effect is absent at elevated pH values. In addition, high pH favors the ionization of the extractant, facilitating metal extraction. The latter statement contradicts the results obtained for the extraction of impurities which would be expected to also increase with increasing pH which is obviously not the case. The explanation lies in Cu loading, as with increasing copper complexation, the competitive reactions of metal impurities do not come to the fore and their loading remains low. This best reflects the separation factor (Figure 5b) which apparently increases after pH 1.5 when the Cu loading reaches about 90%.

According to the experimental results, an increasing trend of Cu extraction is evident, yet at the mentioned upper pH limit the plateau was not achieved. However, additional experiments with an extended pH range were not feasible due to the precipitation of mostly Fe^{3+} hydroxide and co-precipitation of Cu, which led to its loss. Moreover, at pH 2, slight turbidity of aqueous feed occurs, presumably due to iron hydroxide formation, and experiments showed extended phase separation time. Considering the obtained results, pH 1.5 was the optimal choice.

3.2.4. McCabe-Thiele Extraction Diagram

In order to determine the number of the extraction stages, a correlation diagram between metal content in the organic vs. aqueous phase was plotted. The Cu extraction isotherm was constructed according to the experimental results obtained by various O/A ratios maintaining extractant concentration, phase contact time, feed pH, and temperature constant. A McCabe–Thiele diagram (Figure 6) was constructed assuming that the concentration of the Cu in the organic phase was 0 g/L. Results show three theoretical stages needed to achieve quantitative copper extraction.

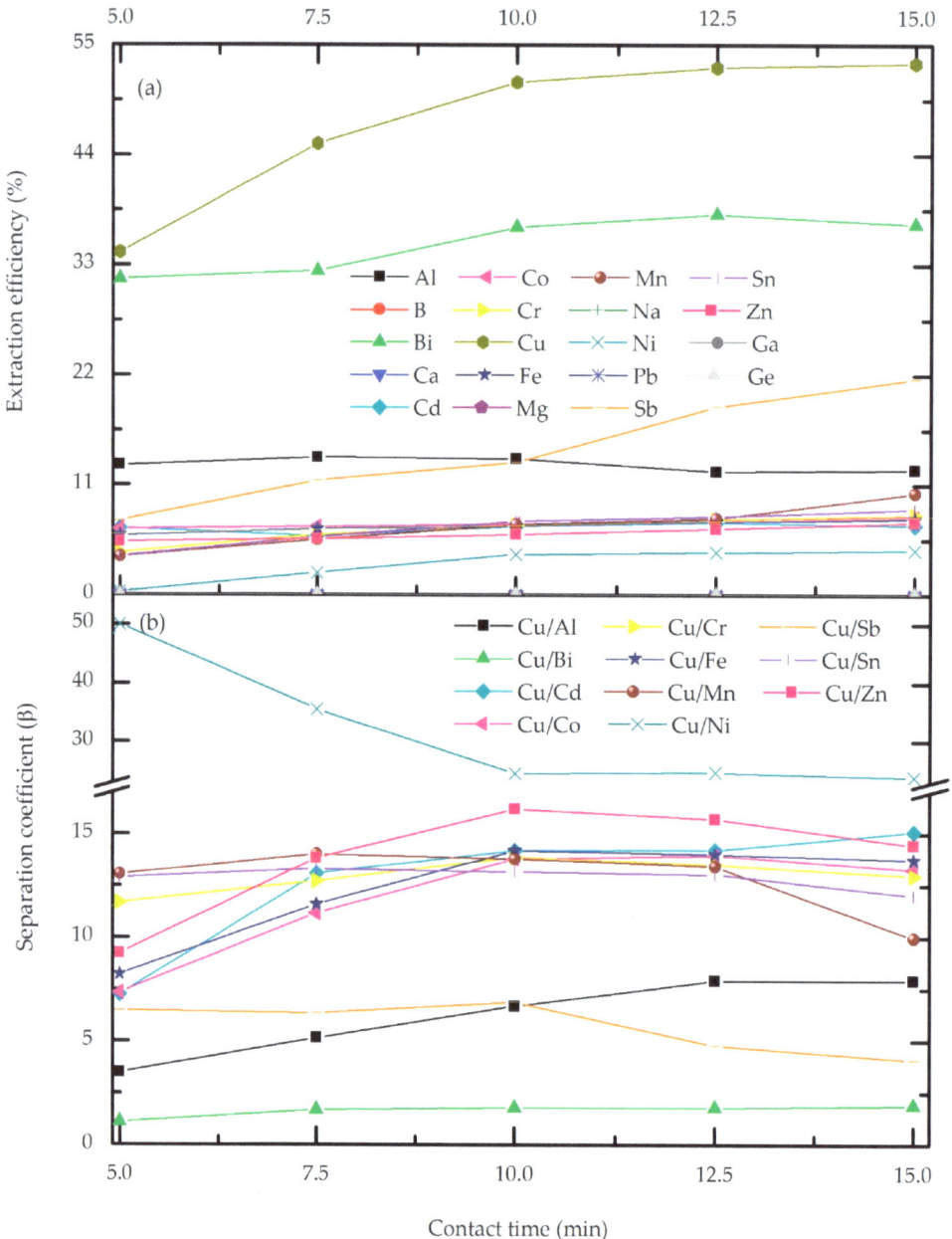

Figure 4. Effect of phase contact time: extraction efficiency (**a**), and separation coefficient (**b**) for the feed pH 0, M5640 20 vol.%, O/A = 3 and temperature 20 °C.

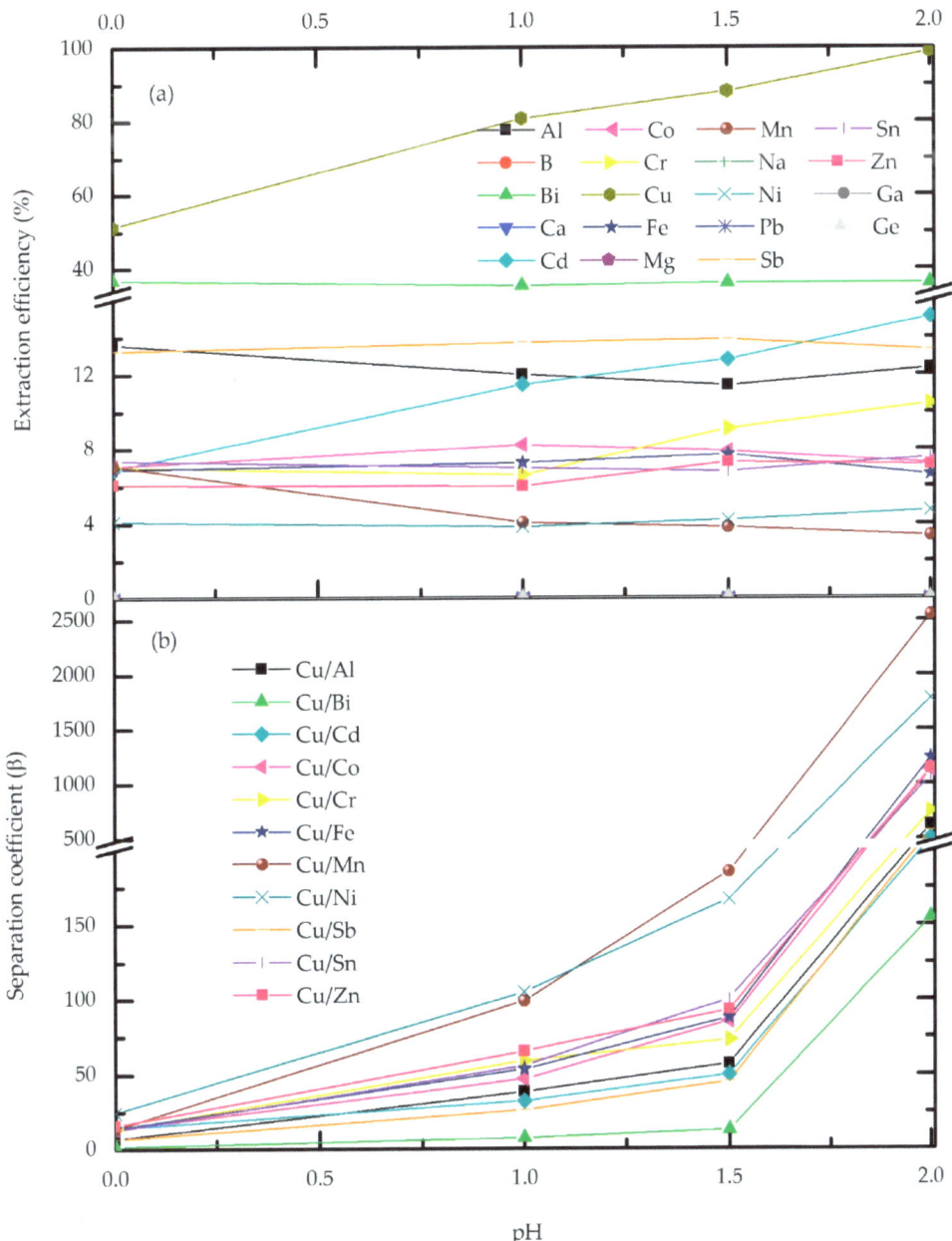

Figure 5. Effect of feed pH on the extraction efficiency (**a**) and Cu-metal separation factor (**b**) for M5640 20 vol.%, O/A = 3, temperature 20 °C and contact time 10 min.

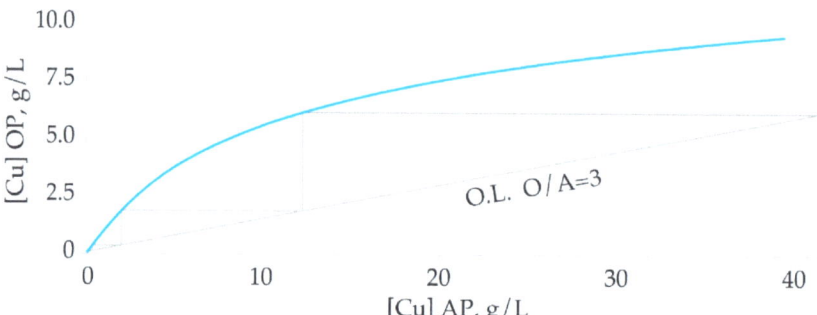

Figure 6. McCabe–Thiele extraction diagram.

3.3. Stripping Optimization

3.3.1. Effect of the Acid Concentration

In order to strip copper from the loaded organic phase, sulfuric acid, as suitable for downstream Cu recycling, i.e., electrowinning, was chosen as a stripping agent. A series of stripping experiments were conducted to determine the effect of the acid concentration on the stripping efficiency and to choose the optimal one. Parameters, such as loaded organic phase extractant concentration (M5649 20 vol.%), temperature 20 °C, O/A ratio 2, and contact time 5 min, were kept constant.

As shown in Figure 7, high efficiency of Cu stripping was achieved in the investigated acid concentration range. By increasing concentration from 100 to 180 g/L H_2SO_4, the stripping efficiency of Cu is increased from 65.2 to 95.3%. A further increase, to 200 g/L H_2SO_4, only slightly increases efficiency, to 96.2%. Higher concentrations of stripping agent were not considered due to the slight effect on Cu stripping and the possible negative effect on the extract, i.e., hydrolytic degradation.

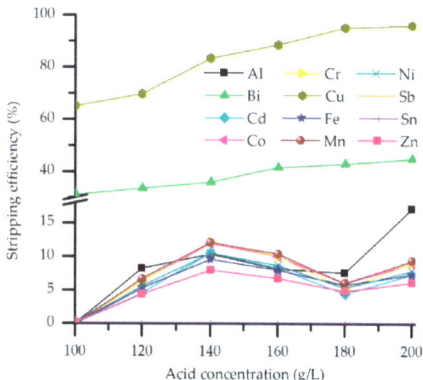

Figure 7. Effect of the H_2SO_4 concentration on the stripping efficiency (M5640 20 vol.% in loaded OP, O/A = 2, temperature 20 °C, and contact time 5 min).

These results are following the extraction mechanism and the influence of H^+ ions on the reaction equilibrium (Equation (6)). Increased concentration of the H^+ ions provides near quantitative metal replacement from its organic complex. Two protons are needed for every CuA_2 complex, while for metals of higher valence, replacement demands more protons (Equations (7) and (8)).

$$CuA_2 + 2H^+ = 2AH + Cu^{2+}, \qquad (7)$$

$$BiA_3 + 3H^+ = 3AH + Bi^{3+}. \tag{8}$$

However, the selectivity of Cu stripping is hindered by Bi, which is co-stripped up to 45%. On the other hand, Sb and Sn cannot be stripped under described experimental conditions and remain in the organic phase. Stripping efficiency of other trace metals and Fe, Zn, and Ni, was under 15%. The predominance of the Cu in the organic phase and affinity in the metal-A_n-H^+ system could explain obtained results. Namely, under optimal extraction conditions, the loaded organic phase contains 36.5 g/L Cu and significantly less metal impurities: 1.6 g/L Zn, 1.9 g/l Fe, while Ni, Sn, and Bi are measured in ppm 415, 31, and 41, respectively. The concentration of other metal impurities is under 30 ppm. According to the results, 180 g/L H_2SO_4 is chosen as the optimal concentration of the stripping agent and thus applied in the subsequent experiments.

3.3.2. Effect of O/A Stripping Ratio

Another set of experiments were conducted to determine the O/A ratio effect on the stripping efficiency. As before, all other process parameters were kept constant. Results are shown in Figure 8. Accordingly, the O/A ratio effect on Cu and Bi stripping efficiency has a negligible effect on trace elements and Fe, Zn, and Ni. Copper stripping efficiency sharply rises to O/A = 2, after which only a slight increase is observed, from 95.3 to 97.2 %S, while Bi co-stripping almost linearly increases in the whole experimental O/A range. Meanwhile, other impurities are stripped up to 8% by the upper O/A range limit.

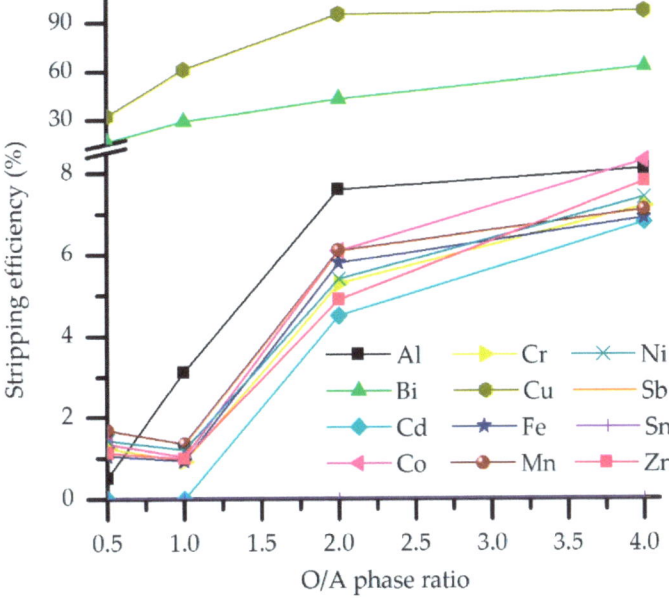

Figure 8. Effect of the phase ratio on the stripping efficiency (M5640 20 vol.% in loaded OP, H_2SO_4 concentration 180 g/L, temperature 20 °C, and contact time 5 min).

The O/A effect can be explained by the amount of metals input in the system through the loaded organic phase. As the amount of metal available for striping increases, so does stripping. However, Cu predominance in the loaded organic phase results in its near quantitative stripping due to the crowding effect of copper on the extraction of metal impurities.

Keeping in mind the optimal extraction ratio (O/A = 3), it would be preferable for the stripping ratio to be greater than 3 to avoid general dilution of the Cu solution. Yet, experimental results showed low selectivity for O/A stripping ratio even above 2, so according to this, O/A = 2 was chosen as the optimal ratio for the Cu stripping from the loaded organic phase.

3.3.3. Effect of Stripping Contact Time

In order to determine the effect of contact time on the stripping efficiency and selectivity, this parameter is varied in the 2.5 to 15 min range. In contrast, other parameters were constant, following previously determined optimal values. Results (Figure 9) indicate that prolonged contact time leads to an increase in the percentage of impurities stripped from the loaded organic phase, especially in the case of Bi, whose stripping efficiency generally rises from 44% to 67% within the studied time range.

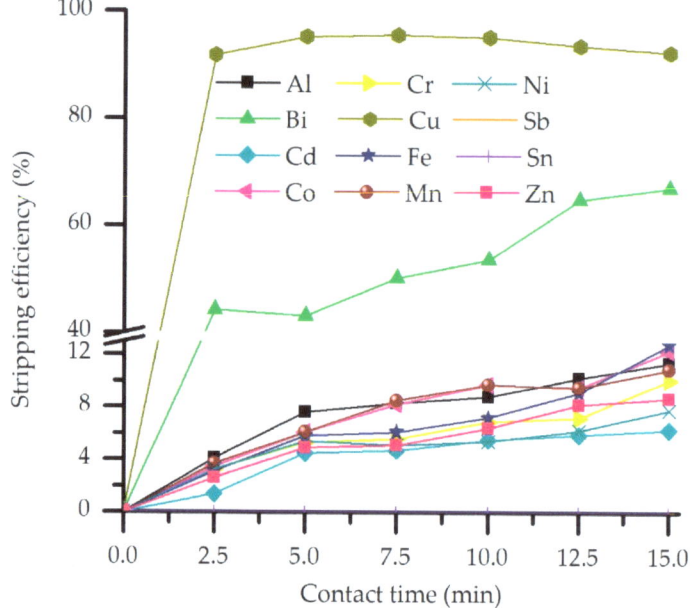

Figure 9. Effect of the phase contact time on the stripping efficiency (M5640 20 vol.% in loaded OP, H_2SO_4 concentration 180 g/L, O/A = 2, and temperature 20 °C).

Accordingly, impurities seem to have slower stripping kinetics than copper, the stripping of which shows a slight decrease after 5 min of phase contacting, achieving the equilibrium at this time point. In addition, almost 92% of Cu is stripped at the lower time range border, indicating fast stripping kinetic under studied process conditions. According to the obtained experimental results, a short contact time would be preferred to achieve high Cu loading while leaving the impurities in the organic phase. Thus, 5 min was chosen as the optimal stripping contact time.

3.3.4. Effect of Extractant Concentration in the Loaded Organic Phase

The loaded organic phase of different extractant concentrations was used as metal feed, and Cu stripping experiments were conducted under previously defined optimal conditions (O/A = 2, 180 g/L H_2SO_4, contact time 5 min, and temperature 20 °C). Results shown in Figure 10 indicate hindered stripping, as reflected in the decrease of stripping efficiency (from 95.3 to 72.6%) for experiments of extractant concentration (M5640) above 20 vol.%. This can be explained by elevated extraction efficiency when the extractant concentration is above the mentioned value due to pronounced Cu-M5640 binding affinity and increased Cu abundance in the organic phase. Such stripping results favor the previous experimentally defined optimal extractant concentration.

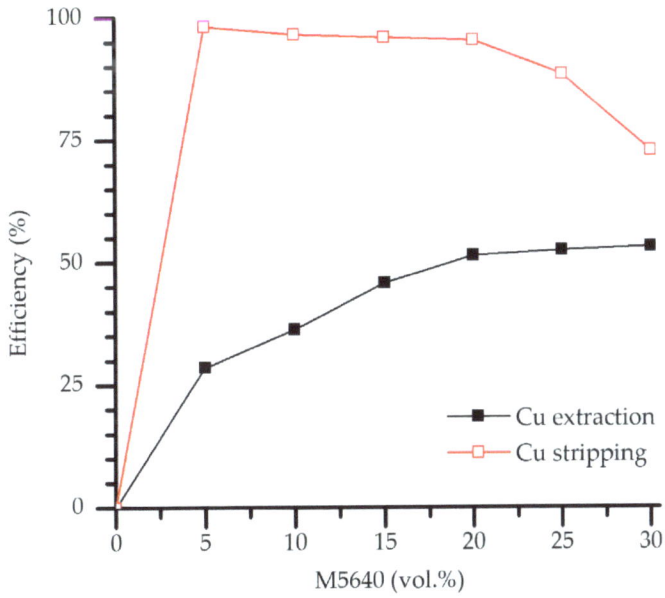

Figure 10. Extraction-stripping efficiency correlation—effect of the extractant concentration in the loaded OP on the stripping efficiency (H_2SO_4 180 g/L, O/A = 2, contact time 5 min).

3.3.5. McCabe-Thiele Stripping Diagram

The distribution isotherm of Cu stripping from the loaded organic phase was plotted according to the results obtained under previously defined optimal experimental conditions: 20 vol.% of extractant concentration in the loaded organic phase, 180 g/L H_2SO_4 as stripping agent, and phase contact time 5 min at 20 °C. Phase ratios were varied from 0.5 to 4. According to the constructed McCabe-Thiele diagram (Figure 11), two theoretical stages are needed to strip Cu from the loaded organic phase quantitatively.

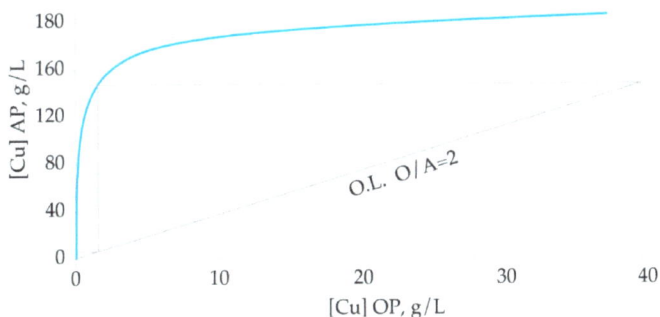

Figure 11. McCabe-Thiele stripping diagram.

3.4. Solvent Extraction with LIX® 984N and OPT 5510

In order to select the most efficient extractant for copper extraction from a complex system, a series of experiments were conducted using, in addition to M5640, two other commercial extractants: LIX® 984N (often used for Cu extraction from sulfate solutions) and OPT 5510 (an extractant modified to facilitate Cu extraction and stripping). According to the obtained results, all three studied reagents show high Cu extraction efficiency (Figure 12). However, although the highest Cu loading was achieved with LIX® 984N, so was impurity co-extraction, especially with Fe, which reaches almost 50%. Co-extraction of Zn, Ni, Bi, and Sn is significant, both with LIX® and OPT. Moreover, selectivity is achieved only by M5640. As for the extraction, the subsequent stripping step also shows favorable Cu results for all three studied reagents, with slightly promoted stripping from loaded M5640. However, the co-stripping of impurities from loaded LIX® 984N and OPT 5510 is elevated, showing poor selectivity towards Cu. In addition, under applied experimental conditions, Sn cannot be stripped from the loaded M5640, yet for the other two extractants it is up to 5%. The trace metals impurities follow a similar trend of increased extraction with LIX® 984N and OPT 5510 for both SX process steps (not shown).

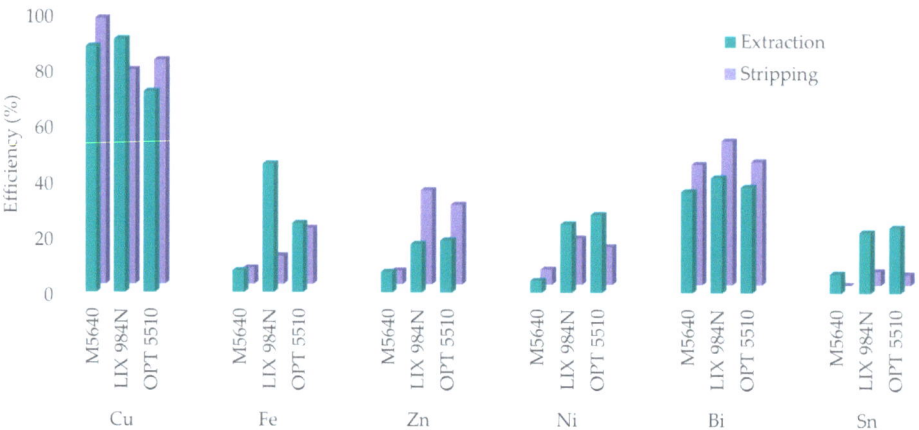

Figure 12. Comparison of extraction (for extractant concentration 20 vol.%, feed pH 1.5, O/A = 3, contact time 10 min) and stripping (H_2SO_4 180 g/L, O/A = 2, contact time 5 min) efficiency of the three studied extractants (one stage SX conducted at 20 °C).

The active species of the extractants could explain the obtained results. Specifically, dissolved Cu cations in sulfuric acid are solvated, $(Cu(H_2O)_m)^{n+}$ and the coordination of these complexes affects CuA_2 stability through the intermediate formation. This depends on the acidity of the extractant active group: the extractants with stronger acidic groups, i.e., aldoxime, facilitate Cu transfer compared with the less acidic active group, i.e., ketoxime one [37]. On the other hand, considering the chemical composition of the studied extractants, due to the weaker MAn bond in ketoxime complexes, stripping is facilitated. In complex solutions, such as the one studied in this paper, aldoxime–ketoxime extractants LIX® 984N and OPT 5510 allow high metal transfer to the organic phase and facilitate stripping, leading to decreased selectivity, both for extraction and stripping steps. In addition, the ketoxime active group, OPT, contains an ester modifier that facilitates Cu extraction and stripping. This OPT feature seems to favor clean systems unburdened by impurities, yet for complex systems, as in this study, it is undesirable, causing low selectivity with high co-extraction percent.

Comparing the obtained experimental results, it is clear that M5640 is the optimal choice for the extractant. The general results for metals behavior using LIX® 984N and OPT 5510 as extractants perhaps could be more efficient under different SX conditions. This remains for further investigation.

3.5. Copper and Metal Impurities Distribution

Based on the experimentally obtained results and calculated values of extraction and stripping efficiencies obtained under optimal process conditions, a diagram of the distribution of the elements through the one-stage SX process (Sankey diagram) is constructed (Figure 13). It should be noted that the width of the lines is proportional to the mass transfer through phases, except for the most trace elements, for which the width of the lines is slightly increased for visualization purposes.

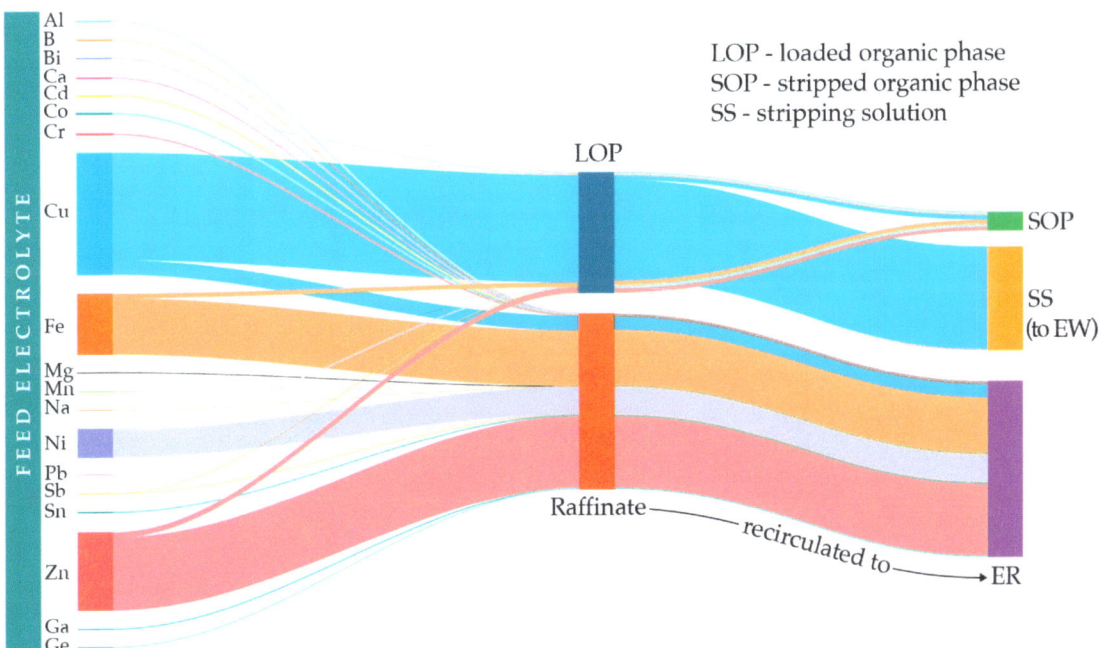

Figure 13. Metal distribution and mass balance of SX process.

By monitoring the impurities transfer through phases, the determined distribution showed that, under the applied conditions, B, Ca, Mg, Na, Pb, Ga, and Ge are not extracted but remain entirely in the raffinate. Moreover, over 90% of each trace element remained in the raffinate, as did the most abundant impurities Fe, Zn, Ni, and Sn. The raffinate is not discarded but returned in the upstream process of ER. Furthermore, after selective stripping of the Cu, most of the previously extracted quantities of the trace elements were retained in the loaded organic phase. Considering their concentration under 30 ppm, a significant effect of the extractant performance is not expected in a repeated process. In addition, reuse of the OP leads to the accumulation of trace metals, allowing their valorization in additional steps, including traditional methods of stripping with a suitable agent. Additionally, implementation of the water wash (scrubbing stage at low acidities) could be an option to remove Fe and other contaminants acting as poison to the organic phase.

Further, the stripping solution obtained under defined process conditions is relieved of impurities, especially Fe: from 20.67 g/L in feed, Fe concentration decreased to 92 ppm in stripping solution, which is 10 times less than the maximal recommended concentration in solution for the downstream Cu EW process, as stated in Section 1.

Having in mind the recirculation of the process solutions, accumulation of the retained elements is inevitable. However, this fact enables effective the extraction of metal values through additional stages. The buildup effect increases the concentration of valuable elements, i.e., Ga and Ge belong to critical technological metals whose extraction is cost-effective and feasible from concentrated solutions [38]. Moreover, increased Sn, Ni, Fe, and Zn concentrations in the Cu unladen raffinate facilitates the separation and valorization of these metals. Fe can be extracted from sulfuric solution by P507 [39] or MOC 55 [40]. Further, Ni and Zn could be separated and extracted with commercial extractants by adjusting the pH of SX feed [29,41]. Sn can be effectively extracted by using PC-88A diluted in EXXSOL D-80, or it can be extracted by coagulation when the optimal concentration in solution is achieved [42].

4. Conclusions

Selected e-waste fractions were used as feed in a specially designed pyro-electrometallurgical processing, which allowed for the generation of an electrolyte of extremely complex chemical composition and high abundance of impurities. Thus, the obtained electrolyte burden with Ni, Sn, Fe, and Zn, and 14 trace metals were used in numerous solvent extraction experiments in order to define optimal conditions for the selective Cu extraction-stripping process.

The results of this study showed that, under optimal conditions of 20 vol.% extractant concentration, feed pH 1.5, O/A ratio 3, and 10-min phase contact time, 88.1% of one stage Cu extraction can be achieved. Simultaneously, the co-extraction of the most abundant impurities (Ni, Sn, Fe, and Zn) was under 8%, and Pb and trace elements co-extraction was negligible. Moreover, it was determined that feed pH and extractant concentration influenced extraction efficiency and selectivity the most through reaction equilibrium shifts and the availability of the active chelating centers. The latter is crucial for selecting the extractant for Cu extraction from a chemically complex system, as can be expected in the future e-waste recycling process. Accordingly, the aldoxime reagent is optimal, with an advantage over the aldoxime/ketoxime blend. Stripping conditions, including H_2SO_4 180 g/L, O/A = 2, and contact time 5 min, showed optimal efficiency from loaded organic phase containing 20 vol.% M5640 which enabled 95.3% Cu stripping and under 6% of the main impurities. Acid concentration and O/A ratio predominantly influenced stripping efficiency and selectivity. Theoretical stages (McCabe Thiele diagram) for quantitative extraction and stripping were determined to be 3 and 2, respectively. Monitoring mass distribution through phases under optimal process conditions allowed the determination of positioning and quantification of each element in the system, enabling the proposal of further treatment and valorization in additional steps.

The methodology of this study and obtained results contribute to the improvement of the existing recycling processes, especially in small and medium-sized enterprises, through the development of both pyro-electro and hydrometallurgical routes capable of effectively separating and valorizing metals detained in e-waste, the valuable secondary raw material product of the increasing waste generation trend.

Author Contributions: Conceptualization, Ž.K.; methodology, Ž.K. and J.D.; formal analysis, Z.N.; data curation, J.D. and D.R.; writing—original draft preparation, J.D.; writing—review & editing, Ž.K., D.R., and Z.A.; supervision, Ž.K. and Z.A. All authors have read and agreed to the published version of the manuscript.

Funding: This research received no external funding.

Institutional Review Board Statement: Not applicable.

Informed Consent Statement: Not applicable.

Acknowledgments: Ministry of Education, Science and Technological Development of Republic of Serbia, Contract number: 451-03-9/2021-14/200288. The expertise on the solvent extraction process provided by Cyril Bourget and Ivan Djurković is highly appreciated.

Conflicts of Interest: The authors declare no conflict of interest.

References

1. Worell, E.; Reuter, M.A. *Handbook of Recycling*; Elsevier: Amsterdam, The Netherlands, 2014; ISBN 978-0-12-396459-5.
2. Dutta, D.; Panda, R.; Kumari, A.; Goel, S.; Jha, M.K. Sustainable Recycling Process for Metals Recovery from Used Printed Circuit Boards (PCBs). *Sustain. Mater. Technol.* **2018**, *17*, e00066. [CrossRef]
3. Wang, L.; Li, Q.; Sun, X.; Wang, L. Separation and Recovery of Copper from Waste Printed Circuit Boards Leach Solution Using Solvent Extraction with Acorga M5640 as Extractant. *Sep. Sci. Technol.* **2019**, *54*, 1302–1311. [CrossRef]
4. Ghosh, B.; Ghosh, M.K.; Parhi, P.; Mukherjee, P.S.; Mishra, B.K. Waste Printed Circuit Boards Recycling: An Extensive Assessment of Current Status. *J. Clean. Prod.* **2015**, *94*, 5–19. [CrossRef]
5. Forti, V.; Baldé, C.P.; Kuehr, R.; Bel, G. *The Global E-Waste Monitor 2020*; United Nations University (UNU)/United Nations Institute for Training and Research (UNITAR)—co-hosted SCYCLE Programme, International Telecommunication Union (ITU) & International Solid Waste Association (ISWA): Bonn, Germany; Geneva, Switzerland; Rotterdam, The Netherlands, 2020; ISBN 978-92-808-9114-0.
6. Cayumil, R.; Khanna, R.; Rajarao, R.; Mukherjee, P.S.; Sahajwalla, V. Concentration of Precious Metals during Their Recovery from Electronic Waste. *Waste Manag.* **2016**, *57*, 121–130. [CrossRef] [PubMed]
7. Wan, X.; Fellman, J.; Jokilaakso, A.; Klemettinen, L.; Marjakoski, M. Behavior of Waste Printed Circuit Board (WPCB) Materials in the Copper Matte Smelting Process. *Metals* **2018**, *8*, 887. [CrossRef]
8. Rai, V.; Liu, D.; Xia, D.; Jayaraman, Y.; Gabriel, J.-C.P. Electrochemical Approaches for the Recovery of Metals from Electronic Waste: A Critical Review. *Recycling* **2021**, *6*, 53. [CrossRef]
9. Hagelüken, C. Recycling of Electronic Scrap at Umicore's Integrated Metals Smelter and Refinery. *World Metall ERZMETALL* **2006**, *11*, 152–161.
10. Lennartsson, A.; Engström, F.; Samuelsson, C.; Björkman, B.; Pettersson, J. Large-Scale WEEE Recycling Integrated in an Ore-Based Cu-Extraction System. *J. Sustain. Metall.* **2018**, *4*, 222–232. [CrossRef]
11. Ariizumi, M.; Takagi, M.; Inoue, O.; Oguma, N. Integrated processing of e-scrap at Naoshima smelter and refinery. In Proceedings of the Copper 2016, Kobe, Japan, 13–16 November 2016; Volume 6, p. RW 1–2.
12. Kamberović, Ž.; Ranitović, M.; Korać, M.; Jovanović, N.; Tomović, B.; Gajić, N. Pyro-Refining of Mechanically Treated Waste Printed Circuit Boards in a DC Arc-Furnace. *J. Sustain. Metall.* **2018**, *4*, 251–259. [CrossRef]
13. Forsén, O.; Aromaa, J.; Lundström, M. Primary Copper Smelter and Refinery as a Recycling Plant—A System Integrated Approach to Estimate Secondary Raw Material Tolerance. *Recycling* **2017**, *2*, 19. [CrossRef]
14. Cui, J.; Zhang, L. Metallurgical Recovery of Metals from Electronic Waste: A Review. *J. Hazard. Mater.* **2008**, *158*, 228–256. [CrossRef] [PubMed]
15. Djokić, J.; Jovančićević, B.; Brčeski, I.; Ranitović, M.; Gajić, N.; Kamberović, Ž. Leaching of Metastannic Acid from E-Waste by-Products. *J. Mater. Cycles Waste Manag.* **2020**, *22*, 1899–1912. [CrossRef]
16. Vasilyev, F.; Virolainen, S.; Sainio, T. Modeling the Liquid–Liquid Extraction Equilibrium of Iron (III) with Hydroxyoxime Extractant and Equilibrium-Based Simulation of Counter-Current Copper Extraction Circuits. *Chem. Eng. Sci.* **2018**, *175*, 267–277. [CrossRef]
17. Das, S.C.; Gopala Krishna, P. Effect of Fe(III) during Copper Electrowinning at Higher Current Density. *Int. J. Miner. Process.* **1996**, *46*, 91–105. [CrossRef]
18. Kamberović, Ž.; Ranitović, M.; Korać, M.; Andjić, Z.; Gajić, N.; Djokić, J.; Jevtić, S. Hydrometallurgical Process for Selective Metals Recovery from Waste-Printed Circuit Boards. *Metals* **2018**, *8*, 441. [CrossRef]
19. Kamberović, Ž.; Korać, M.; Ranitović, M. Hydrometallurgical process for extraction of metals from electronic waste-part II: Development of the processes for the recovery of copper from printed circuit boards (PCB). *Metalurgija* **2011**, *17*, 139–149.
20. Lister, T.E.; Wang, P.; Anderko, A. Recovery of Critical and Value Metals from Mobile Electronics Enabled by Electrochemical Processing. *Hydrometallurgy* **2014**, *149*, 228–237. [CrossRef]
21. Robinson, B.H. E-Waste: An Assessment of Global Production and Environmental Impacts. *Sci. Total Environ.* **2009**, *408*, 183–191. [CrossRef]
22. Kaya, M. Recovery of Metals and Nonmetals from Electronic Waste by Physical and Chemical Recycling Processes. *Waste Manag.* **2016**, *57*, 64–90. [CrossRef]
23. Jiang, F.; Yin, S.; Zhang, L.; Peng, J.; Ju, S.; Miller, J.D.; Wang, X. Solvent Extraction of Cu(II) from Sulfate Solutions Containing Zn(II) and Fe(III) Using an Interdigital Micromixer. *Hydrometallurgy* **2018**, *177*, 116–122. [CrossRef]
24. Tuncuk, A.; Stazi, V.; Akcil, A.; Yazici, E.Y.; Deveci, H. Aqueous Metal Recovery Techniques from E-Scrap: Hydrometallurgy in Recycling. *Miner. Eng.* **2012**, *25*, 28–37. [CrossRef]
25. Agarwal, S.; Ferreira, A.E.; Santos, S.M.C.; Reis, M.T.A.; Ismael, M.R.C.; Correia, M.J.N.; Carvalho, J.M.R. Separation and Recovery of Copper from Zinc Leach Liquor by Solvent Extraction Using Acorga M5640. *Int. J. Miner. Process.* **2010**, *97*, 85–91. [CrossRef]
26. Ochromowicz, K.; Chmielewski, T. Solvent Extraction of Copper(II) from Concentrated Leach Liquors. *Physicochem. Probl. Miner. Process.* **2013**, *49*, 357–367. [CrossRef]
27. Habashi, F. *A Textbook of Hydrometallurgy*; Métallurgie Extractive Québec: Sainte-Foy, QC, Canada, 1999; ISBN 978-2-9803247-7-2.
28. Jha, M.K.; Gupta, D.; Choubey, P.K.; Kumar, V.; Jeong, J.; Lee, J. Solvent Extraction of Copper, Zinc, Cadmium and Nickel from Sulfate Solution in Mixer Settler Unit (MSU). *Sep. Purif. Technol.* **2014**, *122*, 119–127. [CrossRef]

29. Kumari, A.; Jha, M.K.; Lee, J.; Singh, R.P. Clean Process for Recovery of Metals and Recycling of Acid from the Leach Liquor of PCBs. *J. Clean. Prod.* **2016**, *112*, 4826–4834. [CrossRef]
30. Li, X.; Wei, C.; Deng, Z.; Li, C.; Fan, G.; Rong, H.; Zhang, F. Extraction and Separation of Indium and Copper from Zinc Residue Leach Liquor by Solvent Extraction. *Sep. Purif. Technol.* **2015**, *156*, 348–355. [CrossRef]
31. Deep, A.; Kumar, P.; Carvalho, J.M.R. Recovery of Copper from Zinc Leaching Liquor Using ACORGA M5640. *Sep. Purif. Technol.* **2010**, *76*, 21–25. [CrossRef]
32. Ferreira, A.E.; Agarwal, S.; Machado, R.M.; Gameiro, M.L.F.; Santos, S.M.C.; Reis, M.T.A.; Ismael, M.R.C.; Correia, M.J.N.; Carvalho, J.M.R. Extraction of Copper from Acidic Leach Solution with Acorga M5640 Using a Pulsed Sieve Plate Column. *Hydrometallurgy* **2010**, *104*, 66–75. [CrossRef]
33. Vander Linden, J. Selective Recuperation of Copper by Supported Liquid Membrane (SLM) Extraction. *J. Membr. Sci.* **1998**, *139*, 125–135. [CrossRef]
34. Kongolo, K.; Ngoie, N.; Francis, K.; Patric, T. Improving the efficiency of solvent extraction of copper by combination of hydroxyoximic extractants. In Proceedings of the International Conference on Metal Solvent Extraction 2015, Three Gorges, China, 10–14 November 2015; pp. 117–126.
35. Asghari, H.; Safarzadeh, M.S.; Asghari, G.; Moradkham, D. The Effect of Impurities on the Extraction of Copper from Sulfate Medium Using LIX®984N in Kerosene. *Russ. J. Non-Ferrous Metals* **2009**, *50*, 89–96. [CrossRef]
36. Banza, A.N.; Gock, E.; Kongolo, K. Base Metals Recovery from Copper Smelter Slag by Oxidising Leaching and Solvent Extraction. *Hydrometallurgy* **2002**, *67*, 63–69. [CrossRef]
37. Naveed, S.; Nawaz, Z. Copper Extraction from Copper Rolling Mills Scraps using Solvent "aryl aldoxime, 2-hydroxy-5-nonylbenzaldoxime" (ACORGA-M5640). *J. Chem. Soc. Pak.* **2006**, *28*, 44–50.
38. De la Torre, E.; Vargas, E.; Ron, C.; Gámez, S. Europium, Yttrium, and Indium Recovery from Electronic Wastes. *Metals* **2018**, *8*, 777. [CrossRef]
39. Wang, L.; Wang, Y.; Cui, L.; Gao, J.; Guo, Y.; Cheng, F. A Sustainable Approach for Advanced Removal of Iron from CFA Sulfuric Acid Leach Liquor by Solvent Extraction with P507. *Sep. Purif. Technol.* **2020**, *251*, 117371. [CrossRef]
40. Ocaña, N.; Alguacil, F.J. Solvent Extraction of Iron(III) by MOC-55 TD: Experimental Equilibrium Study and Demonstration of Lack of Influence on Copper(II) Extraction from Sulphate Solutions. *Hydrometallurgy* **1998**, *48*, 239–249. [CrossRef]
41. Sridhar, V.; Verma, J.K. Recovery of Copper, Nickel and Zinc from Sulfate Solutions by Solvent Extraction Using LIX 984N. *E-J. Chem.* **2011**, *8*, S434–S438. [CrossRef]
42. Inoue, K.; Mirvaliev, R.; Yoshizuka, K.; Ohto, K.; Babasaki, S.-Y. Solvent extraction of TIN(IV) with PC-88A from sulfuric acid solutions containing chloride ions. *Solvent. Extr. Res. Dev. Jpn.* **2001**, *8*, 21–25.

MDPI AG
Grosspeteranlage 5
4052 Basel
Switzerland
Tel.: +41 61 683 77 34

Metals Editorial Office
E-mail: metals@mdpi.com
www.mdpi.com/journal/metals

Disclaimer/Publisher's Note: The title and front matter of this reprint are at the discretion of the Guest Editor. The publisher is not responsible for their content or any associated concerns. The statements, opinions and data contained in all individual articles are solely those of the individual Editor and contributors and not of MDPI. MDPI disclaims responsibility for any injury to people or property resulting from any ideas, methods, instructions or products referred to in the content.

www.ingramcontent.com/pod-product-compliance
Lightning Source LLC
LaVergne TN
LVHW072349090526
838202LV00019B/2510